AF902500

VOLUME NINETY SIX

Advances in BOTANICAL RESEARCH

ATP Synthase in Photosynthetic Organisms

ADVANCES IN BOTANICAL RESEARCH

Series Editor

Jean-Pierre Jacquot
Membre de l'Institut Universitaire de France, Unité Mixte de Recherche INRA, UHP 1136 "Interaction Arbres Microorganismes", Université de Lorraine, Faculté des Sciences, Vandoeuvre, France

VOLUME NINETY SIX

ADVANCES IN BOTANICAL RESEARCH

ATP Synthase in Photosynthetic Organisms

Volume Editor

TORU HISABORI
Laboratory for Chemistry and Life Science,
Institute of Innovative Research,
Tokyo Institute of Technology,
Yokohama, Japan

ELSEVIER

ACADEMIC PRESS
An imprint of Elsevier

Academic Press is an imprint of Elsevier
125 London Wall, London, EC2Y 5AS, United Kingdom
The Boulevard, Langford Lane, Kidlington, Oxford OX5 1GB, United Kingdom
50 Hampshire Street, 5th Floor, Cambridge, MA 02139, United States
525 B Street, Suite 1650, San Diego, CA 92101, United States

First edition 2020

Copyright © 2020 Elsevier Ltd. All rights reserved.

No part of this publication may be reproduced or transmitted in any form or by any means, electronic or mechanical, including photocopying, recording, or any information storage and retrieval system, without permission in writing from the publisher. Details on how to seek permission, further information about the Publisher's permissions policies and our arrangements with organizations such as the Copyright Clearance Center and the Copyright Licensing Agency, can be found at our website: www.elsevier.com/permissions.

This book and the individual contributions contained in it are protected under copyright by the Publisher (other than as may be noted herein).

Notices

Knowledge and best practice in this field are constantly changing. As new research and experience broaden our understanding, changes in research methods, professional practices, or medical treatment may become necessary.

Practitioners and researchers must always rely on their own experience and knowledge in evaluating and using any information, methods, compounds, or experiments described herein. In using such information or methods they should be mindful of their own safety and the safety of others, including parties for whom they have a professional responsibility.

To the fullest extent of the law, neither the Publisher nor the authors, contributors, or editors, assume any liability for any injury and/or damage to persons or property as a matter of products liability, negligence or otherwise, or from any use or operation of any methods, products, instructions, or ideas contained in the material herein.

ISBN: 978-0-08-102896-4
ISSN: 0065-2296

For information on all Academic Press publications
visit our website at https://www.elsevier.com/books-and-journals

Publisher: Zoe Kruze
Acquisition Editor: Sam Mahfoudh
Editorial Project Manager: Peter J. Llewellyn
Production Project Manager: James Selvam
Cover Designer: Alan Studholme

Typeset by SPi Global, India

ELSEVIER Book Aid International Working together to grow libraries in developing countries
www.elsevier.com • www.bookaid.org

Contents

Contributors

Roland Bienert
Department of Physical Chemistry, Albert-Ludwigs-University of Freiburg, Freiburg im Breisgau, Germany

Michael Börsch
Jena University Hospital, Friedrich Schiller University Jena, Jena, Germany

Felix E. Buchert
Institute of Plant Biology and Biotechnology, University of Münster, Münster, Germany

Riu Furutani
Department of Applied Biological Science, Faculty of Agriculture, Graduate School for Agricultural Science, Kobe University, Kobe, Japan

Peter Gräber
Department of Physical Chemistry, Albert-Ludwigs-University of Freiburg, Freiburg im Breisgau, Germany

Georg Groth
Institute of Biochemical Plant Physiology, Heinrich Heine University, Düsseldorf, Germany

Toru Hisabori
Laboratory for Chemistry and Life Science, Institute of Innovative Research, Tokyo Institute of Technology, Yokohama, Japan

Kentaro Ifuku
Graduate School of Biostudies, Kyoto University, Kyoto, Japan

Amane Makino
Graduate School of Agricultural Science, Tohoku University, Sendai, Japan

Alexander Minges
Institute of Biochemical Plant Physiology, Heinrich Heine University, Düsseldorf, Germany

Chikahiro Miyake
Department of Applied Biological Science, Faculty of Agriculture, Graduate School for Agricultural Science, Kobe University, Kobe; Core Research for Environmental Science and Technology (CREST), Japan Science and Technology Agency, Tokyo, Japan

Ko Noguchi
School of Life Sciences, Tokyo University of Pharmacy and Life Sciences, Tokyo, Japan

Lianwei Peng
Shanghai Key Laboratory of Plant Molecular Sciences, College of Life Sciences, Shanghai Normal University, Shanghai, China

Jean-David Rochaix
Departments of Molecular Biology and Plant Biology, University of Geneva, Geneva, Switzerland

Toshiharu Shikanai
Department of Botany, Graduate School of Science, Kyoto University, Kyoto, Japan

Ginga Shimakawa
Research Center for Solar Energy Chemistry, Osaka University, Osaka, Japan

Takayuki Sohtome
Department of System Development, Bunkoukeiki Co., Ltd, Tokyo, Japan

Yuji Suzuki
Faculty of Agriculture, Iwate University, Morioka, Japan

Paola Turina
Department of Pharmacy and Biotechnology, University of Bologna, Bologna, Italy

Shinya Wada
Department of Applied Biological Science, Faculty of Agriculture, Graduate School for Agricultural Science, Kobe University, Kobe, Japan

Lin Zhang
Shanghai Key Laboratory of Plant Molecular Sciences, College of Life Sciences, Shanghai Normal University, Shanghai, China

Preface

ATP is known as the energy currency in all living organisms on this planet. Therefore, ATP synthesis is considered the root of life. Biological membranes are the main stage for ATP synthesis, and the star of this drama is ATP synthase, a huge protein complex consisting of the membrane-superficial sphere complex and the membrane-embedded ring structure complex. In the 60 years since the enzyme was discovered, researchers have been working to understand the molecular mechanism of how this enzyme complex is able to supply large amounts of ATP. For example, humans need an amount of ATP roughly equal to their body weights each day to maintain their activity. These ATP molecules are produced in the mitochondria that are specialized intracellular organelles required for energy conversion. In contrast, ATP synthesis for photosynthesis in plants is important for promoting the most important production of organic matter, the carbon fixation reaction, the conversion of CO_2 into sugar. As an energy conversion system, phototrophs are an almost unique system on the earth and their activities are roughly estimated to be 4×10^{18} kJ/year, which can convert 0.1% of whole solar energy coming from the sun into the chemical energy useful for biological organisms. Considering that ATP synthase supports this enormous reaction, it is easy to see that understanding the molecular mechanism of the enzymatic reaction of ATP synthase is a matter of understanding the energy conversion process on the earth.

As indicated in Chapter 1, some of the key findings are obtained in the last two decades of the last century. However, the full molecular structure of ATP synthase from phototrophs was reported only 2 years ago. Coupled with major advances in other research areas concerning ATP synthase of phototrophs, this is the good time to collect some reviews on this enzyme.

Due to the serious crises caused by COVID-19, the preparation of this book has been a challenge, and the contents we have been able to cover regarding the importance of this enzyme is still far from complete. However, we hope that the content of the book will help you understand the significance of this enzyme complex.

I express my sincere thanks to all authors and readers of the book.

Toru Hisabori

CHAPTER ONE

Regulation machineries of ATP synthase from phototroph

Toru Hisabori*
Laboratory for Chemistry and Life Science, Institute of Innovative Research, Tokyo Institute of Technology, Yokohama, Japan
*Corresponding author: e-mail address: thisabor@res.titech.ac.jp

Contents

Abstract

Photosynthetic reactions are, so to speak, the source of energy for all organisms on the earth. Thus, ATP synthase of the phototroph is the key enzyme on the reaction process that converts light energy into chemical energy, and is the grand energy conversion system of all photosynthetic organisms. The activity of ATP synthase is finely regulated to ensure that ATP synthesis is carried out continuously and to provide the energy needed for photosynthesis under constantly changing environmental conditions. Since the molecular structure of ATP synthase was elucidated by X-ray crystallography at the end of the 20th century, much progress has been made in understanding the molecular mechanism of ATP synthase catalytic reaction. The active use of cyanobacterial ATP synthase in research has also led to the understanding of

Advances in Botanical Research, Volume 96
ISSN 0065-2296
https://doi.org/10.1016/bs.abr.2020.07.003
© 2020 Elsevier Ltd
All rights reserved.

photosynthetic ATP synthase at the molecular level. The regulatory mechanism of this enzyme has also been understood at the molecular level over the last 20 years. This review provides an overview of the history of such studies, particularly of the regulation of enzyme activity by the γ subunit and the ε subunit.

1. Introduction

1.1 Brief history of F_1-ATPase research

ATP synthase is the ubiquitous enzyme, which is observed in various energy-transducing membranes such as thylakoid membranes in chloroplasts and cyanobacteria, inner membranes of mitochondria, and plasma membranes of prokaryotic bacteria. This enzyme synthesizes ATP from ADP and inorganic phosphate using the electrochemical proton gradient across these energy-transducing membranes, which is formed by photosynthetic or respiratory electron transfer reaction, as a driving force. Therefore, ATP synthase is known as the key player of the energy transduction in living cells (Boyer, 1997; Yoshida, Muneyuki, & Hisabori, 2001). ATP synthase is composed of the multiple subunits. Its membrane-peripheral portion is called F_1, which has five types of subunits with a stoichiometry of $\alpha_3\beta_3\gamma_1\delta_1\varepsilon_1$ (Yoshida, Sone, Hirata, Kagawa, & Ui, 1979). A membrane-embedded portion is F_o, which consists of three types of subunits with a stoichiometry of $a_1b_2c_{8-15}$ (Ferguson, Cook, Montgomery, Leslie, & Walker, 2016; Jiang, Hermolin, & Fillingame, 2001; Matthey, Kaim, Braun, Wuthrich, & Dimroth, 1999; Mitome, Suzuki, Hayashi, & Yoshida, 2004). Historically, F_o subunits of the chloroplast enzyme were named different from those in bacteria and mitochondria, and consist of $IV_1I_1II_1III_{14}$, in which subunit IV corresponds to subunit *a*, subunits I and II are very similar with subunit *b*, and subunit III is with *c*, respectively (Feng & McCarty, 1990; Nelson, Nelson, & Schatz, 1980), although the names of the subunits in the CF_o portion used in this book are those of bacterial subunits for ease of reading. F_1 portion of ATP synthase is a soluble protein complex and is easily detached from the membrane, and shows generally remarkable ATP hydrolysis activity and thus it is commonly called F_1-ATPase.

Because of the intracellular importance of this enzyme, which provides energy-chemical ATP important for various cellular metabolism, the biochemical profile of this enzyme has been actively studied since its discovery (Avron, 1963; Pullman, Penefsky, Datta, & Racker, 1960). In the late 1970s, genomic analysis of ATP synthase from prokaryotic bacteria *Escherichia coli* was initiated by two research groups in Japan and the United Kingdom (Futai & Kanazawa, 1983; Walker & Gay, 1983). Because of the intense competition between these two groups, the entire genetic structure of the subunits of this enzyme complex in the *E. coli* genome was revealed in a very short time. The subunit composition of this enzyme complex, especially the F_1 part, and each role of F_1 part subunits were studied mainly by other Japanese group using ATP synthase from thermophilic bacterium, thermophilic *Bacillus* PS3 (now it is renamed as *Geobacillus stearothermophilus*), isolated from a hot spring in Japan (Kagawa, Sone, Yoshida, Hirata, & Okamoto, 1976; Yoshida, Sone, Hirata, & Kagawa, 1975b). Thus, by 1990, a complete picture of the catalytic mechanism of this enzyme complex had been established. The next big advance on understanding of this enzyme was provided by the X-ray crystal structure analysis of mitochondrial F_1 part in 1994 (Abrahams, Leslie, Lutter, & Walker, 1994). The solved crystal structure only showed the hexagonal $\alpha_3\beta_3$ ring and the part of the γ subunit. However, this structure clearly showed three equivalent but state-different catalytic sites in the protein complex on the β subunits at the interface with the α subunits. In addition, the γ subunit is composed of two α-helices which form the coiled-coil structure in addition to the domain, which is recognized as a so-called Rossman-fold domain, and is located at the center of the $\alpha_3\beta_3$ hexagon as a central axis.

At that time, most researchers superposed this reported crystal structure with the idea of an enzyme-catalyzed model proposed by Boyer, now called the "rotary catalytic mechanism" (Gresser, Myers, & Boyer, 1982). In fact, several groups soon set out to elucidate the mechanism by which this enzyme complex catalyzes rotation. In 1995, Cross and coworkers confirmed the rotational movement of the γ subunit during ATP hydrolysis reaction by way of the alternate cross-link formation between one of β subunits and the central axis γ subunit (Duncan, Bulygin, Zhou, Hutcheon, & Cross, 1995). Based on the results of this experiment, they concluded that the γ subunit can change the opposing β subunit during catalytic reaction. In 1996, W. Junge and colleagues studied the movement of the γ subunit by elaborate biophysical measurements using polarized fluorescence quenching (Sabbert, Engelbrecht, & Junge, 1996). Chloroplast ATP synthase F_1 was used in their experiment. The γ subunit of CF_1 has a cysteine Cys322 at the C-terminus and this cysteine residue is exposed from the enzyme complex, making it easily labeling with a fluorescent probe. They monitored the fluorescence signal from this attached probe, and concluded that the γ subunit can rotate at least 200° during ATP hydrolysis reaction.

However, the most important evidence of the γ subunit rotation was provided by a very simple direct observation of the movement of the γ subunit itself under the optical microscope. In 1997, Noji, Yasuda, Yoshida, and Kinosita succeeded in observing the smooth and continuous counterclockwise ATP-dependent rotation of actin filaments attached to the γ subunit of the F_1-ATPase complex, which was immobilized on the glass surface for optical fluorescent microscopy (Noji, Yasuda, Yoshida, & Kinosita, 1997). The predicted γ subunit rotation speed estimated from the enzyme activity was 200–500 revolutions per second, whereas the observed rotation of 1 μm long fluorescein-labeled actin filament was 2–3 revolutions per second, which was the optimal speed for direct observation of motion under the optical microscope. In addition, the observed counterclockwise rotation direction was consistent with the direction of the reaction sequence proposed by the solved crystal structure, which contains three different state β subunits; ATP bound, ADP bound, and empty (Abrahams et al., 1994). Therefore, this simple experimental result converged the discussion of whether F_1-ATPase rotates during catalytic reaction, and since then F_1-ATPase was recognized as the rotary motor protein.

Several follow-up researches revealed the stator subunits (α, β, δ, *a*, and *b*) and the rotor subunits (γ, ε, and *c*) (Kato-Yamada, Noji, Yasuda, Kinosita, & Yoshida, 1998; Sambongi et al., 1999; Tsunoda et al., 2000). In addition, the rotation motion itself was analyzed very precisely using high-speed camera and a small probe such as gold nanoparticle, then each of the steps of rotation was directly connected with the chemical reaction steps on the enzyme such as substrate binding, catalytic reaction, and products release (Adachi et al., 2000; Yasuda, Noji, Yoshida, Kinosita, & Itoh, 2001).

The latest advance on the analysis of rotation of this enzyme is now explained by Peter Gräber in chapter "Rotation of the γ-subunit in single membrane-bound H+-ATPsynthases from chloroplasts during ATP synthesis"

1.2 Brief history of CF_1-ATPase research

Chloroplast ATP synthase has historically been a critical enzyme for understanding the mechanism of ATP synthesis in the energy-transducing membranes. In 1950s, Arnon discovered the ATP synthesis during illumination of the isolated chloroplasts obtained from spinach leaves, and he named this phenomena "photophosphorylation" in contrast to the oxidative phosphorylation found in mitochondria (Whatley, Allen, Rosenberg, Capindale, & Arnon, 1956). When Mitchell proposed the famous "chemiosmotic theory" as the basic mechanism of ATP synthesis driven by the proton-motive force

(Mitchell, 1966), Jagendorf envisioned a very simple experiment using the chloroplast thylakoid membranes, acid-base transition (Jagendorf & Uribe, 1966). They attempted to artificially forming pH gradient across the thylakoid membranes, and actually succeeded to synthesize ATP when internal pH was adjusted to 4 and the external pH was 7. By using the liposome containing bacteriorhodopsin, light-promoted proton pump, and ATP synthase complex isolated from thermophilic bacteria, Yoshida and his colleagues subsequently succeeded to synthesize ATP just by illumination in 1975 (Yoshida, Sone, Hirata, & Kagawa, 1975a). This experimental design was the simplest working model for ATP synthesis promoted by the proton gradient. Based on these experimental supports, P. Mitchell won the Nobel prize in chemistry in 1978.

Biochemical analysis of the chloroplast ATP synthase was mainly initiated by E. Racker and his colleagues in the 1960s. Racker contributed greatly on the early studies of F_1-ATPase obtained from the beef heart mitochondria and spinach chloroplasts (Pullman et al., 1960; Vambutas & Racker, 1965). Although the biochemical properties of the enzymes from these different origins were very similar, it was not known very well whether these enzymes had same root at that time. The similarity of the subunit composition of the enzyme complex from chloroplast with those from mitochondria and bacteria were then investigated by Yoshida et al. (1979). We can see the trace of the historical different research process on the name of subunits of these enzymes at the F_o portion as described, *i.e.*, *a*, *b*, and *c* of bacterial and mitochondrial ATP synthase F_o portion, and I, II, III, and IV for the chloroplast F_o portion as mentioned elsewhere. Whole structure of this enzyme complex is explained in chapter "Structure and supramolecular architecture of chloroplast ATP synthase" by Georg Groth and chapter "Chloroplast ATP synthase from green microalgae" by Felix Buchert. In addition, how this very complicated enzyme complex is constructed in chloroplasts are discussed in chapter "Regulation of the biogenesis of chloroplast ATP synthase" by Lin Zhang et al.

In the 1970s, Buchanan discovered that activities of many chloroplast enzymes are regulated in a light-dependent manner, and the chloroplast thioredoxin, the reducing equivalent transducer, plays an important role in this regulation (Buchanan & Wolosiuk, 1976; Wolosiuk & Buchanan, 1977). Shortly after his report, the chloroplast ATP synthase was also reported to be a thioredoxin-regulated enzyme. McCarty and his coworkers discovered that the γ subunit of CF_1-ATPase is the key subunit in this regulation (McKinney, Buchanan, & Wolosiuk, 1978; Mills & Mitchell, 1982), and Miki and Futai eventually identified the key cysteines (Cys199 and Cys 205)

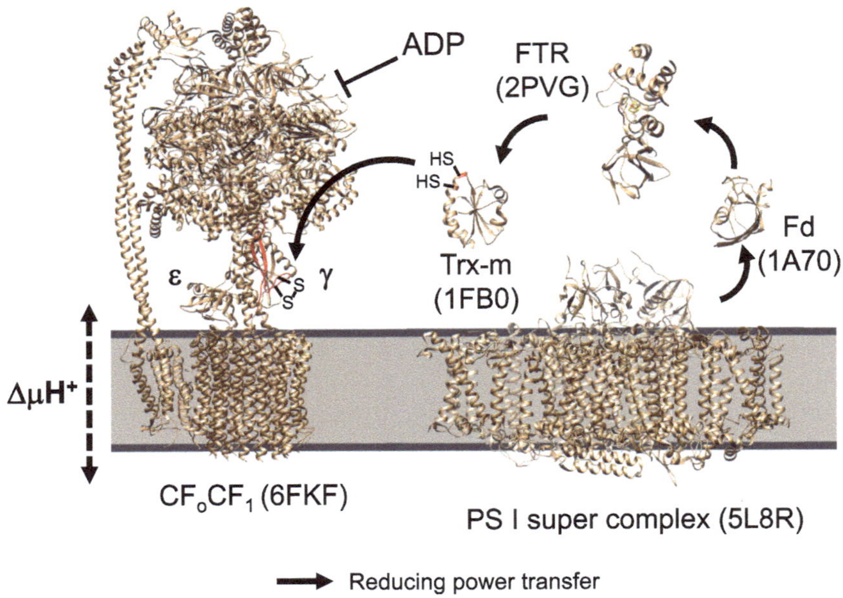

Fig. 1 The regulation of CF_oCF_1 by thioredoxin, which obtains the reducing equivalents from photosynthetic electron transfer system. The molecular structures used were CF_oCF_1 from *Spinacia oleracea*, PDB ID 6FKF (Hahn, Vonck, Mills, Meier, & Kuhlbrandt, 2018); PS I from *Pisum sativum*, PDB ID 5L8R (Mazor, Borovikova, Caspy, & Nelson, 2017); Fd from *S. oleracea*, PDB ID 1A70 (Binda, Coda, Aliverti, Zanetti, & Mattevi, 1998); FTR from *Synechocystis* sp. PCC 6803, PDB ID 2PVD (Dai et al., 2007); and Trx-*m* from *S. oleracea*, PDB ID 1F80 (Capitani et al., 2000). The β-hairpin structure of the γ subunit is shown with *red* color. All structures shown in this figure are prepared using the software "Chimera" by Kentaro Akiyama.

on the γ subunit of spinach CF_1 (Miki, Maeda, Mukohata, & Futai, 1988). Very interestingly, nine amino acid sequences including these two important cysteines of the γ subunit of CF_1 from green plants are not found in the γ subunit of other origins, such as prokaryotes and mitochondria of plants and animals. This means that the redox regulation of ATP synthase is observed only in photosynthetic organisms (Fig. 1).

2. Rotational catalysis mechanism and its regulation

2.1 Rotational catalysis and the related experiments

P. Boyer proposed a conformational change in ATP synthase as the mechanism of ATP synthesis and hydrolysis by this enzyme, but he could not prove it. Thus, after the chemiosmotic theory proposed by Mitchell (1966) was largely established in 1970s, Boyer turned his attention to the catalytic mechanism of ATP synthase itself. The most important information for him was that Yoshida and Kagawa reported subunit stoichiometry of the F_1 portion of thermophilic bacteria by accurate analysis of molecular weights of the subunit proteins using analytical ultracentrifugation measurements (Yoshida et al., 1979). Furthermore, they demonstrated the role of the β subunit as a catalytic subunit by isolating the subunit of the thermophilic F_1-ATPase and reconstituting the protein complex (Yoshida, Okamoto, Sone, Hirata, & Kagawa, 1977). Based on the extensive enzymological knowledge of the enzyme molecules at that time, and the precise biochemical analysis of the catalytic cooperativity of the F_1-ATPases by Boyer's group (Hackney, Rosen, & Boyer, 1979), Boyer ultimately concluded that it was worthwhile to consider the rotational catalytic model (the rotation of three α and β hexagons relative to γ, δ, and ε subunits) (Gresser et al., 1982). This working model was speculative at the time, and many researchers either disagreed or did not consider whether it was a possible molecular mechanism for a catalytic reaction of this enzyme complex. However, several reports have shown that the idea of a rotary catalysis model can easily explain the various experimental results. For example, Penefsky reported in 1981 on the mechanism of a single-site catalytic reaction, showing that hydrolysis of ATP occurs very slowly in the presence of a stoichiometric amount of the substrate ATP, whereas hydrolysis of the first bound ATP to the enzyme molecule is accelerated 10^6 times when the other catalytic sites are occupied by the substrate (Grubmeyer & Penefsky, 1981a, 1981b). This indicates a strong cooperativity between the three catalytic sites in the F_1 complex. In 1996, our group also reported the importance of the γ subunit to catalytic activity (Kaibara, Matsui, Hisabori, & Yoshida, 1996). In this report, we compared the nature of the catalytic reaction catalyzed by the $\alpha_3\beta_3\gamma$ complex with that of the $\alpha_3\beta_3$ complex, and suggested that the structural asymmetry of the F_1-ATPase caused by the γ subunit produced a high-affinity nucleotide-binding site. Furthermore, we have clearly shown that the introduction of one ineligible β subunit into the $\alpha_3\beta_3\gamma$ complex results in a complete loss of catalytic activity, which was also a good evidence that three beta subunits work equivalently and alternately in the enzyme complex (Amano, Hisabori, Muneyuki, & Yoshida, 1996). Finally, Boyer's proposal was directly proven by single-molecule observation experiments reported in 1997, as previously described (Noji et al., 1997).

2.2 ADP inhibition

Biochemical and structural analyses revealed that bacterial and chloroplast F_1-ATPases are composed of five subunits with a stoichiometry of $\alpha_3\beta_3\gamma_1\delta_1\varepsilon_1$ (Yoshida et al., 1979). The α and β subunits show the similarity at the amino acid sequence level and also the molecular structure level (Hisabori, Ueoka-Nakanishi, Konno, & Koyama, 2003). The complex has six nucleotide-binding sites, three on each of three α subunits and three on each of three β subunits. Although the catalytic sites are located on the β subunit at the interface with the α subunit, the function of the nucleotide-binding sites on the α subunits had long been unknown. This question was finally addressed by Matsui et al. using thermophilic enzyme complexes (Matsui et al., 1997). By mutating the nucleotide-binding site on the α subunit, they created an incompetent α subunit that cannot bind nucleotides. When this mutant subunit was introduced into the $\alpha_3\beta_3\gamma$ complex, it was found that the complex could not catalyze successive ATP hydrolysis reactions. The structural effect on the catalytic β subunit caused by the α subunit, when a nucleotide binds to the α subunit and a conformational change is induced, was completely unknown. However, their results clearly showed that binding of a nucleotide to the α subunit releases the enzyme from the inhibitory state caused by the tightly bound ADP at the catalytic site.

Historically, this ADP-dependent inhibitory state was known as "ADP inhibition" (Shoshan & Strotmann, 1980; Strotmann, Bickel-Sandkotter, & Shoshan, 1979) (Fig. 1). In the case of chloroplast ATPases, 1 or 2 ADP molecules were found on the enzyme when the protein complex was purified from chloroplasts (Hisabori & Sakurai, 1985; Sunamura, Konno, Imashimizu-Kobayashi, Sugano, & Hisabori, 2010; Zhou, Xue, Du, Melese, & Boyer, 1988). When the complex was treated with a weak surfactant such as octyl glycoside (Pick & Finel, 1983) or lauryl dimethylamine oxide (LDAO) (Dunn, Tozer, & Zadorozny, 1990; Ren & Allison, 1997), this bound ADP was released from the F_1 complex, and ATP hydrolysis activity was greatly accelerated. Conversely, if the enzyme complex is incubated in the presence of ADP, the enzyme easily drops into the inhibited state.

After the first complete crystal structure information for the mitochondrial F_1 complex was provided by Abrahams et al. (1994), all crystal structures reported for the F_1 complex were found to indicate that the complex was in the ADP-inhibited state (Leslie et al., 1999). In particular, the orientation of the γ subunit in the complex is very important in relation to the three different conformational states of the β subunits. At about the same time, the catalytic process of ATP hydrolysis by this enzyme complex was visualized by the single-molecule observation method as described above. Next, the relationship between the conformations of the three β subunits in different states; empty, ADP bound, and ATP bound, and orientation of the γ subunit in the complex were analyzed (Adachi et al., 2007, 2000; Yasuda et al., 2001). By using a very low concentration of the substrate ATP, the stepwise rotation of the γ subunit was directly visualized and the ATP-waiting position of the γ subunit in the complex was assigned to 0°. Analysis of the rotation of the γ subunit from this position revealed that the ADP-binding position was −80° from the ATP-binding position. Thus, the one-step rotation was divided into an initial 80° rotation followed by a 40° rotation. The first 80° rotation occurs by the hydrolysis of ATP bound to the first catalytic site. At this 80° position, ATP is decomposed to ADP and Pi. Next, for ADP or Pi, the order in which these products are released from the catalytic site, and the order in which the catalytic process moves to the next catalytic site are still discussed. However, the enzyme complex stops at this initial stage when ADP was not released from the binding site on the β subunit. In fact, the γ subunit can be readily observed to stop at this 80° position during rotation analysis. Therefore, this initial 80° stopping is thought to be the position where the enzyme complex is dropped into the ADP inhibition.

2.3 ε Inhibition

In chloroplast and bacterial F_1-ATPases, the release of the ε subunit from the complex greatly enhances the ATP-hydrolyzing activity, and therefore it has been postulated that this subunit is a unique intrinsic inhibitory subunit (Richter & McCarty, 1987; Richter, Patrie, & McCarty, 1984) (Fig. 1). Crystal structure analysis of the bacterial F_1-ATPase complex revealed the molecular structure of the ε subunit; the N-terminal β-sandwich domain and the C-terminal α-helix-loop-α-helix domain are connected by a short loop (Uhlin, Cox, & Guss, 1997; Wilkens, Dahlquist, McIntosh, Donaldson, & Capaldi, 1995). The C-terminal domain showed the extended structure along to the coiled-coil structure of the γ subunit in the $\alpha_3\beta_3\gamma_1\varepsilon_1$ complex (Rodgers & Wilce, 2000; Shirakihara et al., 2015). In contrast, when the structure of the isolated ε subunit was solved by X-ray crystallography and NMR analyses, it was revealed that the C-terminal α-helices domain shows the retracted conformation (Wilkens & Capaldi, 1998; Yagi et al., 2010). In addition, a bound ATP on the C-terminal α-helices domain was found when the structure of the thermophilic F_1-ε subunit was analyzed by X-ray crystallography (Yagi et al., 2007). Based on this finding, the ε subunit was thought to be the sensor to check the intracellular concentrations of ATP in the bacterial cells, and therefore ATP synthesis reaction is controlled by the conformational change of this subunit (Kato-Yamada & Yoshida, 2003). The top of this extended C-terminal helices domain mostly reached to the so-called DELSEED sequence of the β subunit, which is the bottom region of the β subunit, and the possibility of the interaction between this C-terminus domain of the ε subunit and the DELSEED sequence of the β subunit was suggested. Then, the relation between the inhibitory function of the ε subunit and the partial interaction between the ε subunit and the β subunit was studied. This C-terminus domain derived from the thermophilic bacteria contained number of positive-charged residues, Lys and Arg, they were expected to be good candidate amino acids which can interact with the negatively charged residues reside on the DELSEED region of the β subunit. Indeed, the inhibitory effect of the ε subunit was gradually decreased when the multiple positive-charged residues on the C-terminal helices part was stepwisely substituted with noncharged amino acids (Hara, Kato-Yamada, Kikuchi, Hisabori, & Yoshida, 2001).

However, this interaction between the C-terminus domain of the ε subunit and the DELSEED sequence on the β subunit is not feasible in the case of cyanobacterial F_1. When the NMR structure of the ε subunit of thermophilic cyanobacterial F_1 was solved, it showed the so-called retracted conformation (Yagi et al., 2010). In addition, this retracted conformation of the C-terminus domain of the ε subunit was hardly released because many hydrogen bonds within two α-helices and also between the α-helices and the N-terminal domain were observed. This structure was maintained even in the F_1 complex of green plants and cyanobacteria as mentioned elsewhere.

3. Redox regulation via the redox switch on the γ subunit

3.1 Activation of CF_1 by reduction

It was well studied that the ATP hydrolysis activity (ATPase activity) of the isolated CF_1 from spinach chloroplasts is greatly accelerated by the incubation with the high concentrations (certain mM) of dithiothreitol (DTT) (Ketcham, Davenport, Warncke, & McCarty, 1984; Nalin & McCarty, 1984). The significance of the disulfide bond switch on the control of the enzyme activity was then investigated. For instance, Possmayer and Gräber studied the ATP synthesis by using CF_oCF_1 reconstituted in the proteoliposome (Possmayer & Gräber, 1994). Threshold of the size of the pH gradient required for the ATP synthesis was smaller when the enzyme was reduced in advance, suggesting the importance of the reduction of the redox switch on CF_1-γ (-EICDINGXC-) for the enzyme activity (Fig. 1). However, the physiological impact of this redox switch was not studied very well for long time. Ort et al. then performed the mutation analysis of this redox switch. For the purpose, they first randomly introduced the mutation into the seeds of *Arabidopsis thaliana* by using ethylmethanesulfonate and surveyed the mutant plants which could survive under the normal growth conditions, but not grow well under the weak light intensity conditions. By this way, they checked 100,000 seedlings and obtained 117 mutant plants, which showed the above tendencies (Gabrys, Kramer, Crofts, & Ort, 1994). Finally, they selected three mutant plants, which contain a certain mutation at around the redox switch of the γ subunit based on the electrochromic shift measurement. These mutations were supposed to disturb the reduction of the redox switch (Wu & Ort, 2008).

In contrast, Selman and his coworkers succeeded to prepare the mutant strain of the green algae *Chlamydomonas reinhardtii,* in which the γ subunit lacks the redox-regulation responsible cysteines (Ross, Zhang, & Selman, 1995). By the introduced mutation, the enzyme complex became insensitive to the environmental redox changes. However, they could not find any physiological disadvantages of this mutant strain. Hence, the physiological significance of the redox-regulation switch was not conclusively clarified.

In the case of cyanobacteria, the γ subunit originally lacks CF_1-specific 9-amino acid sequence including two redox-regulation responsible cysteines (-EICDINGXC-), which are located at the end of Rossman-fold structure (Hisabori et al., 2003). However, they still contain about 25 amino acid sequence at this redox-responsive region and this region is not observed in the γ subunit of non-photosynthetic prokaryotes and that of the mitochondrial F_1. To reveal the role of this additional region, we have prepared the mutant cyanobacterium *Synechocystis* sp. PCC6803, which lacks this additional 25 amino acid sequence from the γ subunit, and measured the change in the intracellular amounts of ATP during light/dark transition (Sunamura et al., 2010). In the dark conditions, the amounts of the intracellular ATP of the wild-type cells reached to almost the half of those in the cells kept in the light. In contrast, the dark-induced decrease of ATP was quite obvious when the γ subunit of *Synechocystis* cells lacked the unique 25 amino acids region as mentioned. Because the ATP hydrolysis activity of the ATP synthase complex of *Synechocystis* is greatly accelerated by the deletion of this phototroph-unique amino acid sequence region, the above results must imply that the lower intracellular amounts of ATP is the consequence of the vigorous ATP hydrolysis activity in the dark. Because the cyanobacterial thylakoid membrane possesses the respiratory electron transport system as well, the intracellular ATP level once dropped by the transition into the dark has recovered gradually even in the dark. In addition, ATP hydrolysis activity of the CF_1 complex which lacks the unique region was much higher than that of the wild-type CF_1 complex, and was not accelerated by the surfactant such as LDAO, indicating that the mutant enzyme is relieved from ADP inhibition state from the beginning. This suggests that the function of the unique 25 amino acids region is related to the induction of the ADP inhibition.

3.2 Redox regulation of rotation of the γ subunit of CF_1

As described above, direct observation of continuous counterclockwise rotation under an optical microscope using fluorescein-labeled actin filaments bound to the γ subunit of the thermophilic bacterial F_1 complex revealed that the γ subunit is the rotational axis of the F_1 complex (Hisabori, Kondoh, & Yoshida, 1999; Noji et al., 1997). Since oxidation of CF_1-γ results in a decrease in ATP hydrolysis activity, analysis of the change in rotation of the enzyme by oxidation and reduction is considered to be crucial for understanding the redox-regulation mechanism of this enzyme. However, there was a significant problem with the use of green plant CF_1 in such rotation experiments, in that it was not possible to prepare a recombinant CF_1 complex in, for example, *E. coli*. This was considered to be a prerequisite for introducing some of the amino acid mutations required for rotation experiments. In fact, our group has attempted to construct expression systems in *E. coli* for CF_1 complexes derived from green plants such as *Spinacia oleracea* and *Arabidopsis thaliana*, but all attempts have been unsuccessful. Furthermore, we could not construct even a CF_1 complex expression system of the most primitive eukaryotic photosynthetic organism, *Cyanidioschyzon merolae*.

To overcome this critical problem, we introduced the amino acid sequence containing the redox switch region of spinach CF_1 into the expression system of the $\alpha_3\beta_3\gamma$ complex of *G. stearothermophilus* (Bald, Noji, Stumpp, Yoshida, & Hisabori, 2000), previously constructed by Matsui and Yoshida (Matsui & Yoshida, 1995). After preparing the protein and confirming the redox regulation of ATP hydrolysis activity, we attempted to observe the redox regulation of rotation of this mutated γ subunit (Bald, Noji, Yoshida, Hirono-Hara, & Hisabori, 2001; Ueoka-Nakanishi et al., 2004). Indeed, the hybrid complexes showed smooth rotation under reducing conditions, but frequently stopped rotation under oxidizing conditions. However, the hybrid enzyme complex was unstable and its characteristic rotation properties could not be fully characterized.

The conserved redox switch region consisting of nine amino acids was then introduced into the counter part of the γ subunit of the cyanobacterial F_1 (Kim, Konno, Sugano, & Hisabori, 2011), since the expression system of the F_1 complex of the cyanobacterium was successfully constructed in *E. coli* using the gene obtained from the thermophilic cyanobacteria *Thermosynechococcus elongatus* BP-1, and the obtained complex was more stable (Konno et al., 2006). This way, we could confer the redox-regulation properties into the cyanobacterial ATPase complex. The mutant complex was then subjected to rotation experiments to determine its rotational properties under reducing or oxidizing conditions. The results showed that the low enzymatic activity of the oxidized complex was due to pauses during rotation of the γ subunit. The frequency of pauses was found to be more frequent than the rotation of the reducing enzyme, and the length of each pause was found to be longer whereas the period of continuous rotation was not affected very much by the change in redox situation. In addition, it was clarified that the rotational pause occurs at the position, where the enzyme complex falls into the ADP-inhibited state. Thus, the inactivation caused by the oxidation of the enzyme is thought to be the same situation as ADP inhibition.

3.3 Redox regulation of CF_1 by thioredoxin

Redox regulation of CF_1 in chloroplasts was reported in the late 1970, when Bob Buchanan suggested that thioredoxin is functioning in chloroplasts for the regulation of various enzyme activities, and mediates to transfer the reducing equivalents from the photosynthetic electron transport system to the target proteins (Buchanan, 1980; McKinney et al., 1978). Because the chloroplasts of green plants possess two major isoforms of thioredoxin, *f*-type and *m*-type, their specificities were the target of interest for long time. The *f*-type thioredoxin was named because this type of thioredoxin specifically reduces fluctose-1,6-bis phosphatase in the Calvin-Benson cycle, and the *m*-type was named because it was working for reduction of malate dehydrogenase (Buchanan, 1980). Although these names are simply derived based on the results of the in vitro experiments, the specificities of chloroplast thioredoxin isoforms were not studied very well in vivo. Interestingly, the *f*-type thioredoxin must be the eukaryotic origin and *m*-type is the prokaryotic origin based on their sequence similarities.

Schwarz et al. studied thioredoxin-dependent reduction of CF_1 isolated from spinach chloroplasts (Schwarz, Schurmann, & Strotmann, 1997). Based on the kinetic analysis of the activation process of CF_1, they concluded that the *f*-type thioredoxin must be the major reductant. In contrast, we recently studied the in vivo reduction process of CF_1-γ by using the CF_1-γ specific antibody and the direct redox status observation method using AMS as described below (Sekiguchi et al., 2020). We prepared both the *f*-type thioredoxin-deficient mutant *Arabidopsis* strain and the *m*-type thioredoxin-deficient strain. Both strains showed similar reduction kinetics of CF_1-γ as that of the wild-type plants in the light, implying that both thioredoxin isoforms can reduce CF_1-γ on the thylakoid membranes. This implies that CF_1 does not show so critical specificity against thioredoxin isoforms on the thylakoid membranes (Fig. 1).

3.4 Redox regulation of CF_1 in vivo

As mentioned above, the physiological significance of redox regulation of CF_1 is very mysterious. In addition, there were no direct biochemical evidence, which shows oxidation of CF_1 in the dark or reduction in the light in plant leaves. Kramer et al. then estimated the reduction level of CF_1 in sunflower leaves using the flash-induced electrochromic shift measurement at 515 nm (Kramer et al., 1990), which was found as a linear indicator of the $\Delta\psi$ component of the proton-motive force (Kramer & Crofts, 1989). By using the decay of this flash-induced electrochromic shift value, they could estimate the activation ratio of CF_1, namely the reduced form of CF_1 under different light conditions. Consequently, they reported the immediate reduction of CF_1 with the very weak light at around $22\,\mu E\,m^{-2}\,s^{-1}$ by the flash-induced electrochromic shift measurement, though they did not show the reduction of the disulfide bond on the γ subunit directly.

To observe oxidation or reduction of thiol residues on the protein, thiol-specific chemical modification method was recognized to be the very powerful tool. Kobayashi et al. used 4-acetamido-4′-maleimidylstilbene-

2,2′-disulfonate (AMS) for thiol modification and successfully separated the reduced form protein containing two thiol residues, which were labeled with AMS, and the oxidized one with the disulfide bond by the nonreducing SDS-PAGE (Kobayashi et al., 1997). The method was successfully applied to the observation of the redox state of CF_1-γ in the leaves. By using this method, we have observed the redox change in the γ subunit in the spinach leaves at the outdoor from 2:30 am for 24 h (Konno et al., 2012). Indeed, the reduction of the γ subunit was observed just after the sunrise at 6:00 am, and it reached to the plateau level of about 80% reduction in total already at around 9:00 am, whereas the reduction of FBPase, well-characterized thiol-regulated enzyme working at Calvin-Benson cycle, was linearly proceeded according to the increase of the light intensity after sunrise. The light intensity was gradually decreased after 14:00 pm on that day, and the sunset time was 17:30 pm. Interestingly, the oxidation of FBPase was already initiated at around 15:00 pm, when the light intensity was still enough strong, and reached to the full-oxidation level before the sunset. In contrast, the oxidation of CF_1-γ subunit was not observed until the very last period before the sunset (Konno et al., 2012). These tendencies were also confirmed by the leaves of *A. thaliana* under the laboratory light conditions (Yoshida, Matsuoka, Hara, Konno, & Hisabori, 2014). Again, this implies that the reduction and the oxidation of CF_1-γ show very different manner from that of FBPase and other redox-regulated enzymes.

In addition, we realized that the reduction of CF_1-γ requested the formation of the proton gradient across thylakoid membranes (Konno et al., 2012) (Fig. 1). When the uncoupler FCCP was applied to the illuminated thylakoid membranes, the reduction of CF_1-γ was not observed even under the light conditions. When FCCP was added to the thylakoid membranes after when the part of CF_1-γ was reduced by the light illumination, immediate oxidation of CF_1-γ was observed. These results strongly suggest that the mode of the reduction and oxidation process of the CF_1-γ is very different from those of the thioredoxin-regulated stroma enzymes, such as the member of Calvin-Benson cycle. Furthermore, the existence of an oxidation factor in chloroplasts even under the light conditions was strongly suggested from this observation, though we could not have any idea for this factor at that time. Therefore, reduction/oxidation dynamics of CF_1-γ in chloroplasts must be studied in light of the relation with the membrane potential formation in future. In this book, the physiological significance of the cyclic electron transport system, which largely responsible for the ATP synthase activity, is discussed by Chikahiro Miyake et al. in chapter entitled "P700 oxidation suppresses the production of reactive oxygen species in photosystem I" and by Toshiharu Shikanai in chapter entitled "Regulation of photosynthesis by cyclic electron transport around photosystem I."

We recently found the important protein which can mediate the oxidation of the reduced proteins in chloroplasts, which are reduced by thioredoxin under the light conditions (Yoshida, Hara, Sugiura, Fukaya, & Hisabori, 2018; Yoshida, Uchikoshi, Hara, & Hisabori, 2019). Thioredoxin-like protein 2, Trx-L2, was revealed as one of the important proteins for this oxidation process. The midpoint redox potential of Trx-L2 was much higher than those of other redox-mediator proteins in chloroplasts such as various Trx isoforms, and therefore this protein cannot assist the reduction of Trx-target proteins in the presence of DTT, for example. However, we clearly showed the oxidation of the reduced forms of chloroplast localized, thiol-regulated proteins by Trx-L2, although the experiments were carried out in vitro. When the reduced forms of these proteins are oxidized by the transfer of reducing equivalents to Trx-L2, the reduced form of Trx-L2 then transfers them to H_2O_2 via 2-Cys peroxiredoxin. Because H_2O_2 is always produced in chloroplasts as a consequence of the excess electron transfer and superoxide dismutase reaction, it will be a good final acceptor of the reducing equivalents mediated by the redox-regulation system in chloroplasts (Ojeda, Perez-Ruiz, & Cejudo, 2018; Vaseghi et al., 2018; Yoshida et al., 2018). Because there might be several proteins in chloroplasts which may have a similar function with Trx-L2, we need more information to describe the whole image of oxidation process in chloroplasts (Yokochi et al., 2019).

3.5 Structure analysis of the γ subunit of CF_1

As mentioned, the partial structure of ATP synthase was reported for the mitochondrial ATP synthase by the X-ray crystallography analysis in 1994 (Abrahams et al., 1994). The structure of the αβ-εsubcomplex of CF_1 was then reported in 2001 by Groth and Pohl (2001). However, they could not observe the structure of the γ subunit at that time due to the asymmetric positioning of the γ subunit in the crystal lattice. Since then, a lot of the ATP synthase structures were reported mainly for bacterial and mitochondrial enzymes. Advances of the structural analyses of CF_1 and $CFoCF_1$ are well summarized in Chapters 2 and 3. In 2004, we have reported the NMR structure of the ε subunit of the cyanobacterial ATP synthase (Yagi et al., 2010). This report was the very rare case of the report on the structure of the ATP synthase subunit of phototrophs so far. Because we have succeeded to construct the expression system of the $\alpha_3\beta_3\gamma$ complex of the thermophilic cyanobacterial ATP synthase in 2006 (Konno et al., 2006), we thought this protein complex will be a good resource for the crystal structure analysis because of its higher stability. Indeed, our group gave a lot of efforts for this work for more than10years. However, we could not obtain the suitable crystals from this protein complex at all, and finally we changed the target for the crystallization analysis to the partial subunit complex of the γ- of cyanobacterial CF_1, which was once solved by Rodgers et al. for *E. coli* F_1. Finally, we could obtain the atomic resolution structure of the co-complex of the γ subunit and the ε subunit at 1.98Å (Murakami et al., 2018). The most important structure in the γ subunit was thought to be the β-hairpin structure of the so-called inserted sequence region only observed in phototrophs. Mostly at the similar timing, the German group reported the whole structure of CF_oCF_1 using the cryo-electron microscopy method with the resolution of 2.4Å for CF_1 and 3.6Å for whole CF_oCF_1 (Hahn et al., 2018) (Fig. 1). In this structure, the β-hairpin structure (Fig. 1, shown as red) is clearly indicated with the second loop structure containing the disulfide bond at the root position of the γ subunit.

3.6 Regulation mechanism governed by β-hairpin structure of the γ subunit

Although our structure only revealed the part of the γ subunit and the whole ε subunit, interface between the γ and ε subunits was very well revealed (Murakami et al., 2018). Again, the C-terminal α-helices region of the ε subunit was shown as a retracted conformation, and the ε subunit structure in the CF_oCF_1 shown by the cryo-EM structure was very similar with ours (Hahn et al., 2018). Very interesting finding from our structure was that the β-strands located at the bottom of the γ subunit form the mixed β-sheet structure with the two β-strands of the N-terminal β-sandwich structure of the ε subunit. This implies that the interaction between these two subunits must be tightly coupled and the binding of the ε subunit to the γ subunit may induce the conformational change of the γ subunit simultaneously (Murakami et al., 2018). Another interaction position of the γ subunit in the complex was suggested to be the top of the β-hairpin structure, which must reach to the so-called DELSEED region of the open form of the β subunit. Because the contribution of this DELSEED region for the ε inhibition (Hara et al., 2001) and for the ADP inhibition is already reported (Sunamura, Konno, Imashimizu, Mochimaru, & Hisabori, 2012), we examined how the interaction between the top of the β-hairpin structure of the γ subunit and the DELSEED region of the β subunit affects the activity or the regulation of this enzyme complex. To confirm the significance of this interaction, we have deleted the top of the β-hairpin structure and studied the change in the ATP hydrolysis activity thoroughly. As a result, we found only two amino acids deletion from the top of the β-hairpin structure has already released the enzyme from ADP inhibition state (Murakami et al., 2018). This result is in harmony with our former report that the deletion of the inserted sequence region of the γ subunit of *Synechocystis* ATP synthase results the higher ATP hydrolysis activity and then the amounts of the intracellular ATP drastically decreased in the dark (Sunamura et al., 2010).

From the solved crystal structure of the γ-ε co-complex, we realized that the thermal factor of the lower part of the β-hairpin structure is rather high, indicating the flexibility of this structure in the complex. We, therefore, investigated the possible movement of the structure in the enzyme complex (Akiyama et al., 2019). We then have chosen two positions where β-hairpin structure of the γ subunit can be fixed to the C-terminal side α-helix region of the central helices structure by the introduction of Cys residue for formation of the disulfide bond. In addition, two Cys residues were introduced at the root position of the β-hairpin structure, which may fix the potential movement of this structure.

This way, we have successfully introduced the intramolecular disulfide bond under the oxidizing conditions to suppress the possible movement of the β-hairpin structure. When the β-hairpin structure was fixed to the central axis of the γ subunit, the level of ADP inhibition was largely suppressed implying the certain interaction between the β-hairpin structure possibly with the DELSEED region of the β subunit induces ADP inhibition. In contrast, the formation of the disulfide bond at the bottom part of the β-hairpin structure was disturbed in the presence of the ε subunit, implying that the ε binding somehow induce the conformational change at the bottom part of the γ subunit as mentioned.

4. Function of the CF_1 ε subunit

4.1 Intrinsic inhibitor protein for ATP hydrolysis activity

In CF_1, the ε subunit is known as an intrinsic inhibitor protein (Richter et al., 1984). Amino acid sequences of the bacterial and chloroplast ATP synthase ε subunits are well conserved and both of them consist of the N-terminus β-sandwich structure and the C-terminus two α-helices domain connected by loop region as mentioned. Although the large conformational change of the C-terminus two α-helices domain was reported in the case of prokaryotic bacterial F_1 complex (Kato-Yamada, Yoshida, & Hisabori, 2000; Yagi et al., 2007), the NMR structure of the CF_1 ε subunit was thought to be very stable as a retracted conformation caused by the formation of multiple hydrogen bonds within two α-helices (Yagi et al., 2010). In addition, the retracted form α-helices part is also fixed on the surface of the β-sandwich structure by hydrogen bonds.

By using the recombinant cyanobacterial ATP synthase complex $\alpha_3\beta_3\gamma$ of *T. elongatus* BP-1, we have studied the function of the ε subunit on the ATP hydrolysis activity (Konno et al., 2006, 2011). The ε subunit strongly inhibited the ATP hydrolysis activity with the K_d at around 2nM. Interestingly, this inhibitory function was not so remarkable when the insertion sequence of the γ subunit (from 198 to 222 in the case of the γ subunit of *T. elongatus* BP-1) was deleted, implying the inhibition caused by the ε subunit binding may induce a certain conformational change of the γ subunit as well (Konno et al., 2006). In addition, this recombinant cyanobacterial ATP synthase complex enabled us to analyze the inhibition phenomena at the single-molecule level. When the rotation of the γ subunit was inhibited by the ε subunit, the γ subunit stopped at the similar position where the enzyme was inhibited by the tight binding of ADP, 80° apart from the ATP-waiting position as mentioned. This implies that the ADP inhibition state is the very stable inhibited form of the ATP synthase complex, and it is often observed when rotation of this enzyme complex stopped.

We, therefore, studied the difference of the inhibition states caused by tightly ADP binding (ADP inhibition) and by the ε subunit binding (ε inhibition) (Konno et al., 2011). When stiffness of the inhibition was assessed by the magnetic tweezer method, which can forcedly rotate the inhibited γ subunit to the certain direction, it was revealed that the ε inhibition was much tighter than ADP inhibition. We, therefore, concluded that these two inhibition modes are structurally different from each other although the stop position of the γ subunit in the $\alpha_3\beta_3$ hexagon is almost same.

4.2 The interaction between the γ subunit and the ε subunit

The relevance between the redox state of the γ subunit and the binding of the ε subunit on CF_1 is well summarized in the review (Richter, Hein, &

Huchzermeyer, 2000). To understand the mode of the ε binding to the γ subunit thoroughly, we have prepared a couple of the ε subunit mutant proteins, and applied them to examine if they can inhibit the ATP hydrolysis activity of the $\alpha_3\beta_3\gamma$ complex of thermophilic cyanobacteria (Inabe, Kondo, Yoshida, Wakabayashis, & Hisabori, 2019). We found that only the N-terminal β-sandwich structural domain of the ε subunit can inhibit ATP hydrolysis activity whereas the binding affinity of this partial domain of the ε subunit was much lower than the wild-type protein. To confirm the inhibitory effect of the binding of this N-terminal β-sandwich structural domain, we then prepared two mutant ε subunits, in which the movement or the conformational change of the C-terminal α-helices domain is stabilized by the formation of the disulfide bond within C-terminal two α-helices, or by the formation of the disulfide bond between the C-terminal α-helix part and the β-sandwich structure. Again, both mutant ε subunits inhibit the ATP hydrolysis activity with the almost same extent and the similar affinity with those of the wild-type ε subunit. This means that only the interaction between the N-terminal region of the ε subunit and the γ subunit is critical to induce the ε subunit-assisted inhibition of the ATP hydrolysis activity. As mentioned, we have found the β-sheet structure at the bottom of the γ subunit, which is formed by the two β-strands from the ε subunit and others from the γ subunit, implying the γ-ε interaction certainly induces the conformational change of the whole complex.

5. Conclusion and future perspective

We are now mainly focusing on the two subunits of CF_1, the γ and ε subunits as the regulatory apparatus of this enzyme complex. Both subunits are located at the interface between CF_1 and CF_o. In addition, it is revealed that the disulfide switch of the γ subunit is positioning at the very bottom of the γ subunit, and the γ-ε interaction is also strengthened by the formation of the mixed β-sheet formation as mentioned. Because these areas are also the interface between CF_1 complex and the *c*-ring structure of CF_o, these regulation apparatuses must be strongly influenced by the conformation or the movement of *c*-ring as well. Although we still do not understand the molecular dynamics of this interesting complex during rotation movement caused by the proton transfer at the *c*-ring and the *a* subunit at all, understanding of the molecular dynamics at this area must be very important to clarify the regulation of this enzyme complex under the fluctuating light conditions, which might produce various electrochemical conditions at thylakoid membranes.

Acknowledgments

T.H. is supported by JSPS KAKENHI Grant Number 16H06556 and by Dynamic Alliance for Open Innovation Bridging Human, Environment and Materials.

References

Abrahams, J., Leslie, A., Lutter, R., & Walker, J. (1994). Structure at 2.8 Å resolution of F_1-ATPase from bovine heart mitochondria. *Nature*, *370*, 621–628.

Adachi, K., Oiwa, K., Nishizaka, T., Furuike, S., Noji, H., Itoh, H., et al. (2007). Coupling of rotation and catalysis in F_1-ATPase revealed by single-molecule imaging and manipulation. *Cell*, *130*, 309–321.

Adachi, K., Yasuda, R., Noji, H., Itoh, H., Harada, Y., Yoshida, M., et al. (2000). Stepping rotation of F_1-ATPase visualized through angle-resolved single-fluorophore imaging. *Proceedings of the National Academy of Sciences of the United States of America*, *97*, 7243–7247.

Akiyama, K., Kondo, K., Inabe, K., Murakami, S., Wakabayashi, K., & Hisabori, T. (2019). The beta-hairpin region of the cyanobacterial F_1-ATPase γ-subunit plays a regulatory role in the enzyme activity. *Biochemical Journal*, *476*, 1771–1780.

Amano, T., Hisabori, T., Muneyuki, E., & Yoshida, M. (1996). Catalytic activities of $\alpha_3\beta_3\gamma$ complexes of F_1-ATPase with 1, 2, or 3 incompetent catalytic sites. *The Journal of Biological Chemistry*, *271*, 18128–18133.

Avron, M. (1963). A coupling factor in photophosphorylation. *Biochimica et Biophysica Acta*, *77*, 699–702.

Bald, D., Noji, H., Stumpp, M. T., Yoshida, M., & Hisabori, T. (2000). ATPase activity of a highly stable $\alpha_3\beta_3\gamma$ subcomplex of thermophilic F_1 can be regulated by the introduced regulatory region of γ subunit of chloroplast F_1. *The Journal of Biological Chemistry*, *275*, 12757–12762.

Bald, D., Noji, H., Yoshida, M., Hirono-Hara, Y., & Hisabori, T. (2001). Redox regulation of the rotation of F_1-ATP synthase. *The Journal of Biological Chemistry*, *276*, 39505–39507.

Binda, C., Coda, A., Aliverti, A., Zanetti, G., & Mattevi, A. (1998). Structure of the mutant E92K of [2Fe-2S] ferredoxin I from Spinacia oleracea at 1.7 Å resolution. *Acta Crystallographica Section D, Biological Crystallography*, *54*, 1353–1358.

Boyer, P. D. (1997). The ATP synthase–a splendid molecular machine. *Annual Review of Biochemistry*, *66*, 717–749.

Buchanan, B. B. (1980). Role of light in the regulation of chloroplast enzymes. *Annual Review of Plant Physiology and Plant Molecular Biology*, *31*, 341–374.

Buchanan, B. B., & Wolosiuk, R. A. (1976). Photosynthetic regulatory protein found in animal and bacterial cells. *Nature*, *264*, 669–670.

Capitani, G., Markovic-Housley, Z., DelVal, G., Morris, M., Jansonius, J. N., & Schurmann, P. (2000). Crystal structures of two functionally different thioredoxins in spinach chloroplasts. *Journal of Molecular Biology*, *302*, 135–154.

Dai, S., Friemann, R., Glauser, D. A., Bourquin, F., Manieri, W., Schurmann, P., et al. (2007). Structural snapshots along the reaction pathway of ferredoxin-thioredoxin reductase. *Nature*, *448*, 92–96.

Duncan, T. M., Bulygin, V. V., Zhou, Y., Hutcheon, M. L., & Cross, R. L. (1995). Rotation of subunits during catalysis by *Escherichia coli* F_1-ATPase. *Proceedings of the National Academy of Sciences of the United States of America*, *92*, 10964–10968.

Dunn, S. D., Tozer, R. G., & Zadorozny, V. D. (1990). Activation of *Escherichia coli* F_1-ATPase by lauryldimethylamine oxide and ethylene glycol: Relationship of ATPase activity to the interaction of the ε and β subunits. *Biochemistry*, *29*, 4335–4340.

Feng, Y., & McCarty, R. E. (1990). Purification and reconstitution of active chloroplast F_0. *The Journal of Biological Chemistry*, *265*, 5104–5109.

Ferguson, S. A., Cook, G. M., Montgomery, M. G., Leslie, A. G., & Walker, J. E. (2016). Regulation of the thermoalkaliphilic F_1-ATPase from *Caldalkalibacillus thermarum*. *Proceedings of the National Academy of Sciences of the United States of America*, *113*, 10860–10865.

Futai, M., & Kanazawa, H. (1983). Structure and function of proton-translocating adenosine triphosphatase (F_0F_1): Biochemical and molecular biological approaches. *Microbiological Reviews*, *47*, 285–312.

Gabrys, H., Kramer, D. M., Crofts, A. R., & Ort, D. R. (1994). Mutants of chloroplast coupling factor reduction in *Arabidopsis*. *Plant Physiology*, *104*, 769–776.

Gresser, M. J., Myers, J. A., & Boyer, P. D. (1982). Catalytic site cooperativity of beef heart mitochondrial F_1 adenosine triphosphatase. Correlations of initial velocity, bound intermediate, and oxygen exchange measurements with an alternating three-site model. *The Journal of Biological Chemistry*, *257*, 12030–12038.

Groth, G., & Pohl, E. (2001). The structure of the chloroplast F_1-ATPase at 3.2 Å resolution. *The Journal of Biological Chemistry*, *276*, 1345–1352.

Grubmeyer, C., & Penefsky, H. S. (1981a). Cooperatively between catalytic sites in the mechanism of action of beef heart mitochondrial adenosine triphosphatase. *The Journal of Biological Chemistry*, *256*, 3728–3734.

Grubmeyer, C., & Penefsky, H. S. (1981b). The presence of two hydrolytic sites on beef heart mitochondrial adenosine triphosphatase. *The Journal of Biological Chemistry*, *256*, 3718–3727.

Hackney, D. D., Rosen, G., & Boyer, P. D. (1979). Subunit interaction during catalysis: Alternating site cooperativity in photophosphorylation shown by substrate modulation of [^{18}O]ATP species formation. *Proceedings of the National Academy of Sciences of the United States of America*, *76*, 3646–3650.

Hahn, A., Vonck, J., Mills, D. J., Meier, T., & Kuhlbrandt, W. (2018). Structure, mechanism, and regulation of the chloroplast ATP synthase. *Science*, *360*, eaat4318.

Hara, K. Y., Kato-Yamada, Y., Kikuchi, Y., Hisabori, T., & Yoshida, M. (2001). The role of the βDELSEED motif of F_1-ATPase: Propagation of the inhibitory effect of the ε subunit. *The Journal of Biological Chemistry*, *276*, 23969–23973.

Hisabori, T., Kondoh, A., & Yoshida, M. (1999). The γ subunit in chloroplast F_1-ATPase can rotate in a unidirectional and counter-clockwise manner. *FEBS Letters*, *463*, 35–38.

Hisabori, T., & Sakurai, H. (1985). Effect of methanol on the nucleotide binding to high-affinity sites on chloroplast coupling factor 1. *Plant and Cell Physiology*, *26*, 505–514.

Hisabori, T., Ueoka-Nakanishi, H., Konno, H., & Koyama, F. (2003). Molecular evolution of the modulator of chloroplast ATP synthase: Origin of the conformational change dependent regulation. *FEBS Letters*, *545*, 71–75.

Inabe, K., Kondo, K., Yoshida, K., Wakabayashis, K., & Hisabori, T. (2019). The N-terminal region of the ε subunit from cyanobacterial ATP synthase alone can inhibit ATPase activity. *Journal of Biological Chemistry*, *294*, 10094–10103.

Jagendorf, A. T., & Uribe, E. (1966). ATP formation caused by acid-base transition of spinach chloroplasts. *Proceedings of the National Academy of Sciences of the United States of America*, *55*, 170–177.

Jiang, W., Hermolin, J., & Fillingame, R. H. (2001). The preferred stoichiometry of c subunits in the rotary motor sector of *Escherichia coli* ATP synthase is 10. *Proceedings of the National Academy of Sciences of the United States of America*, *98*, 4966–4971.

Kagawa, Y., Sone, N., Yoshida, M., Hirata, H., & Okamoto, H. (1976). Proton translocating ATPase of a thermophilic bacterium. Morphology, subunits, and chemical composition. *Journal of Biochemistry*, *80*, 141–151.

Kaibara, C., Matsui, T., Hisabori, T., & Yoshida, M. (1996). Structural asymmetry of F_1-ATPase caused by the γ subunit generates a high affinity nucleotide binding site. *The Journal of Biological Chemistry*, *271*, 2433–2438.

Kato-Yamada, Y., Noji, H., Yasuda, R., Kinosita, K., Jr., & Yoshida, M. (1998). Direct observation of the rotation of ε subunit in F_1-ATPase. *The Journal of Biological Chemistry*, *273*, 19375–19377.

Kato-Yamada, Y., & Yoshida, M. (2003). Isolated ε subunit of thermophilic F_1-ATPase binds ATP. *The Journal of Biological Chemistry*, *278*, 36013–36016.

Kato-Yamada, Y., Yoshida, M., & Hisabori, T. (2000). Movement of the helical domain of the ε subunit is required for the activation of thermophilic F_1-ATPase. *The Journal of Biological Chemistry*, *275*, 35746–35750.

Ketcham, S. R., Davenport, J. W., Warncke, K., & McCarty, R. E. (1984). Role of the γ subunit of chloroplast coupling factor 1 in the light-dependent activation of photophosphorylation and ATPase activity by dithiothreitol. *The Journal of Biological Chemistry*, *259*, 7286–7293.

Kim, Y., Konno, H., Sugano, Y., & Hisabori, T. (2011). Redox regulation of rotation of the cyanobacterial F_1-ATPase containing thiol regulation switch. *The Journal of Biological Chemistry*, *286*, 9071–9078.

Kobayashi, T., Kishigami, S., Sone, M., Inokuchi, H., Mogi, T., & Ito, K. (1997). Respiratory chain is required to maintain oxidized states of the DsbA-DsbB disulfide bond formation system in aerobically growing *Escherichia coli* cells. *Proceedings of the National Academy of Sciences of the United States of America*, *94*, 11857–11862.

Konno, H., Isu, A., Kim, Y., Murakami-Fuse, T., Sugano, Y., & Hisabori, T. (2011). Characterization of the relationship between ADP- and ε-induced inhibition in cyanobacterial F_1-ATPase. *The Journal of Biological Chemistry*, *286*, 13423–13429.

Konno, H., Murakami-Fuse, T., Fujii, F., Koyama, F., Ueoka-Nakanishi, H., Pack, C. G., et al. (2006). The regulator of the F_1 motor: Inhibition of rotation of cyanobacterial F_1-ATPase by the ε subunit. *The EMBO Journal*, *25*, 4596–4604.

Konno, H., Nakane, T., Yoshida, M., Ueoka-Nakanishi, H., Hara, S., & Hisabori, T. (2012). Thiol modulation of the chloroplast ATP synthase is dependent on the energization of thylakoid membranes. *Plant & Cell Physiology*, *53*, 626–634.

Kramer, D. M., & Crofts, A. R. (1989). Activation of the chloroplast ATPase measured by the electrochromic change in leaves of intact plants. *Biochimica et Biophysica Acta*, *976*, 28–41.

Kramer, D. M., Wise, R. R., Frederick, J. R., Alm, D. M., Hesketh, J. D., Ort, D. R., et al. (1990). Regulation of coupling factor in field-grown sunflower: A redox model relating coupling factor activity to the activities of other thioredoxin-dependent chloroplast enzymes. *Photosynthesis Research*, *26*, 213–222.

Leslie, A. G., Abrahams, J. P., Braig, K., Lutter, R., Menz, R. I., Orriss, G. L., et al. (1999). The structure of bovine mitochondrial F_1-ATPase: An example of rotary catalysis. *Biochemical Society Transactions*, *27*, 37–42.

Matsui, T., Muneyuki, E., Honda, M., Allison, W. S., Dou, C., & Yoshida, M. (1997). Catalytic activity of the $\alpha_3\beta_3\gamma$ complex of F_1-ATPase without noncatalytic nucleotide binding site. *The Journal of Biological Chemistry*, *272*, 8215–8221.

Matsui, T., & Yoshida, M. (1995). Expression of the wild-type and the Cys-/Trp-less $\alpha_3\beta_3\gamma$ complex of thermophilic F_1-ATPase in *Escherichia coli*. *Biochimica et Biophysica Acta*, *1231*, 139–146.

Matthey, U., Kaim, G., Braun, D., Wuthrich, K., & Dimroth, P. (1999). NMR studies of subunit c of the ATP synthase from *Propionigenium modestum* in dodecylsulphate micelles. *European Journal of Biochemistry*, *261*, 459–467.

Mazor, Y., Borovikova, A., Caspy, I., & Nelson, N. (2017). Structure of the plant photosystem I supercomplex at 2.6 Å resolution. *Nature Plants*, *3*, 17014.

McKinney, D. W., Buchanan, B. B., & Wolosiuk, R. A. (1978). Activation of chloroplast ATPase by reduced thioredoxin. *Phytochemistry*, *17*, 794–795.

Miki, J., Maeda, M., Mukohata, Y., & Futai, M. (1988). The γ-subunit of ATP synthase from spinach chloroplasts. Primary structure deduced from the cloned cDNA sequence. *FEBS Letters*, *232*, 221–226.

Mills, J. D., & Mitchell, P. (1982). Modulation of coupling factor ATPase activity in intact chloroplasts. Reversal of thiol modulation in the dark. *Biochimica et Biophysica Acta, 679*, 75–83.

Mitchell, P. (1966). Chemiosmotic coupling in oxidative and photosynthetic phosphorylation. *Biological Reviews of the Cambridge Philosophical Society, 41*, 445–502.

Mitome, N., Suzuki, T., Hayashi, S., & Yoshida, M. (2004). Thermophilic ATP synthase has a decamer c-ring: Indication of noninteger 10:3 H^+/ATP ratio and permissive elastic coupling. *Proceedings of the National Academy of Sciences of the United States of America, 101*, 12159–12164.

Murakami, S., Kondo, K., Katayama, S., Hara, S., Sunamura, E. I., Yamashita, E., et al. (2018). Structure of the γ-ε complex of cyanobacterial F_1-ATPase reveals a suppression mechanism of the γ subunit on ATP hydrolysis in phototrophs. *Biochemical Journal, 475*, 2925–2939.

Nalin, C. M., & McCarty, R. E. (1984). Role of a disulfide bond in the γ subunit in activation of the ATPase of chloroplast coupling factor 1. *The Journal of Biological Chemistry, 259*, 7275–7280.

Nelson, N., Nelson, H., & Schatz, G. (1980). Biosynthesis and assembly of the proton-translocating adenosine triphosphatase complex from chloroplasts. *Proceedings of the National Academy of Sciences of the United States of America*, 77, 1361–1364.

Noji, H., Yasuda, R., Yoshida, M., & Kinosita, K. Jr. (1997). Direct observation of the rotation of F_1-ATPase. *Nature, 386*, 299–302.

Ojeda, V., Perez-Ruiz, J. M., & Cejudo, F. J. (2018). 2-cys peroxiredoxins participate in the oxidation of chloroplast enzymes in the dark. *Molecular Plant, 11*, 1377–1388.

Pick, U., & Finel, M. (1983). Detection of conformational changes in chloroplast coupling factor 1 by 8-anilino-1-naphthalene-sulphonate fluorescence changes. *European Journal of Biochemistry, 135*, 559–567.

Possmayer, F. E., & Graber, P. (1994). The pH_{in} and pH_{out} dependence of the rate of ATP synthesis catalyzed by the chloroplast H^+-ATPase, CF_0F_1, in proteoliposomes. *The Journal of Biological Chemistry, 269*, 1896–1904.

Pullman, M. E., Penefsky, H. S., Datta, A., & Racker, E. (1960). Partial resolution of the enzymes catalyzing oxidative phosphorylation. I. Purification and properties of soluble dinitrophenol-stimulated adenosine triphosphatase. *The Journal of Biological Chemistry, 235*, 3322–3329.

Ren, H. M., & Allison, W. S. (1997). Photoinactivation of the F_1-ATPase from spinach chloroplasts by dequalinium is accompanied by derivatization of methionine β183. *The Journal of Biological Chemistry, 272*, 32294–32300.

Richter, M. L., Hein, R., & Huchzermeyer, B. (2000). Important subunit interactions in the chloroplast ATP synthase. *Biochimica et Biophysica Acta, 1458*, 326–342.

Richter, M. L., & McCarty, R. E. (1987). Energy-dependent changes in the conformation of the ε subunit of the chloroplast ATP synthase. *The Journal of Biological Chemistry, 262*, 15037–15040.

Richter, M. L., Patrie, W. J., & McCarty, R. E. (1984). Preparation of the ε subunit and ε subunit-deficient chloroplast coupling factor 1 in reconstitutively active forms. *The Journal of Biological Chemistry, 259*, 7371–7373.

Rodgers, A. J., & Wilce, M. C. (2000). Structure of the γ-ε complex of ATP synthase. *Nature Structural Biology*, 7, 1051–1054.

Ross, S. A., Zhang, M. X., & Selman, B. R. (1995). Role of the *Chlamydomonas reinhardtii* coupling factor 1 γ-subunit cysteine bridge in the regulation of ATP synthase. *The Journal of Biological Chemistry, 270*, 9813–9818.

Sabbert, D., Engelbrecht, S., & Junge, W. (1996). Intersubunit rotation in active F-ATPase. *Nature, 381*, 623–625.

Sambongi, Y., Iko, Y., Tanabe, M., Omote, H., Iwamoto-Kihara, A., Ueda, I., et al. (1999). Mechanical rotation of the c subunit oligomer in ATP synthase (F_0F_1): Direct observation. *Science, 286*, 1722–1724.

Schwarz, O., Schurmann, P., & Strotmann, H. (1997). Kinetics and thioredoxin specificity of thiol modulation of the chloroplast H^+-ATPase. *The Journal of Biological Chemistry, 272*, 16924–16927.

Sekiguchi, T., Yoshida, K., Okegawa, Y., Motohashi, K., Wakabayashi, K., & Hisabori, T. (2020). Chloroplast ATP synthase is reduced by both *f*-type and *m*-type thioredoxins. *Biochimica et Biophysica Acta-Bioenergetics, 1861*, 148261.

Shirakihara, Y., Shiratori, A., Tanikawa, H., Nakasako, M., Yoshida, M., & Suzuki, T. (2015). Structure of a thermophilic F_1-ATPase inhibited by an ε-subunit: Deeper insight into the ε-inhibition mechanism. *The FEBS Journal, 282*, 2895–2913.

Shoshan, V., & Strotmann, H. (1980). The effect of phosphate on light-induced exchange of ADP at the tight nucleotide binding site of CF_1. *The Journal of Biological Chemistry, 255*, 996–999.

Strotmann, H., Bickel-Sandkotter, S., & Shoshan, V. (1979). Kinetic analysis of light-dependent exchange of adenine nucleotides on chloroplast coupling factor CF_1. *FEBS Letters, 101*, 316–320.

Sunamura, E., Konno, H., Imashimizu, M., Mochimaru, M., & Hisabori, T. (2012). A conformational change of the γ subunit indirectly regulates the activity of cyanobacterial F_1-ATPase. *The Journal of Biological Chemistry, 287*, 38695–38704.

Sunamura, E., Konno, H., Imashimizu-Kobayashi, M., Sugano, Y., & Hisabori, T. (2010). Physiological impact of intrinsic ADP inhibition of cyanobacterial F_oF_1 conferred by the inherent sequence inserted into the γ subunit. *Plant & Cell Physiology, 51*, 855–865.

Tsunoda, S. P., Aggeler, R., Noji, H., Kinosita, K., Jr., Yoshida, M., & Capaldi, R. A. (2000). Observations of rotation within the F_oF_1-ATP synthase: Deciding between rotation of the F_oc subunit ring and artifact. *FEBS Letters, 470*, 244–248.

Ueoka-Nakanishi, H., Nakanishi, Y., Konno, H., Motohashi, K., Bald, D., & Hisabori, T. (2004). Inverse regulation of rotation of F_1-ATPase by the mutation at the regulatory region on the γ subunit of chloroplast ATP synthase. *The Journal of Biological Chemistry, 279*, 16272–16277.

Uhlin, U., Cox, G. B., & Guss, J. M. (1997). Crystal structure of the ε subunit of the proton-translocating ATP synthase from *Escherichia coli*. *Structure, 5*, 1219–1230.

Vambutas, V., & Racker, E. (1965). Partial resolution of enzymes catalyzing photophosphorylation .I. Stimulation of photophosphorylation by a preparation of a latent Ca^{++}-dependent adenosine triphosphatase from chloroplasts. *Journal of Biological Chemistry, 240*, 2660–2667.

Vaseghi, M. J., Chibani, K., Telman, W., Liebthal, M. F., Gerken, M., Schnitzer, H., et al. (2018). The chloroplast 2-cysteine peroxiredoxin functions as thioredoxin oxidase in redox regulation of chloroplast metabolism. *eLife*, 7, e38194.

Walker, J. E., & Gay, N. J. (1983). Analysis of *Escherichia coli* ATP synthase subunits by DNA and protein sequencing. *Methods in Enzymology, 97*, 195–218.

Whatley, F. R., Allen, M. B., Rosenberg, L. L., Capindale, J. B., & Arnon, D. I. (1956). Photosynthesis by isolated chloroplasts. V. Phosphorylation and carbon dioxide fixation by broken chloroplasts. *Biochimica et Biophysica Acta, 20*, 462–468.

Wilkens, S., & Capaldi, R. A. (1998). Solution structure of the ε subunit of the F_1-ATPase from *Escherichia coli* and interactions of this subunit with β subunits in the complex. *The Journal of Biological Chemistry, 273*, 26645–26651.

Wilkens, S., Dahlquist, F. W., McIntosh, L. P., Donaldson, L. W., & Capaldi, R. A. (1995). Structural features of the ε subunit of the *Escherichia coli* ATP synthase determined by NMR spectroscopy. *Nature Structural Biology, 2*, 961–967.

Wolosiuk, R. A., & Buchanan, B. B. (1977). Thioredoxin and glutathione regulate photosynthesis in chloroplasts. *Nature*, *266*, 565–567.
Wu, G., & Ort, D. R. (2008). Mutation in the cysteine bridge domain of the γ-subunit affects light regulation of the ATP synthase but not photosynthesis or growth in *Arabidopsis*. *Photosynthesis Research*, *97*, 185–193.
Yagi, H., Kajiwara, N., Tanaka, H., Tsukihara, T., Kato-Yamada, Y., Yoshida, M., et al. (2007). Structures of the thermophilic F_1-ATPase ε subunit suggesting ATP-regulated arm motion of its C-terminal domain in F_1. *Proceedings of the National Academy of Sciences of the United States of America*, *104*, 11233–11238.
Yagi, H., Konno, H., Murakami-Fuse, T., Isu, A., Oroguchi, T., Akutsu, H., et al. (2010). Structural and functional analysis of the intrinsic inhibitor subunit ε of F_1-ATPase from photosynthetic organisms. *Biochemical Journal*, *425*, 85–94.
Yasuda, R., Noji, H., Yoshida, M., Kinosita, K., Jr., & Itoh, H. (2001). Resolution of distinct rotational substeps by submillisecond kinetic analysis of F_1-ATPase. *Nature*, *410*, 898–904.
Yokochi, Y., Sugiura, K., Takemura, K., Yoshida, K., Hara, S., Wakabayashi, K. I., et al. (2019). Impact of key residues within chloroplast thioredoxin-*f* on recognition for reduction and oxidation of target proteins. *The Journal of Biological Chemistry*, *294*, 17437–17450.
Yoshida, K., Hara, A., Sugiura, K., Fukaya, Y., & Hisabori, T. (2018). Thioredoxin-like2/2-Cys peroxiredoxin redox cascade supports oxidative thiol modulation in chloroplasts. *Proceedings of the National Academy of Sciences of the United States of America*, *115*, E8296–E8304.
Yoshida, K., Matsuoka, Y., Hara, S., Konno, H., & Hisabori, T. (2014). Distinct redox behaviors of chloroplast thiol enzymes and their relationships with photosynthetic electron transport in *Arabidopsis thaliana*. *Plant & Cell Physiology*, *55*, 1415–1425.
Yoshida, K., Uchikoshi, E., Hara, S., & Hisabori, T. (2019). Thioredoxin-like2/2-Cys peroxiredoxin redox cascade acts as oxidative activator of glucose-6-phosphate dehydrogenase in chloroplasts. *Biochemical Journal*, *476*, 1781–1790.
Yoshida, M., Muneyuki, E., & Hisabori, T. (2001). ATP synthase—A marvellous rotary engine of the cell. *Nature Reviews. Molecular Cell Biology*, *2*, 669–677.
Yoshida, M., Okamoto, H., Sone, N., Hirata, H., & Kagawa, Y. (1977). Reconstitution of thermostable ATPase capable of energy coupling from its purified subunits. *Proceedings of the National Academy of Sciences of the United States of America*, *74*, 936–940.
Yoshida, M., Sone, N., Hirata, H., & Kagawa, Y. (1975a). ATP synthesis catalyzed by purified DCCD-sensitive ATPase incorporated into reconstituted purple membrane vesicles. *Biochemical and Biophysical Research Communications*, *67*, 1295–1300.
Yoshida, M., Sone, N., Hirata, H., & Kagawa, Y. (1975b). A highly stable adenosine triphosphatase from a thermophillie bacterium. Purification, properties, and reconstitution. *The Journal of Biological Chemistry*, *250*, 7910–7916.
Yoshida, M., Sone, N., Hirata, H., Kagawa, Y., & Ui, N. (1979). Subunit structure of adenosine triphosphatase. Comparison of the structure in thermophilic bacterium PS3 with those in mitochondria, chloroplasts, and *Escherichia coli*. *The Journal of Biological Chemistry*, *254*, 9525–9533.
Zhou, J. M., Xue, Z. X., Du, Z. Y., Melese, T., & Boyer, P. D. (1988). Relationship of tightly bound ADP and ATP to control and catalysis by chloroplast ATP synthase. *Biochemistry*, *27*, 5129–5135.

CHAPTER TWO

Structure and supramolecular architecture of chloroplast ATP synthase☆

Alexander Minges and Georg Groth*
Institute of Biochemical Plant Physiology, Heinrich Heine University, Düsseldorf, Germany
*Corresponding author: e-mail address: georg.groth@uni-duesseldorf.de

Contents

Abstract

The chloroplast F_oF_1 ATP synthase—a multi-subunit enzyme located in the plant thylakoid membrane—couples the electrochemical proton gradient built up by photosynthetic electron transport reactions to the synthesis of ATP. Synthesis of this universal cellular energy source in the extra-membranous F_1 module is driven by rotary motions in the F_oF_1 holoenzyme that are powered by proton transport across the membrane-embedded F_o-submodule. Our mechanistic understanding of the ATP synthase has tremendously benefited from detailed insights into the structure and supramolecular architecture of this marvelous molecular machine which have enormously grown in the last decades. This started with low-resolution electron microscopic images of the F_oF_1 complex, went on with atomic resolution structures of individual subunits or subcomplexes resolved by X-ray crystallography or nuclear magnetic resonance and eventually culminated in a high-resolution molecular model of the complete chloroplast F_oF_1 holoenzyme obtained by cryo-electron microscopy.

☆This article is dedicated to Wolfgang Junge (Osnabrück) at the occasion of his 80th birthday.

Advances in Botanical Research, Volume 96
ISSN 0065-2296
https://doi.org/10.1016/bs.abr.2020.07.004
© 2020 Elsevier Ltd
All rights reserved.

1. Introduction

ATP synthases localized in the bioenergetic membranes of bacteria, mitochondria and chloroplasts produce the majority of ATP in aerobic cells at the expense of an electrochemical gradient generated by electron transfer reactions across the membrane. The enzymes are multi-subunit macromolecular machines composed of two physically and functionally separable modules. The membrane peripheral F_1 module protrudes outside the membrane and catalyzes synthesis/hydrolysis of ATP, whereas the membrane-embedded F_o module serves as a proton channel and catalyzes proton transfer across the lipid bilayer membrane. The tightly-coupled interplay of these two modules interconverts the chemical energy of ATP hydrolysis and the proton electrochemical gradient via mechanical rotation. Accordingly, ATP synthases are considered mechano-enzymes or mechanical rotary motors and the related rotor and stator elements in the F_o and F_1 module have been identified. The overall architecture, core structure and basic catalytic mechanism are the same for all F_oF_1 ATP synthases. Consequently, data from different sources have been merged to obtain a comprehensive picture of this remarkable bioenergetic molecular machines. Over the past decades ATP synthases have been extensively studied and research has furnished many exciting insights into the molecular structure and mechanism. In this review we will focus on the molecular architecture and structure of the plant chloroplast ATP synthase which is located in the stroma lamella of the thylakoid membranes. The primary reaction of this enzyme is ATP synthesis driven by the transmembrane proton gradient that is formed in the photosynthetic electron transport reactions. The current picture on this enzyme at near-atomic resolution was achieved by advances in detergent-solubilization, membrane reconstitution, X-ray crystallography and single-particle cryo-electron microscopy.

2. Structure determination of macromolecular protein complexes

2.1 Challenges for structure determination of membrane proteins

About 20–30% of sequenced genomes encode for membrane proteins (Krogh, Larsson, von Heijne, & Sonnhammer, 2001; Wallin & Heijne, 2008). However, the vast majority of solved protein structures deposited within the Protein Data Bank (PDB) are water-soluble proteins whereas only 15% are annotated as membrane proteins. If this analysis is restricted to those proteins containing at least one transmembrane region, this fraction further decreases to about 3% of all deposited structures (Shimizu, Cao, Saad, Shoji, & Terada, 2018). In absolute numbers, only 999 structures of unique membrane proteins have been deposited to date (White, 2020). The main reason for this under-representation is that production and isolation of membrane proteins from their natural source or heterologous host in sufficient quality and quantity is usually more difficult than for their soluble counterparts. However, even if these requirements are met, the actual structure determination process is often more complicated and cumbersome. Historically the most prominent and successful method for structure determination of proteins was—and still is—X-ray crystallography which has evolved to the high-throughput work horse of structural biology due to the rise of modern synchrotron radiation sources, fast photon-counting detectors and automated sample handling (Owen, Juanhuix, & Fuchs, 2016). While it now is possible to screen thousands of crystals and collect data sets in mere hours, the real rate-limiting step in structure determination has never been data collection, but protein crystallization. For water-soluble proteins, the procedure is usually straight forward: First the protein has to be extracted from natural sources or from a suitable expression system used for recombinant over expression of the target protein. For proteins of eukaryotic origin, a eukaryotic expression system (e.g., yeast) might be necessary to provide post-translational modifications needed for proper folding and/or function of the protein. When produced in sufficient amounts—which for most structure determination processes means in the range of several milligrams—the protein of interest then is purified using standard biochemical methods such as ion exchange or affinity chromatography. In addition, size exclusion chromatography may be used to ensure high sample homogeneity that is mandatory for a number of structural biology methods which will be briefly introduced later in this chapter. Although soluble proteins are kept in their preferred environment—an aqueous solution—during purification still special care has to be taken to optimize the conditions for protein stability, e.g., by adjusting the pH of the purification protocol or by supplementing certain additives needed for conformational stability and/or protein functionality. Finally, the purified concentrated protein solution is used to set up crystallization trials employing a variety of vapor diffusion methods. Furthermore, additives and precipitants are added in varying concentrations to induce and optimize crystal growth.

In contrast to water-soluble proteins which never leave their preferred environment throughout purification and crystallization, structure determination of membrane proteins poses a number of challenges. These are related to (i) their low natural abundance and low expression yields, (ii) their poor solubility and tendency to form inclusion bodies, (iii) their instability once solubilized from their membranes and (iv) their preparation required for structure determination (e.g., crystal growth) (Birch et al., 2018).

In the past the low abundance of membrane proteins in their natural hosts often hampered comprehensive and systematic structural studies. Hence, the few naturally-abundant representatives in this family were among the first membrane proteins that have been successfully crystallized and their structures being solved such as the matrix protein from *E. coli* (Garavito & Rosenbusch, 1980) or the photosynthetic reaction center from *Blastochloris viridis* (Bill et al., 2011; Deisenhofer, Epp, Miki, Huber, & Michel, 1985; Deisenhofer, Epp, Sinning, & Michel, 1995). Nowadays, heterologous protein expression is routinely used to produce the quantities of a target protein required for structural studies. About a third of all reported structures in the PDB have been determined with recombinant proteins and over 70% of these have been obtained by expression in a bacterial host. Among these, *E. coli* is the most popular choice (Dilworth et al., 2018; Wagner, Bader, Drew, & de Gier, 2006). Nonetheless, as for water-soluble proteins, post-translational modifications such as specific glycosylation may require the use of a eukaryotic expression system such as yeast (*Pichia pastoris* or *Saccharomyces cerevisiae*) and mammalian or insect cell lines (Junge et al., 2008; Pandey, Shin, Patterson, Liu, & Rainey, 2016).

Although being the most used system, additional issues compared to the expression of water-soluble proteins are frequently encountered when *E. coli* or other bacteria are employed as expression host for membrane proteins. These include: (i) a potential toxicity of the over expressed membrane protein (Gubellini et al., 2011; Wagner et al., 2006), (ii) the overloading of translation and/or secretion pathways (Geertsma, Groeneveld, Slotboom, & Poolman, 2008; Schlegel et al., 2012; Wagner et al., 2007, 2008), (iii) the capacity limit of the host cell membrane which when exceeded leads to the formation of inclusion bodies (Zorman, Botte, Jiang, Collinson, & Schaffitzel, 2015) or (iv) the requirement of a particular lipid composition of the host membrane (Kanonenberg et al., 2019; Wagner et al., 2006). These issues often can be solved by fine-tuning the transcription levels of the target gene (Gubellini et al., 2011; Wagner et al., 2008) or by the use of specific host strains suitable for the expression of membrane proteins such as *E. coli* C41(DE3) or C43(DE3) (Arechaga, Miroux,

Runswick, & Walker, 2003; Miroux & Walker, 1996). Along these lines, the deletion of two transporters associated with fatty acid metabolism in *E. coli* strain C43(DE3) has been reported in a recent study to lead to a change in lipid composition with favorable characteristics for membrane protein expression and purification (Kanonenberg et al., 2019).

Once sufficient expression levels are obtained, the protein of interest can be extracted from the host cell membrane with detergents. However, the choice of the detergent is a crucial step since not all these molecules are equally suited to solubilize a given protein from the membrane in sufficient amounts (Birch et al., 2018; Moraes, Evans, Sanchez-Weatherby, Newstead, & Stewart, 2014; Stetsenko & Guskov, 2017). Generally, detergents can be categorized according to their respective head groups in: (i) ionic (ii) non-ionic and (iii) zwitterionic detergents. Positively or negatively charged ionic detergents are considered harsh as they are not only disrupting interactions between the protein and membrane, but also interfere with intramolecular hydrophobic interactions within the protein which may lead to rapid unfolding or denaturing (Michel, 1983). Sodium dodecylsulfate (SDS) is an example for such a harsh ionic detergent. Non-ionic detergents feature uncharged hydrophilic head groups and are considered mild as they tend to disturb only the interactions between membrane lipids and protein. Examples of non-ionic detergents are Triton X-100 and *n*-dodecyl-β-D-maltoside (DDM). Zwitterionic detergents such as 3-[(3-cholamidopropyl)dimethylammonio]-1-propanesulfonate (CHAPS) house both positively and negatively charged groups and are usually less harsh than ionic detergents which make them suitable for solubilization of membrane proteins. Special care in detergent choice has to be taken, as some detergents may not be compatible with projected downstream applications. Ionic detergents for example are detrimental for ion exchange chromatography, while detergents forming large micelles may prevent protein interactions and thus impede crystal formation. (Bamber, Harding, Butler, & Kunji, 2006; Birch et al., 2018; Chipot et al., 2018). In recent years, novel surfactants such as the sugar-based neopentenyl glycosides or tripod amphiphiles have been developed for effective but gentle solubilization of membrane proteins from their host membranes (Birch et al., 2018; Breibeck & Rompel, 2019).

2.2 Current methods for structure determination of membrane proteins

A number of biophysical methods are available for structure determination of membrane proteins all of which have been successfully applied to ATP synthases. Traditionally, X-ray crystallography has been the most widely used method. Nonetheless, structures of smaller membrane proteins or isolated domains of larger membrane proteins and complexes have also been solved by nuclear magnetic resonance spectroscopy (NMR) (Gautier, Mott, Bostock, Kirkpatrick & Nietlispach, 2010, Gautier 2014). In recent years, cryo-electron microscopy (cryo-EM) has gained much attention in the field of structural biology since advances in instrumentation and data processing boosted the obtainable resolution to the atomic range, competing with medium to high-resolution X-ray structures (Egelman, 2016; Kühlbrandt, 2014a,b).

In spite of these new developments in cryo-EM, X-ray crystallography still remains the dominant method for structure determination of membrane proteins. It has been successfully applied to gain structural insights into F-type ATPases beginning with the first atomic structure of the bovine mitochondrial F_1 module (Abrahams, Leslie, Lutter, & Walker, 1994; van Raaij, Abrahams, Leslie & Walker, 1996), the first structure of chloroplast F_1 ATPase (Groth & Pohl, 2001) and of the membrane-integral *c*-ring from yeast, bacteria and spinach chloroplast (Meier, Polzer, Diederichs, Welte, & Dimroth, 2005; Stock, Leslie, & Walker, 1999; Vollmar, Schlieper, Winn, Buchner, & Groth, 2009). Besides sample preparation sufficient diffraction quality of protein crystals is still the main bottleneck for structure solution of membrane proteins by X-ray crystallography (Bill et al., 2011; Birch et al., 2018). Macromolecular X-ray crystallography (MX) has benefited from major advances in automated sample handling and miniaturization within the last decade which now allow to explore a large number of different crystallization conditions and methods. Furthermore, the increasing number of solved membrane protein structures and related crystallization conditions available in the PDB, has fostered the development of specific crystallization kits for membrane proteins (Parker & Newstead, 2012). Several crystallization methods are available to produce three-dimensional crystals suitable for structure solution (for more details the reader is referred to excellent reviews by Birch et al., 2018; Caffrey, 2015; Pandey et al., 2016).

Crystallization of detergent-solubilized membrane protein (*in surfo*) follows the same principles and methods as for their water-soluble counterparts. Crystallization trials can be set up using vapor diffusion methods (hanging or sitting drop), microbatch or microdialysis. However, the type and concentration of the added detergent has to be taken into account in order to maximize crystallization success. Detergents forming large micelles—beyond their intended shielding of hydrophobic surfaces—tend to interrupt or prevent crystal contacts by masking available hydrophilic contact sites. Additionally, high detergent concentrations lead to an excess of unbound detergent micelles that may inhibit crystal formation and promote the formation of detergent crystals or phase separation. Hence, it is necessary to carefully screen the detergents used during the purification process for their compatibility with protein crystallization. If needed, their concentration can be reduced to a range just slightly above their respective critical micelle concentration (CMC) by dialysis or chromatographic techniques such as size exclusion, ion exchange or affinity chromatography. Similarly, unsuitable detergents can be exchanged prior to crystallization against other detergents that are more compatible with crystal formation. In the latter case, it is necessary to carefully validate the stability and—if possible—functionality of the target protein in its new detergent environment. Crystals of membrane proteins are usually very fragile due to their large solvent content which hampers safe harvesting of the crystals. Additionally, crystals tend to be rather small which further impedes visualization and harvesting, but also data collection and screening on common in-house X-ray sources.

In addition to direct so-called *in surfo* crystallization of solubilized protein-detergent complexes, membrane proteins have also been crystallized in lipidic phases by a method usually referred to as *in meso* crystallization. The most prominent type of *in meso* crystallization is the lipidic cubic phase (LCP) (Landau & Rosenbusch, 1996) which has become popular in recent years and led to structure determination of numerous G protein coupled receptors (GPCR) (Grisshammer, 2017). Although GPCRs form the majority of structures solved with the *in meso* approach, the method has been successfully applied for other protein families such as photosynthetic or outer membrane proteins (Caffrey, 2015; Cherezov, 2011). Lipids from the monoacylglycerol family (MAG)—specifically monoolein (MAG 9.9)—are the most commonly used matrices for cubic phase crystallization. The protein is reconstituted from its solubilized state into the continuous interconnected bilayer of the cubic phase that forms spontaneously by a vigorous mixing of lipid and protein at room temperature. Addition of a precipitant then triggers the formation of a lamellar phase into which the protein diffuses and concentrates (Caffrey, 2008; Caffrey, Li, & Dukkipati, 2012; Caffrey, 2015). Special care has to be taken regarding additives. Organic molecules and solvents such as polyethylene glycols (PEGs) may lower the bilayer interfacial curvature of the cubic phase, cause swelling and eventually result in a transformation of the lamellar phase into a so-called lipidic sponge phase (LSP). Still, this LSP retains the ability to support crystallogenesis (Wadsten et al., 2006) and is particular useful for membrane proteins with relatively large extramembrane domains (Caffrey, 2015; Wadsten et al., 2006). With the advent of microfocus beamlines, the overall small size of crystals formed by *in meso* methods has become less critical. Recent developments in this approach include techniques to inject a continuous stream of nanocrystals embedded in the LCP matrix into X-ray free electron lasers (XFEL) or microfocus beamlines equivalent to serial X-ray crystallography (SX) on soluble proteins (Fromme & Spence, 2011; Martin-Garcia, Conrad, Coe, Roy-Chowdhury, & Fromme, 2016; Weierstall et al., 2014). Another recently emerging technique is the *in meso in situ* X-ray crystallography (IMISX) (Cheng, Huang, Hennig, Nar, & Schnapp, 2019; Huang et al., 2015). Crystals grown under *in meso* conditions are measured within their crystallization well (*in situ*) without further manipulation of the crystal prior to data collection which circumvents the often difficult process of harvesting crystals from the viscous mesophase. IMISX is compatible with SX at room temperature or as IMISX-cryo for data collection at cryogenic conditions (Huang et al., 2016). Crystallization from lipidic bicelles (Poulos, Morgan, Zimmer, & Faham, 2015; Sanders & Prestegard, 1990)—small disks of lipid bilayers surrounded by a ring of amphiphiles—is another mesophase-based method that in contrast to LCP/LSP does not require special instrumentation and has been successfully applied to solve a number of membrane protein structures including bacteriorhodopsin (Faham & Bowie, 2002) or the β2-andreneric GPCR (Rasmussen et al., 2007). Similarly, crystallization assisted by high lipid and detergent concentrations (HiLiDe) is a simple method compatible with classic vapor diffusion or microbatch methods (Gourdon et al., 2011). Here, the protein at a concentration suitable for crystallization is added to a series of highly concentrated lipid supplemented with high concentrations of detergent. After overnight incubation insoluble material is separated by centrifugation and the supernatant is further used for crystallization.

Solution NMR is another useful technique successfully employed to determine the structure in particular of individual subunits of larger protein complexes. Examples include subunits δ, ε and *c* of bacterial (Girvin, Rastogi, Abildgaard, Markley, & Fillingame, 1998; Wilkens, Borchardt, Weber, & Senior, 2005; Wilkens & Capaldi, 1998; Wilkens, Dahlquist, McIntosh, Donaldson, & Capaldi, 1995; Wilkens, Dunn, Chandler, Dahlquist, & Capaldi, 1997) or subunit F_6 of mitochondrial F_oF_1 ATP synthase (Carbajo, Silvester, Runswick, Walker, & Neuhaus, 2004). In contrast to other structure determination techniques a principal advantage of NMR spectroscopy is the ability to study membrane proteins in their native lipid environment, e.g., by incorporating them in artificial membrane systems such as nanodisks (Bayburt & Sligar, 2010) or bicelles (Dürr, Gildenberg, & Ramamoorthy, 2012; Poulos et al., 2015; Sanders & Prestegard, 1990). Recently, additional membrane mimetics such as Saposin A, amphiphols or amphiphilic peptides have been explored for structural studies by solution NMR (Puthenveetil & Vinogradova, 2019; Sim et al., 2017). In addition to the potential to study membrane proteins in a lipid or lipid-like environment, solution NMR provides valuable data on structural dynamics and conformational flexibility of the protein which are hardly accessible by other techniques. Previously, these benefits were limited to proteins below 25 kDa size. However, high-field magnets and novel isotope labeling schemes helped to partly overcome this limit (Puthenveetil & Vinogradova, 2019). Still, the requirement of isotope labeling, relatively large sample volumes and the reduced sensitivity for larger proteins are relevant drawbacks of this technique. Solid-state NMR which is theoretically not limited in protein size has become a powerful addition to solution NMR and a number of membrane protein structures have been solved by this technique in recent years (Opella & Marassi, 2017).

Significant technological breakthroughs in the last years enabled a so-called resolution revolution in cryo-EM (Kühlbrandt, 2014a,b). The most remarkable are the introduction of direct electron detection cameras and techniques to correct for beam induced image motion. Notably, advances in image processing also contributed to a large extend to this outstanding improvement of resolution (Cheng, 2018). Based on these improvements, single-particle cryo-EM has gained traction in high-resolution structure determination of large complexes and membrane proteins. Since the first structure was solved by single-particle cryo-EM in 2013 (Cao, Liao, Cheng, & Julius, 2013), depositions of cryo-EM structures to the PDB have grown exponentially with 1422 new structures in 2019 alone and a total of 4437 entries (RCSB PDB, 2020). Newly deposited cryo-EM structures reached almost 50% of annually deposited X-ray structures (9666 depositions in 2019). For some families of integral membrane proteins even the majority of new structures was solved by single-particle cryo-EM. A major advantage of single-particle cryo-EM is that this technique does not require prior crystallization of the protein of interest which still is the main bottleneck in X-ray crystallography as already discussed in this chapter. By contrast, a sample which otherwise meets all criteria for crystallization in terms of purity, homogeneity and concentration, but fails to produce diffraction quality crystals routinely fulfills all requirements for cryo-EM. Though some degree of sample heterogeneity may be tolerated and, in the end, may resolve different conformational states, a homogeneous sample composition is preferable and usually probed by negative-stain EM before single-particle analysis. For both EM, the sample is put onto a grid consisting of a carbon film on a support structure (usually copper) (Dobro, Melanson, Jensen, & McDowall, 2010). While this support is continuous for negative-stain EM, a porous grid is used for single-particle cryo-EM to vitrify samples. Vitrification is obtained by plunge freezing of the sample in liquid ethane or propane before EM imaging (Dobro et al., 2010). Despite of the advantages for structure analysis of complex membrane proteins, cryo-EM has its own bottlenecks and challenges. While atomic resolution suitable for structure building can readily be achieved in many cases using modern instruments, the map quality is often still insufficient to properly identify non-protein densities such as bound ligands or lipids. This problem can be partly addressed by a comparison of EM maps obtained from the bound and unbound state of the protein which may help to locate the molecule in question (Cao et al., 2013). An even greater challenge is the identification and placement of metal ions in the structure which are crucial for proper protein function or folding for many proteins considering the fact that about 30% of all known proteins are metalloproteins. Again, difference maps of the bound and unbound state may help as previously described for the identification of bound lipids and ligands providing that the protein is stable in the absence of the metal ion. Metal ion densities along symmetry axes are furthermore difficult to assign as they may reflect either bound genuine metal ions, i.e., along a permeation pathway or simply noise which is amplified by symmetry during image processing (Autzen et al., 2018). Larger proteins and protein complexes tend to yield higher quality data and higher resolution structures. Additionally, well-ordered secondary structure elements and internal symmetry help in image alignment and further boost data quality. In contrast, large amounts of unstructured mass due to detergent micelles, lipids or unstructured protein regions contribute to the low-resolution signal only, produce high-resolution noise and are thus harmful for proper image alignment and data quality (Cheng, 2018).

3. Supramolecular organization of chloroplast thylakoid membranes

3.1 Architecture and ultrastructure of the thylakoid bioenergetic membrane

In the chloroplast, the organelle where the ATP synthase is found, the thylakoid membrane is one of three membrane systems—the other ones are the outer and inner envelope membranes—defining the structure of the mature chloroplast (Kirchhoff, 2019; Staehelin, 1986). The combination of these membrane systems leads to the formation of related aqueous compartments: (i) the intermembrane space that is located between the outer and inner envelope membrane, (ii) the chloroplast stroma that is confined by the inner envelope membrane and the thylakoid membrane housing the light-independent metabolic reactions and (iii) the thylakoid lumen that is enclosed by the thylakoid membrane containing the light-dependent reactions (Staehelin, 1986). Since the invention of electron microscopy (EM), this technique has been applied to study the structure of the chloroplast. The most striking feature revealed by these studies is the stacking or stapling of membranes, forming so-called grana thylakoids. These are connected by a continuous network of unstacked stroma lamellae. Individual grana cylinders are composed of stacks of flat membrane disks with a diameter of about 300–600 nm and are enwrapped in the stroma lamellae. As previously discussed, advances in single-particle cryo-EM lead to the availability of high-resolution structures of protein complexes embedded in the thylakoid membrane. Cryo-electron tomography (cryo-ET) as a closely-related technique to cryo-EM—although at a lower resolution—helped to understand the overall composition and structural features of the thylakoid membrane system and incorporated protein complexes on a mesoscopic level (Austin & Staehelin, 2011; Kirchhoff, 2019). However, the precise three-dimensional organization of the thylakoid membrane system is still under debate. Two fundamentally different types of models exist (extensively reviewed in Daum & Kuhlbrandt, 2011; Mustardy, 2003; Nevo et al., 2009) trying to explain the multitude of EM data gathered on thylakoids over the last decades. These are (i) a helical model (Mustárdy, Buttle, Steinbach, & Garab, 2008; Paolillo, 1970; Wehrmeyer, 1963, 1964) and (ii) different types of a forking or bifurcation model (Arvidsson & Sundby, 1999; Shimoni, Rav-Hon, Ohad, Brumfeld, & Reich, 2005). The helical model (Fig. 1C) assumes that thylakoids are built of a fretwork of stroma lamellae wrapped around grana stacks in form of a right-handed helix which connects individual grana disks. In contrast, the forking or bifurcation models (Fig. 1A/B) postulate that grana are formed by bifurcations of stroma lamellae, either as repeating units formed by symmetrical invaginations (Fig. 1A) of a thylakoid pair (Arvidsson & Sundby, 1999) or as paired layers of simple bifurcations originating from stroma lamellar sheets (Fig. 1B). In these later models, the layers are thought to be interconnected by membrane bridges (Shimoni et al., 2005). Although light induced changes at the thylakoid system such as stacking and unstacking of grana

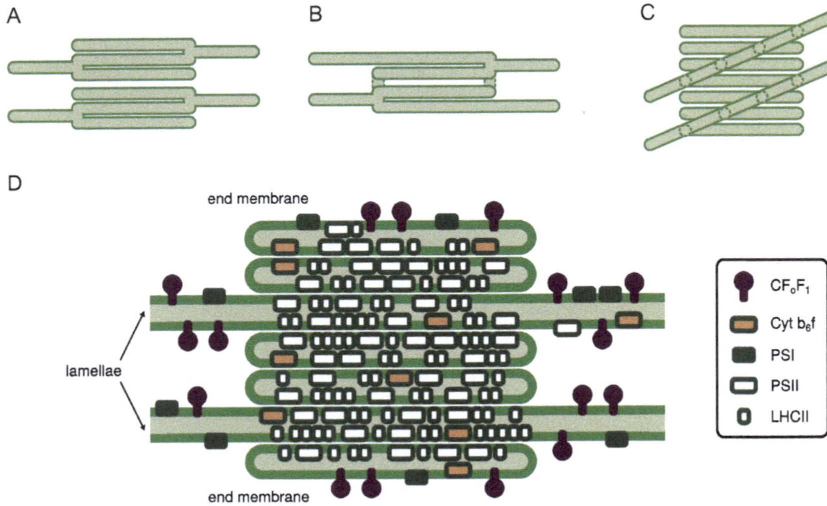

Fig. 1 Different models of thylakoid organization. Forking models according to (A) Arvidsson and Sundby (1999), suggesting a repetitive grana architecture built from units consisting of three grana disks formed by symmetrical invaginations of a thylakoid pair. (B) Bifurcating model proposed by Shimoni et al. (2005) where grana are composed of layers which are the result of bifurcations of stroma lamellar sheets. (C) Helical model as originally proposed by Wehrmeyer (1963) and supported by recent EM data (Austin & Staehelin, 2011; Daum & Kuhlbrandt, 2011; Pribil, Labs, & Leister, 2014) with fretwork of lamellae wrapping around stacks of grana disks in a right-handed helix. (D) Illustration of the thylakoid lateral heterogeneity. Light harvesting complex II (LHCII) and photosystem II (PSII) are mostly found in appressed membranes (grana) whereas Photosystem I (PSI) and chloroplast ATP synthase (CF_oF_1) are located in stroma lamellae and grana end membranes due to steric restrictions. Cytochrome b_6f complex (Cyt b_6f) can be found in both appressed membranes and stroma lamellae. PSII that is undergoing repair is also found in stroma lamellae.

are better explained by the forking models, the helical model is still widely-accepted and is sustained by recent experimental data from cryo-ET studies (Austin & Staehelin, 2011; Daum & Kuhlbrandt, 2011; Daum, Nicastro, Austin, McIntosh, & Kühlbrandt, 2010; Pribil et al., 2014).

Another hallmark of the thylakoid ultrastructure is its so-called lateral heterogeneity (Fig. 1D) referring to the different protein composition of stroma lamellae and grana (Andersson & Anderson, 1980; Dekker & Boekema, 2005; Nevo, Charuvi, Tsabari, & Reich, 2012). Photosystem II (PSII) and its related light harvesting complexes II (LHCII) are concentrated in the grana where they form super- and megacomplexes, whereas photosystem I (PSI), light harvesting complex I (LHCI) and the chloroplast ATP synthase are located in unstacked thylakoid regions, the stroma lamellae and grana end membranes. In these regions the ATP synthase covers about 20% of the surface area (Daum et al., 2010). The physicochemical rationale for membrane stacking and lateral heterogeneity are still not fully understood (Pribil et al., 2014) although they are most likely closely connected to the spatial separation of the different types of photosynthetic protein complexes (Nevo et al., 2012). The absence of PSI and the chloroplast ATP synthase from stacked grana is readily explained by steric hindrance as both include large extramembrane regions protruding from the membrane. Consequently, these protein complexes are spatially incompatible with the tightly stacked grana thylakoids where the distance between neighboring membranes is only in the range of a few nanometers (Dekker & Boekema, 2005; Kirchhoff, 2013). In contrast, LHCII and PSII both include relatively flat surfaces which allow them to localize in the tightly stacked grana membranes (Pribil et al., 2014).

4. Molecular architecture and structure of the ATP synthase molecular machine

4.1 Structure of the membrane peripheral catalytic F_1 module

The F_1 module of the chloroplast ATP synthase is comprised of five different subunits with a stoichiometry of $\alpha_3\beta_3\gamma\delta\varepsilon$ which is conserved across organism. The multi-subunit complex is connected to the chloroplast thylakoid membrane by the membrane-embedded F_o module of the enzyme and faces the chloroplast stroma. The overall F_1 shape was revealed by electron microscopic studies (Akey, Crepeau, Dunn, McCarty, & Edelstein, 1983; Boekema, Schmidt, Gräber, & Berden, 1988; Böttcher & Gräber, 2000; Böttcher, Schwarz, & Gräber, 1998; Gräber, Fromme, Schmidt, & Boekema, 1988). These studies showed in the first instance that the F_1 has a globular spherical structure, but further indicated that this part of the enzyme carrying the catalytic sites for ATP turnover resembles a pseudohexagonal, symmetrical ring of alternating large subunits α (507 aa, corresponding to 55.3 kDa) and β (498 aa, corresponding to 53.9 kDa). The central cavity of the $\alpha_3\beta_3$ core is partially filled by minor subunits γ (354 aa after processing of the 15 aa chloroplast transit peptide, corresponding to 38.5 kDa) and ε (141 aa, corresponding to 15.4 kDa). Further studies on subunit deficient F_1 implied that subunit ε is essential to preserve the central arrangement of the γ-subunit. Deletion of the peripheral δ-subunit (187 aa after processing of the 70 aa transit peptide, corresponding to 20.5 kDa) affected the integrity of the F_oF_1 holo-complex but not of the $\alpha_3\beta_3$ core.

Crystals of the F_1 from spinach chloroplasts (CF_1) gave a more detailed insight into the structure of the membrane peripheral module of the ATP synthase (Groth, 2002; Groth & Pohl, 2001). Structure analysis disclosed that the α- and β-subunit have almost identical folds with a central nucleotide binding site and are arranged alternately along a central axis in the $\alpha_3\beta_3$ hexagon (see Fig. 2A and B). Location and orientation of the nucleotide binding sites in CF_1 have been addressed prior to X-ray analysis by fluorescence mapping studies (Engelbrecht, Giakas, Marx, & Lill, 1998; McCarty & Hammes, 1987; Richter, Snyder, McCarty, & Hammes, 1985; Snyder & Hammes, 1984). These studies indicated structural asymmetry among the in total six nucleotide binding sites. Further labeling studies with substrate analogues also pointed to differences between conformational states of the isolated and membrane bond CF_1 (Groth, Mills, Christiansen, Richter, & Huchzermeyer, 2000).

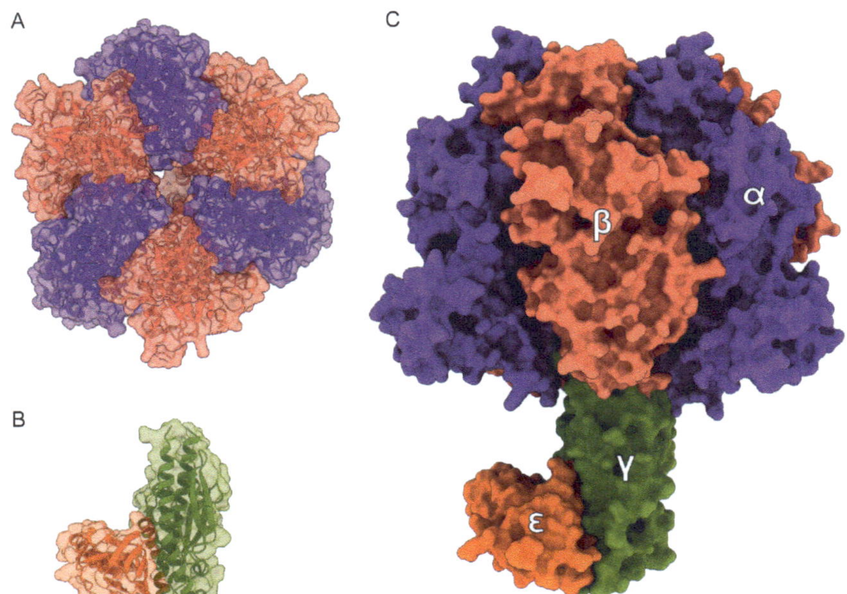

Fig. 2 (A) Top view of the chloroplast F_1 core complex composed of subunits α (blue) and β *(salmon). For illustration of the* $\alpha_3\beta_3$ *hetero-hexameric assembly in the* F_1 *module the crystal structure of chloroplast subunits* α *and* β *(PDB* 1fx0*) was aligned on the bovine mitochondrial* F_1 *structure (PDB* 2jdi*). (B) Side view of the* $\gamma\varepsilon$ *complex (*γ*: green;* ε*: orange; PDB* 5zwl*). (C) Reconstruction of the chloroplast* $\alpha_3\beta_3\gamma\varepsilon$ *submodule based on (A) and (B).*

In view of these studies, the crystal structure of the spinach chloroplast enzyme unexpectedly revealed no conformational differences among the six nucleotide binding sites which are located at the $\alpha\beta$- and $\beta\alpha$-interfaces, respectively. In agreement with previous biochemical and structural data the catalytic sites are predominantly formed by the β-subunits, whereas the non-catalytic sites reside mainly on the α-subunits. The overall fold of the α- and β-subunits is highly similar to the homologous structures from mitochondrial and bacterial ATPases (Abrahams et al., 1994; Bianchet, Hullihen, Pedersen, & Amzel, 1998; Shirakihara et al., 1997) which have been solved beforehand by X-ray crystallography. However, in contrast to these structures no bound nucleotides were detected in the nucleotide binding sites which seem to represent a closed state of the binding site probably reflecting a unique latent state of the chloroplast enzyme that is not found in bacterial or mitochondrial ATPases. In this latent form the enzyme shows no ATP synthesis or hydrolysis in the absence of an activating membrane potential (Richter, 2004). In order to overcome the latent state and to catalyze ATP hydrolysis membrane-bound F_1 has to be activated either by the reduction of a disulfide in the central γ-subunit mediated by the physiological activator thioredoxin or by a continuous low electrochemical gradient created by ATP-driven backward proton pumping across the membrane-embedded F_o module (Arana & Vallejos, 1982; Ketcham, Davenport, Warncke, & McCarty, 1984). The catalytic latency of the membrane-bound form is also present in CF_1 isolated from the thylakoid membranes. However, various treatments have been established to overcome this latency such as the addition of reducing agents or the exposure to proteases which ultimately result in the removal of the ε-subunit and/or reduction of the γ-disulfide (Richter & McCarty, 1987; Schumann, Richter, & McCarty, 1985).

The chloroplast γ and ε-subunit which control the activation of the enzyme were not resolved in the spinach chloroplast F_1 crystal structure (Groth & Pohl, 2001). However, a recent crystal structure of the $\gamma\varepsilon$-subcomplex from the photosynthetic cyanobacterium *Thermosynechococcus elongatus* BP-1 (Murakami et al., 2018) which like other photosynthetic organism contains a unique regulatory sequence of 35–40 aa inserted in the γ-subunit that is not found in bacterial or mitochondrial ATPases (Hisabori, Sunamura, Kim, & Konno, 2013; Miki, Maeda, Mukohata, & Futai, 1988) has bridged this gap. This structure indicated that the overall architecture of the cyanobacterial $\gamma\varepsilon$-complex (see Fig. 2B and C) is similar to those of the corresponding bacterial and mitochondrial complexes.

Common to the homologous complexes the α-helical N- and C-termini of the γ-subunit are connected by a Rossman fold and form an antiparallel coiled coil structure (Gibbons, Montgomery, Leslie, & Walker, 2000). In addition to this common structural element, the structure of the cyanobacterial $\gamma\varepsilon$-complex uncovered that the insertion unique to photosynthetic organism is located almost at the end of the N-terminal α-helix where it forms a β-hairpin structure along the central stalk. In green algae and green plants this insertion contains additional nine aa with two cysteines (C199 and C205) responsible for the redox regulation of chloroplast ATPase (Miki et al., 1988) which is provided by the plant thioredoxin system (Mills, Mitchell, & Schürmann, 1980; Schwarz, Schürmann, & Strotmann, 1997). Reduction of this regulatory C199-C205 disulfide was shown to reduce the activation energy for ATP hydrolysis and to increase ATP synthesis at limiting light conditions (Ketcham et al., 1984).

Superimposing the structure of the isolated $\gamma\varepsilon$-complex on the corresponding region in the structure of the *Caldalkalibacillus thermarum* F_1 ATPase (Ferguson, Cook, Montgomery, Leslie, & Walker, 2016) demonstrated that the β-turn located at the top of the inserted β-hairpin extends into the central cavity of the $\alpha_3\beta_3$ core and might function as a catch between the γ-subunit and the conserved DELSEED region in the β-subunit which is thought to play a critical role in energy transmission in the F_1 module (Furuike et al., 2008; Koga & Takada, 2006; Mnatsakanyan, Krishnakumar, Suzuki, & Weber, 2009; Pu & Karplus, 2008). Further biochemical analyses of truncations of the γ-insertion disclosed that the β-hairpin element plays a critical regulatory role and accounts for significant interaction with the $\alpha_3\beta_3$-core to induce ADP inhibition during ATP hydrolysis (Murakami et al., 2018). Structural and biochemical analyses in this work further identified that the β-hairpin element forms critical interactions with the ε-subunit which binds on the opposite site of the γ-subunit from the hairpin structure in the cyanobacterial $\gamma\varepsilon$-crystal structure.

The overall structure of the ε-subunit in the cyanobacterial $\gamma\varepsilon$-complex turned out similar to the previous solved NMR structure of the isolated ε-subunit (Yagi et al., 2010). With the N-terminal β-sandwich structure and the C-terminal two-helix hairpin the ε-subunit of a photosynthetic organism shows the characteristic domains that have been also observed with the homologous bacterial equivalents (Uhlin, Cox, & Guss, 1997; Wilkens et al., 1995; Yagi et al., 2007).

In both organism, chloroplast and bacteria, the ε-subunits acts an endogenous inhibitor of the F_1 module. The inhibitory effect was attributed to the interaction of the C-terminal α-helical region with the γ-subunit and the catalytic β-subunit. A series of structural and biochemical analyses have identified two distinct conformations of the ε-subunit in the catalytic cycle (Iino et al., 2005; Suzuki et al., 2003; Tsunoda et al., 2001). Both structures differ in the orientation of the C-terminal helix hairpin which is associated with the β-sandwich domain in the retracted form, but protrudes from this structural element in the extended conformation. Whereas these conformational changes depend on ATP in the bacterial ε-subunit which has the ability to bind ATP, the chloroplast ε-subunit has no such nucleotide binding site and works as stable nucleotide-independent intrinsic inhibitor of the ATPase activity. In this case, transformation of the conformational state from the retracted to the extended form blocking catalysis is controlled by the electrochemical gradient across the thylakoid membrane (Johnson & McCarty, 2002). This coupling of transmembrane potential and inhibitory conformation of the ε-subunit ensures that ATP consumption is avoided at limiting transmembrane potential in the dark. Still, the molecular principle is the same for both, the bacterial and chloroplast ε-subunits. As for the β -hairpin element in the regulatory sequence of photosynthetic organism the α-helical C-terminal domain of the ε-subunit mechanistically acts as a plug-in of an α-helical element to the catalytic β-subunit. This steric hindrance in turn then blocks the catalysis-related conformational changes in the γ-subunit (Iino, Hasegawa, Tabata, & Noji, 2009; Iino et al., 2005; Saita et al., 2010; Suzuki et al., 2003). The close relative arrangement of the γ- and ε-subunit indicated by the early electron microscopic studies on chloroplast F_1 is resolved more precisely in the crystal structure of the cyanobacterial $\gamma\varepsilon$-complex (see Fig. 2B) which pinpoints that the ε-subunit binds to the lower part of the γ-subunit (Murakami et al., 2018). Furthermore, the high-resolution crystal structure exposed possible contact sites on the γ- and ε-subunit with the oligomeric *c*-ring of the F_o module. Interaction of these subunits, i.e., the bottom of the F_1 region with the top of the chloroplast *c*-ring in the intact F_oF_1 complex in the thylakoid membrane is further supported by electrostatic surface potential calculations of the $\gamma\varepsilon$-complex and the *c*-ring (Murakami et al., 2018). Such connection of the F_o and F_1 module by the $\gamma\varepsilon$-entity protruding the $\alpha_3\beta_3$ core was already projected in previous studies (Kato et al., 1997; Sternweis & Smith, 1980).

The remaining subunit in the F_1 complex, subunit δ, that was shown to be located in the periphery of the $\alpha_3\beta_3$ core, is missing in the crystallized spinach F_1 complex as it had turned out in these studies that removal of this subunit before crystallization was indispensable to improve crystal diffraction and structure resolution (Groth & Pohl, 2001). Though, structural information on this part of the F_1 module was obtained by solution NMR on the homologous bacterial δ-subunit and its relative arrangement to the N-terminal part of the α-subunits (Wilkens et al., 2005, 1997). These studies specified that the intact δ-subunit is a two-domain protein consisting of a well-structured N-terminal, compact six α-helix bundle that provides the contact side to the F_1, and a less-well characterized C-terminal domain. Later structural analysis by cryo-EM (Hahn, Vonck, Mills, Meier, & Kuhlbrandt, 2018) uncovered that the chloroplast δ-subunit binds to all three α-subunits ensuring that only one peripheral stalk (consisting of the δ-subunit and the *b* and *b'* subunits in the F_o module) can attach to the $\alpha_3\beta_3$ core (for more details see Section 4.3 on the F_oF_1 holoenzyme later in this chapter). Together, all structural approaches support that the stator function is the main task of the δ-subunit in the intact F_oF_1 complex. Precisely, the δ-subunit keeps the $\alpha_3\beta_3$ core stationary relative to the *c*-ring in the F_o module and prevents sliding off the catalytic core from the central stalk formed by the γ- and ε-subunits.

A major breakthrough in the understanding of the molecular mechanism of the F_1 ATPase was made with the structure determination of the homologous F_1 from bovine mitochondria complex by the group of John Walker (Stock et al., 1999). In contrast to the CF_1 structure this structure resolved three different conformations of the β-subunit in the $\alpha_3\beta_3$ core and supported the binding change mechanism which has been proposed based on kinetic data and isotope exchange studies earlier by Boyer (1975, 1989).

The resolved conformations of the β-subunits in the MF_1 complex differ in particular in their nucleotide binding site, their bound nucleotide and their position with respect to the asymmetrical γ-subunit in the center of the $\alpha_3\beta_3$ core. One of them was empty (β_E), another contained ADP (β_{DP}-subunit), and a third had bound the ATP analogue AMP-PNP (β_{TP}-subunit). The β_{TP}- and β_{DP}-subunit were in a closed conformation, whereas the empty subunit adopted an open conformation. In addition to these three distinct states of the catalytic β-subunit later structural studies identified a half-closed conformation of the β-subunit which was interpreted as an intermediate state shortly before product release (Menz, Walker, & Leslie, 2001). The different conformational states of the β-subunit identified in the crystal structures are thought to reflect the different binding affinities—*tight* corresponding to high ATP affinity, *loose* reflecting medium affinity and *open* for the nucleotide-free state—of the three catalytic centers that have been proposed by the alternating binding change mechanism. Each of the three β-subunits is supposed to pass sequentially through these affinities or conformational states in a cooperative manner. The related changes in the binding affinities were supposed to occur by rotational catalysis that is powered by the rotating central $\gamma\varepsilon$-subcomplex of the enzyme which in turn is driven by proton transport across the bioenergetic membrane in the F_o module. Hence, the central $\gamma\varepsilon$-stalk transmits the energy from the F_o motor to the F_1 module resulting in the synthesis of 3 ATP per revolution. Molecular dynamics simulation, biochemical and biophysical studies confirmed this mechanism (Bockmann & Grubmuller, 2002; Duncan, Bulygin, Zhou, Hutcheon, & Cross, 1995; Sabbert, Engelbrecht, & Junge, 1996). Direct visualization of the γ-rotation in the $\alpha_3\beta_3$ catalytic core in response to ATP reflecting ATP hydrolysis in the isolated F_1 complex as a matter of fact was accomplished in single molecule experiments where a fluorescent actin filament was attached as a marker to the γ-subunit and the $\alpha_3\beta_3$ core fixed upside-down on a cover slip (Noji, Yasuda, Yoshida, & Kinosita Jr., 1997).

A selective modulator of this motion in the chloroplast F_1 complex is the cyclic peptide tentoxin produced by phytopathogenic fungi from the *Alternaria* species which was shown to strongly reduce γ-rotation at low and reinstall γ-motion similar to the non-inhibited enzyme at high concentration (Meiss, Konno, Groth, & Hisabori, 2008). Peculiarly, this inhibitor affects only the chloroplast but not homologous bacterial or mitochondrial enzymes. Structural analyses on crystals of a tentoxin-CF_1 inhibitor complex revealed that the binding site is located at the $\alpha\beta$-interface in the N-terminal β-barrel part of the subunits stabilizing the $\alpha_3\beta_3$ hexagon (Groth, 2002). The closed conformation of the catalytic β-subunit that is associated with the tentoxin binding at the N-terminal interface suggested that tentoxin inhibits the activity of the enzyme by blocking conformational shifts at the catalytic sites. Later single molecule experiments confirmed that the compound affects ADP release at the catalytic sites (Meiss et al., 2008). Although the main part of the binding site is located in the non-catalytic α-subunit, a single aspartate in the β-subunit (D83 in the chloroplast enzyme) was identified critical for tentoxin binding and response. Based on this knowledge the naturally non-sensitive enzymes from *E. coli* and *Bacillus PS3* were turned into tentoxin-sensitive bacterial ATPases by site-directed mutagenesis affecting the related aspartate (Groth, Hisabori, Lill, & Bald, 2002; Schnick, Kortgen, & Groth, 2002).

4.2 Structure of the membrane-embedded F_o proton transport module

The membrane-embedded F_o part of the chloroplast ATP synthase which catalyzes proton translocation across the thylakoid bioenergetic membrane resembles the basic architecture of the bacterial and mitochondrial F_o complex. However, in contrast to bacterial F_o which is composed of only three different types of subunits (*a*, *b* and *c*) arranged in a stoichiometry of ab_2c_{10-17} (Jiang, Hermolin, & Fillingame, 2001; Schulz, Wilkes, Mills, Kuhlbrandt, & Meier, 2017; Sobti et al., 2016) and the more complex composition of the mitochondrial F_o sector which is an assembly of single copies of subunits *a* (or ATP6), *b*, *c* (or ATP9), *e*, *f*, *g*, and *A6L* (or ATP8) (Fearnley & Walker, 1986; Walker, Lutter, Dupuis, & Runswick, 1991) chloroplast F_o is composed of four different subunits (*a*, *b*, *b′* and *c*—originally named IV, I, II and III) which in green plants occur with a stoichiometry of $a_1b_1b'_1c_{14}$ (Seelert et al., 2000; Vollmar et al., 2009). Evolutionary analysis suggests that subunit *b′* which is the only nuclear-encoded gene product of the F_o chloroplast membrane complex arose by gene duplication from subunit *b* (Hennig & Herrmann, 1986). Based on this evolutionary background a similar function of the chloroplast *bb′*-heterodimer and the bacterial b_2-homodimer in their respective F_o and F_oF_1 complexes was assumed. Along these lines, cross-linking studies disclosed that subunits *c*, *b* (and *b′*) provide the sites for the attachment of the membrane peripheral catalytic F_1 (Beckers, Berzborn, & Strotmann, 1992; Bulygin, Duncan, & Cross, 2004; McLachlin, Coveny, Clark, & Dunn, 2000). Subunit *c* is the only subunit present in multiple copies in the membrane-embedded F_o module.

Low-resolution electron spectroscopic imaging resolved the shape, dimensions and relative orientation of subunits in the F_o complex and indicated that the *c*-subunits form a ring-shaped complex in the membrane (Fromme, Boekema, & Gräber, 1987). Topological studies by atomic force microscopy (Neff et al., 1997) and mutagenesis studies (Groth, Tilg, & Schirwitz, 1998) further confirmed the annular arrangement of the *c*-oligomer in the membrane. Further electron microscopic analysis, site-directed mutagenesis and cross-linking studies revealed that subunits *a*, *b* and *b′* are located on the periphery of the homo-oligomeric *c*-ring cylinder (Birkenhäger, Hoppert, Deckers-Hebestreit, Mayer, & Altendorf, 1995; Jiang & Fillingame, 1998; Jones, Jiang, & Fillingame, 1998; Singh, Turina, Bustamante, Keller, & Capaldi, 1996; Takeyasu et al., 1996).

High-resolution structure determination of an isolated intact F_o complex by X-ray crystallography turned out difficult and has not been successful yet maybe because of the dynamic nature of the complex. Therefore, solving the structure of individual subunits or subcomplexes at higher resolution and to combine these information with the existing low-resolution information on the complete F_o complex was considered an appropriate approach to obtain high-resolution structural information on the membrane-embedded part of the ATP synthase. To this end, the structure of individual F_o subunits was addressed by NMR spectroscopy. Pioneering work along these lines was performed on the bacterial homolog (79 aa, corresponding to 8.3 kDa) of the chloroplast *c*-subunit (81 aa, corresponding to 8.0 kDa) which was analyzed by NMR spectroscopy in a mixed solvent suggested to reflect the typical membrane environment and preserve the native structure of the subunit (Dmitriev, Altendorf, & Fillingame, 2004; Dmitriev, Jones, Jiang, & Fillingame, 1999; Girvin et al., 1998). In these studies, Fillingame and co-workers uncovered that the *c*-monomer forms a hairpin-like structure consisting of two hydrophobic membrane spans (TMH1 and THM2) connected by a polar loop (Girvin et al., 1998). Later studies disclosed that conserved residues in the connecting loop form part of the physical link between the membrane-embedded F_o and the $\gamma\varepsilon$-subunits in the peripheral F_1 catalytic complex. Because earlier mutagenesis studies had identified the critical importance of a conserved central carboxylate (D61 in bacteria, E61 in chloroplasts) in TMH2 of subunit *c* for proton transport, the NMR structure of the *c*-monomer was studied at different pH (Girvin et al., 1998; Rastogi & Girvin, 1999a). Structural changes in the *c*-monomer conformation were observed in these studies at pH 5 and pH 8 reflecting different protonation state of the central carboxylate. Based on the observed conformational changes rotation of TMH2 in the *c*-monomer was proposed as a potential mechanism by which proton translocation in the F_o may drive ATP synthesis in the F_1 complex (Rastogi & Girvin, 1999b).

Based on their sequence homology similar conformational changes are assumed for the chloroplast *c*-monomer although no high-resolution structural information on the individual *c*-subunit is available. However, the oligomeric assembly of *c*-monomers which forms the core of the membrane-embedded F_o in all organism has been crystallized from spinach and pea chloroplast and the related structure has been solved by X-ray crystallography at a resolution of 3.4–3.8 Å, respectively (Saroussi, Schushan, Ben-Tal, Junge, & Nelson, 2012; Vollmar et al., 2009). These structures show that the 14 *c*-subunits in the chloroplast F_o which represent the key part of the membrane rotor of the enzyme are arranged in a circle as a cylindrical palisade around a central hydrophobic cavity (see Fig. 3B and D).

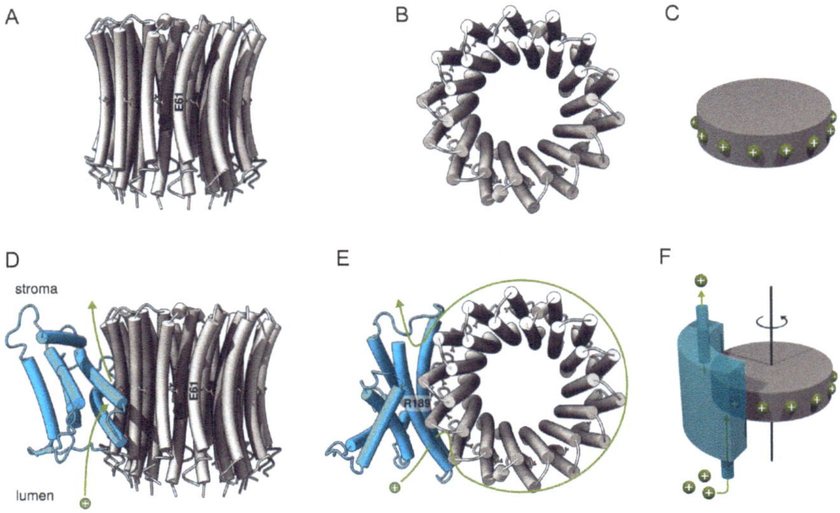

Fig. 3 Side (D) and top (E) view of the chloroplast F_o module composed of subunit *a* (blue) and the *c*-ring (gray) extracted from PDB 6fkf. The path of protons passing from the luminal entry channel to the stromal exit channel driven by the rotation of the *c*-ring is depicted by green arrows. The isolated *c*-ring without subunit *a* bound is shown in side (A) and top view (B), highlighting its radial assembly and hourglass-like shape. Schematic representation (F) of subunit *a* and the chloroplast *c*-ring, demonstrating the proton translocating mechanism as proposed by Junge, Lill, and Engelbrecht (1997). Individual protons are illustrated as green spheres. Schematic representation (C) of the chloroplast ATP synthase *c*-ring with its 14 protonation sites.

The N-terminal (TMH1) helices of the *c*-protomers pack tightly on the inside of the ring, while the C-terminal (TMH2) helices locate at the periphery. Based on the crystallization conditions the spinach and pea c_{14}-structures probably reflect the protonated state of the chloroplast *c*-ring rotor. When viewed from the membrane the overall shape of the *c*-oligomer resembles that of an hourglass (see Fig. 3B). In this hourglass figure the conserved E61 carboxylate is located approximately in the middle of TMH2 at the central waist of the complex. The side chain of this carboxylate fundamental for proton transport in the F_o sector is located at the periphery of the cylinder close to the outer surface of the *c*-ring in an environment shielded from the hydrophobic membrane. Additional density in the central cavity of the *c*-ring was attributed to lipid or detergent molecules bound to the complex although clear interpretation of this density was not possible at the obtained low-medium resolution. Recent studies by Gordeliy and co-workers (Vlasov et al., 2019) provided a higher resolution picture of the spinach chloroplast c_{14}-ring resolving more details on the intersubunit contacts in the ring. Their data support a tight interaction of the inner helices (TMH1) which form a dense hydrogen bonding network in particular in the region close to the stroma side. In contrast, TMH2-helices of the *c*-protomers interact only at their E61 center. Comparison to other *c*-ring structures supports that the conserved carboxylate is part of the oligomerization interface in bacterial, mitochondrial and plant ATPases. However, it should be noted that the side chain of the conserved carboxylate is directly bond to the backbone of neighboring protomers only in large *c*-rings, whereas the contact between adjacent protomers is mediated by water molecules in smaller rings. Apart from that, the high-resolution structure shows additional electron density inside the *c*-ring which form circles parallel to the membrane plane. Based on further spectroscopic analysis the authors suggested that these densities which are also found in other high-resolution structures of *c*-rings correspond to isoprenoid quinones such as coenzyme Q or plastoquinone. In contrast to this interpretation of the additional density in the *c*-ring, previous crystallization studies implied that carotenoids and chlorophylls are associated with the *c*-ring (Varco-Merth, Fromme, Wang, & Fromme, 2008). Possible functions of the isoprenoid quinones proposed from the high-resolution structure of the chloroplast c_{14}-rotor would be to stabilize the *c*-ring, to prevent ion leakage through the protomer cylinder and/or to protect proton uptake/release in the *c*-subunits against reactive oxygen species (ROS) (Vlasov et al., 2019).

Importantly, similar hourglass figures as observed for the chloroplast *c*-ring were obtained in crystallization studies with bacterial and mitochondrial ATPases (Meier et al., 2005; Morales-Rios, Montgomery, Leslie, & Walker, 2015; Pogoryelov et al., 2005; Preiss et al., 2015; Stock et al., 1999) which further support the chloroplast c_{14}-assemby. However, due to the different stoichiometry in these complexes ranging from 8 to 17 protomers (Schulz et al., 2017; Watt, Montgomery, Runswick, Leslie, & Walker, 2010) the diameter of the *c*-ring complexes in the different organisms varies substantially, but is constant for a given species. Nesci and co-workers propose that the distinct subunit *c* stoichiometry reflects the different proton to ATP ratios and thus the species specific bioenergetic cost of ATP in the diverse organisms (Nesci, Trombetti, Ventrella, & Pagliarani, 2016).

Subunit *a* (247 aa, corresponding to 27.1 kDa) is the largest polypeptide in the membrane-embedded F_o sector of the ATP synthase. As for the *c*-subunit most of the primary structural information on this part of the F_o module was obtained with the bacterial *a*-subunit. Though, solution NMR studies in the membrane-mimicking methanol:chloroform:water mixed solvent used for structure determination of the monomeric *c*-subunit were not as successful in this case and did not provide high-resolution structural information on the *a*-subunit. Thus, for long time most structural information on this subunit was based on genetic, physiological and biochemical studies on mutants of the bacterial enzyme. First, these studies identified critical functional and disruptive residues and second mapping studies provided topological information on the number, orientation and arrangement of the transmembrane spans. Cysteine mapping was used to probe and identify surface-accessible regions, whereas cysteine cross-linking was applied to gain information on the relative arrangement of the transmembrane segments (for details see Vik, Long, Wada, & Zhang, 2000 and references therein).

These experiments identified a critical arginine (R210 corresponding to R189 in the chloroplast enzyme) as the only residue which is strictly conserved among all species and which tolerates no substitution. Moreover, these studies suggested that the *a*-subunit contains four or five transmembrane spans, though both topologies are roughly identical after residue 65 (Long, Wang, & Vik, 1998; Valiyaveetil & Fillingame, 1998; Wada, Long, Zhang, & Vik, 1999).

Based upon the derived topology and results from suppressor screens Vik and colleagues proposed a partial structural model of the *a*-subunit assuming a uniform length of 18-residues for each of the 5 transmembrane α-helices. An essential feature of this model is the surface-exposure of the previously identified essential R210 which is pointing outward the *a*-subunit helix bundle and contacts the *c*-ring oligomer. Such contact of both subunits, precisely the interaction of transmembrane helix (TMH) 2 of the *c*-subunit with TMH4 and TMH5 of subunit *a* was determined in preceding cross-linking studies (Jiang & Fillingame, 1998).

A similar, but more detailed model of the ac_{12}-complex has been proposed by Rastogi and Girvin based on distance constraints from cross-linking studies, the solution NMR structure of the *c*-monomer and a structural model of the four consensus transmembrane helices of subunit *a* (Jäger, Birkenhäger, Stalz, Altendorf, & Deckers-Hebestreit, 1998; Valiyaveetil & Fillingame, 1998; Wada et al., 1999; Yamada, Moriyama, Maeda, & Futai, 1996) obtained by standard NMR structure calculation methods and further cross-linking constraints (Rastogi & Girvin, 1999b). Similar to the model by Vik and colleagues this model was used to explain and further explore possible proton translocation pathways in the F_o on the molecular level. Together, both models refer to a basic concept of proton transfer in the

F_o module and rotary catalysis outlined by Wolfgang Junge (Junge et al., 1997). In this initial mechanical approach to mechanistically explain proton transport across the F_o module, the *c*-ring was identified as the rotor and the *a*-subunit as the stator element of the membrane sector of the ATP synthase which was considered to work as a molecular motor (see Fig. 3E and F). This model which had been proposed in similar form by Vik and Antonio (1994) is based on the premise that the strictly conserved membrane-embedded arginine in the *a*-subunit couples proton translocation to *c*-ring rotation across in the *ac*-core complex. In more detail, these mechanistic models assume direct interaction of protonated R210 (R179 in chloroplasts) and a single deprotonated carboxylate (D61 in bacteria, E61 in chloroplasts) in the *c*-oligomeric rotor ring. In contrast to the Junge model Vik and Antonio assume that a second *c*-subunit adjacent to this contact is also unprotonated and interacts with the *a*-subunit. All remaining carboxylates at the waist of the hourglass shaped *c*-ring are protonated and are located in the hydrophobic lipid environment in contrast to those facing the *a*-stator. Protonation of the carboxylate in the *c*-subunit interacting with the conserved arginine in subunit *a* will move the newly protonated *c*-protomer to the membrane environment. While the associated rotation of the *c*-ring is driven by random Brownian motion in the Junge model, *c*-ring rotation is electrostatically driven in the Vik and Antonio version as in their setup the second still unprotonated *c*-protomer will interact with the positive charged arginine. As a consequence of the *c*-ring rotation the (n-1) or (n-2) *c*-protomer will move from the membrane environment coming in contact with the *a*-subunit. In consequence of the resulting polar protein environment, the conserved carboxylate in the *c*-protomer experiences a pK-shift and proton release to the negative site of the membrane (chloroplast stroma). In the ATP-driven proton pumping mode the process runs in reverse sequence. In consequence of these events, no matter whether ATP synthesis or hydrolysis is considered, a *c*-ring protomer travels by an almost full rotation through the hydrophobic membrane environment before releasing the proton to the other side of the membrane. In any case this rotation of the *c*-ring is coupled to the rotation of the F_1 rotor subunits which in turn drive sequential conformational changes at the three β-subunits resulting in ATP catalysis.

Structural data of the *c*-ring rotor further elaborate the idea that the *ac*-complex works as a torque-generation, single file ion channel (Junge et al., 1997) and propose that the protonated carboxylate switches from an open to a locked conformation that is energetically favored to penetrate into the membrane (Meier et al., 2005; Vollmar et al., 2009).

To pinpoint the exact mechanism of proton transport in the membrane-embedded F_O module and to discriminate critical features in the construction and mechanism of rotor function high-resolution information is crucial. However, location and identity of the proposed two half-channels at the interface of the *c*-ring and subunit *a* have not been resolved by crystallization of a related *ac*-core complex yet.

Nonetheless, several residues which contribute to the half-channel controlling proton access from the P-side of the membrane have been recognized in the *ac*-modeling approach by Vik et al. (2000). In this context, E219, H245, D44 and W241 in the center of the modeled subunit *a* bundle were considered as critical residues that contribute to the proton transport channel toward the conserved carboxylate in the *c*-subunit.

Furthermore, based on the crystal structure of the chloroplast *c*-ring (Vollmar et al., 2009) Gohlke et al. identified and mapped by molecular dynamic simulations in an explicit lipid bilayer a water-channel as part of the proton path and as half-channel from the N-side of the membrane in the chloroplast F_O membrane module (Gohlke, Schlieper, & Groth, 2012). In contrast to all previous models assuming a vertical alignment of the two half-channels in subunit *a* with the *c*-ring rotor, the N-side half-channel identified in the MD studies is inclined against the sense of rotation of the *c*-ring—an arrangement that according to Gohlke et al. would minimize futile proton flow (Gohlke et al., 2012). Recent high-resolution structural data on the *a*-subunit of the mitochondrial ATP synthase from the green alga *Polytomella* sp. (for further details see Section 4.3) suggested indeed that the N-side half-channel has horizontal helices, while the P-side half-channel is arranged vertical (Allegretti et al., 2015).

In contrast to subunits *a* and *c* which constitute the core complex of the F_O module (see Fig. 3A and B) subunit *b* (and subunit *b′* in the chloroplast ATP synthase) are amphipathic polypeptides containing both integral membrane and peripheral membrane stretches. In the chloroplast F_O module both subunits are present in single copies. The nuclear-encoded *b′*-subunit (222 aa, corresponding to 24.5 kDa) contains a transit peptide of 75 aa which is naturally lacking in the chloroplast-encoded *b*-subunit (184 aa, corresponding to 21 kDa). Together, both subunits have a single highly hydrophobic transmembrane span (residues 27–49 in subunit *b*, residues 83–103 in subunit *b′*) which is supposed to anchor them in the membrane, followed by a tether domain, a dimerization domain and a C-terminal domain linking them to the peripheral F_1. Cross-linking studies on the bacterial *b*-subunit revealed a direct interaction of both copies in the transmembrane domain. Based upon NMR studies and distance constraints from further cross-linking a structural model has been proposed for the N-terminal membrane-embedded part of the *b*-homodimer (Dmitriev et al., 1999). Further analysis by analytical ultracentrifugation (AUC) and small angle scattering (SAXS) propose that the *b*-module forms an elongated dimer and support an α-helical coiled coil structure of the membrane peripheral stretch (Del Rizzo, Bi, & Dunn, 2006). A high-resolution structure of subunit *b* corresponding to residues 62–122 was solved by X-ray crystallography. Combination of this partial structure of the subunit *b* monomer with SAXS data from the *b*-dimer suggests that the two helices form a coiled coil with a right hand super helical twist (Del Rizzo, Bi, Dunn, & Shilton, 2002). Cross-linking data propose that the loosely unwinded α-helices of subunit *b* interact with the *a*-subunit and clamp this central part of the F_o stator in its position next to the *c*-ring rotor. This interaction reflects a critical function of the *b*-dimer namely connecting the membrane intrinsic F_o and the membrane peripheral F_1. In this connection, subunit *b* works as peripheral stator that avoids unproductive rotation in the *ac*-core complex. Remarkably, mutagenesis studies demonstrated that the membrane peripheral stretch of subunit *b* in spite of its critical function as stator element in the F_oF_1 complex is robust against insertion and deletion suggesting that the *b*-dimer acts as an elastic connector for smooth torque generation (Sorgen, Bubb, & Cain, 1999; Sorgen, Caviston, Perry, & Cain, 1998). Structural studies on the related *Thermus thermophilus* A-type ATP synthase support the concept that the peripheral stator holds lateral rigidity, but at the same time conformational flexibility to store transient elastic energy.

4.3 Structure of the intact complete F_oF_1 holoenzyme

In order to understand the detailed molecular mechanism of the intact ATP synthase and the coupling of the membrane-embedded F_o motor module with the membrane peripheral F_1 catalytic module a high-resolution structure of the complete F_oF_1 holoenzyme seemed indispensable. However, crystallization of the entire protein complex turned out difficult particular due to the dynamic nature and conformational flexibility of the protein as a direct consequence of the rotary catalytic mechanism employed by the enzyme. Eventually single-particle cryo-EM allowing classification and structural resolution of distinct states of a biological macromolecule in combination with recent technological and methodological advances in this technique turned out the right approach to solve the structure of full F_oF_1 at crystallographic resolution.

Initial low resolution cryo-EM of the detergent-solubilized chloroplast F_oF_1 holoenzyme indicated that the membrane extrinsic F_1 domain is connected to the membrane-embedded F_o domain by two stalks (Böttcher, Lücken, & Gräber, 1995; Böttcher et al., 1998). The central stem was attributed to the $\gamma\varepsilon$-rotor, whereas the second peripheral stalk was assigned to subunits *b*, *b′* and δ. Further studies were able to discriminate two different conformational classes of the enzyme which were characterized by different tilts of the peripheral F_1 with respect to the membrane F_o module. These classes were interpreted as different protonation states of the F_o module (Mellwig & Bottcher, 2001). From these initial efforts to get a high-resolution picture of the complete F_oF_1 by cryo-EM it took another about 15 years before this objective was finally achieved. Then, a number of medium to high-resolution cryo-EM structures from ATP synthases of various organisms revealed important new details about the structure and the catalytic mechanism and allowed a new level of molecular understanding of the enzyme (Allegretti et al., 2015; Hahn et al., 2016; Sobti et al., 2016; Zhou et al., 2015).

A major breakthrough in structural resolution of the chloroplast ATP synthase was the structure determination of the complete F_oF_1 complex from spinach chloroplasts by single-particle cryo-EM (Hahn et al., 2018). The structure (see Fig. 4A–D) was resolved in three different conformations with the central $\gamma\varepsilon$- and *c*-ring rotor stalled at different resting positions in the rotary catalytic cycle which provide new insights into the catalytic mechanism and the proton pathway of the ATP synthase. Detailed structural analyses revealed that the position of the *c*-subunits relative to the *a*-subunit differed in the resolved conformational classes corresponding to discrete rotation steps of 103°, 112° and 145°. Notably, this uneven distribution is in contrast to previous FRET experiments on detergent-solubilized bacterial ATPase showing multi-stepping rotation which was interpreted as discrete 36° sub-step rotations based on the 10-fold symmetry of the bacterial *c*-ring rotor (Duser et al., 2009). Similar 36° sub-rotation steps at hydrolysis conditions were also observed with the bacterial F_oF_1 embedded in lipid bilayer nanodisks (Ishmukhametov, Hornung, Spetzler, & Frasch, 2010).

In accordance to other cryo-EM studies on homologous ATP synthases the CF_oF_1 cryo-EM structure resolved the peripheral stalk connecting the F_o and the F_1 module over the entire length and identified that the δ-subunit simultaneously interacts with all three α-subunits in the F_1 head although

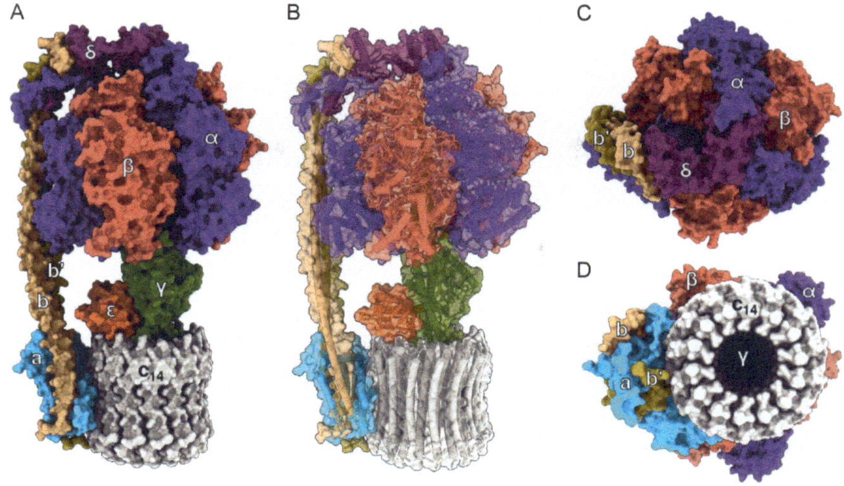

Fig. 4 Surface representation in side view (A) of the complete chloroplast F_oF_1 ATP synthase (PDB 6fkf), colored by subunit identity. Secondary structure composition is outlined in (B). Top view of complete F_oF_1 ATP synthase with the stromal side toward the thylakoid membrane (C) and bottom view (D) of the holoenzyme with the luminal side toward the membrane.

at different contact sites (see Fig. 4A and B). The remaining *b* and *b′* subunits which connect the *δ*-subunit in the peripheral stalk to the membrane-embedded F_o module showed up as loose right-handed coiled coil that ends just above the membrane surface. At the site where the helices enter the membrane they bifurcate, attach to the *a*-subunit and clamp it to the *c*-ring rotor. Further analysis of the CF_oF_1 cryo-EM structure indicated that the peripheral stalk bends relative to central axis and couples F_1 elastically to F_o. That way, the peripheral stalk works like an elastic spring and evens out energy differences from the three unequal rotation steps in the full *c*-ring rotation cycle. In contrast to this interpretation, previous single molecule experiments with the bacterial enzyme have implied opposite torsional elasticity and proposed a stiff peripheral stalk with a flexible *γ*-rotor that stores the energy differences caused by F_o-F_1 symmetry mismatches as torsional force (Wachter et al., 2011).

Essentially, the chloroplast ATP synthase cryo-EM structure provided further detailed knowledge on the proton transport pathways to and from the *c*-ring in the membrane-embedded F_o module. The corresponding map confirms the previously noted horizontal orientation of the membrane intrinsic *α*-helices in the *a*-subunit (Allegretti et al., 2015; Zhou et al., 2015) which are arranged at a certain inclination to the *c*-ring rotor protomers (for illustration see Fig. 3A and C). In total, six α-helices were identified in the chloroplast *a*-subunit. Among these the long membrane intrinsic hairpin of helices H5 and H6 that follows the curvature of the *c*-ring forms the major part of the proton transport path in the *a*-subunit leading to the conserved E61 in the *c*-protomers. Moreover, this hairpin provides the luminal proton entry and the stromal proton exit channel (Hahn et al., 2018). The access channel was attributed to a deep cavity lined by charged and polar residues in helices H5, H6 and the H3-H4 loop which funnels protons to the conserved E61 carboxylate in the *c*-ring located almost in the middle of the thylakoid membrane. The wedge-shaped release channel ranging from the stomal side to a narrow pocket next to the conserved R189 (corresponding to R210 in the bacterial *a*-subunit) is defined by charged and polar residues in H5, H6 and the proline-rich H4-H5 loop. In the cryo-EM map of the chloroplast complex the position of the conserved arginine that is essential for proton transport in the F_o module was well defined facing the release channel as reported for the mitochondrial enzyme (Guo, Bueler, & Rubinstein, 2017; Guo & Rubinstein, 2018; Zhou et al., 2015). This positional assignment further supports previous notions that the strictly conserved arginine separates access and release half-channels in order to prevent proton leakage through the membrane. Further, the chloroplast F_oF_1 cryo-EM structure disclosed that the nearest distance observed for the conserved arginine in the *a*-subunit and the conserved carboxylate in the *c*-protomers prevents salt bridge formation of these residues which would impede ring rotation and arrest both subunits at a fixed position.

A recent high-resolution cryo-EM map of mitochondrial F_oF_1 from the unicellular green algae *Polytomella* did not only reveal two different rotational positions of the *c*-ring, but also recognized a metal ion and ordered water molecules in the F_o proton access and release channel. These exciting results opened up a new avenue of mechanistic understanding of proton transport and rotational coupling in the F_o module. In particular, the proposed role of the metal ion in synchronizing *c*-ring protonation with rotation that was drawn based on the changes observed in the coordination environment of the metal ion with *c*-ring position is ground-breaking, but waits for approval in other F_oF_1 structures (Murphy et al., 2019).

In addition to the membrane-embedded F_o module the cryo-EM structure of the chloroplast F_oF_1 further provided substantial information on the membrane peripheral F_1 module. In contrast to the CF_1 crystal structure (Groth, 2002) all F_1 subunits were resolved. More importantly, the F_1 head appeared asymmetric due to three different conformations of the three

$\alpha\beta$-pairs. In total five bound nucleotides were resolved, three in the non-catalytic a α-subunits and two in the catalytic β-subunits. As both, βDP and βTP, contain Mg-ADP the structure probably reflected the ADP-inhibited state of the chloroplast ATP synthase. Though, the CF_oF_1 cryo-EM structure reflects an inhibited state of the enzyme none of the structural classes resolved by the cryo-EM study shows the extended inhibitory conformation of the ε-subunit previously reported in structural studies on the isolated ε-subunit. Still, such extended inhibitory conformation was observed in the bacterial F_oF_1 holoenzyme (Sobti et al., 2016).

Ultimately, the CF_oF_1 cryo-EM structure pinpointed the exact mechanism of the γ-redox switch that inhibits ATPase activity in the dark. As previously noted, the chloroplast γ-subunit contains an insertion including two regulatory cysteines which are redox-modulated by the plant thioredoxin system. Similar to what was observed with the cyanobacterial γ-subunit which also has an insertion although lacking the regulatory cysteines, the inserting residues in the chloroplast γ-subunit adopt a β-hairpin structure. However, in contrast to the cyanobacterial enzyme where the insertion forms a single β-hairpin element, the residues inserted into the chloroplast γ-subunit form two β-hairpins that are arranged in an L-shaped loop located in between the nucleotide-free β-subunit (βE) and the γ-stator. The shorter N-terminal hairpin (hp1) contains the regulatory cysteines that form a disulfide bride in the cryo-EM structure, whereas the larger hairpin (hp2) interacts with the spindle domain in the γ-subunit similar to the situation in the cyanobacterial ATPase. At the tip of this hairpin residue E261 forms a salt bridge with R397 in the empty catalytic β-subunit that blocks rotation as hp2 would otherwise clash with the highly conserved DELSEED motif in the β-subunit upon counter-clockwise γ-rotation required for ATP hydrolysis. Reduction of the disulfide would destabilize the hp1 and allow for hp2 to push aside when the rotor turns in ATP synthesis direction.

4.4 Superassemblies of ATP synthase

Scaffolding of enzymes into stoichiometric complexes driven by specific protein-protein interactions is a well-known phenomenon in a number of metabolic pathways such as protein or lipid synthesis. This higher-order quinary structures enable direct transfer of substrates and metabolic products between the involved enzymes thereby increasing the efficiency of enzymatic reactions (Cohen & Pielak, 2017; Wirth & Gruebele, 2013). In case of membrane proteins, the relatively high local protein concentration in biological membranes (Engelman, 2005) facilitates such protein complex formation that is further supported by the preferential orientation of proteins in the lipid bilayer (Helms, 2002). Another factor in the recruitment of membrane-embedded protein complexes are lipid molecules and the general lipid composition of the given biological membrane. Since the highest ratio of membrane proteins compared to lipids is found in energy-transducing membranes such as the inner membrane of mitochondria or the thylakoid membrane of chloroplasts, a high level of protein interactions is expected in these membrane systems (Seelert & Dencher, 2011).

First evidence of a supramolecular organization of the ATP synthase was provided by early electron microscopy (EM) studies on mitochondria where a regular arrangement of ATP synthase particles was observed (Sjöstrand, 1963). Later studies further refined the quinary structure of the mitochondrial ATP synthase observed in these initial EM studies and unraveled a double-rowed formation of F_1 complexes in the inner mitochondrial membrane. Moreover, these studies indicated a preferred localization of the ATP synthase supramolecular complexes at the convex side in the curved tubular regions of the inner mitochondrial membrane (Seelert & Dencher, 2011). Later, electron cryo-tomography (cryo-ET) that allows for the three-dimensional reconstruction of microstructural features within tissue slices or organelles has provided new insights into the supramolecular arrangement of eukaryotic ATP synthases. The three-dimensional tomograms (Davies et al., 2011) allowed for the first time a direct visualization of the spatial organization of the ATP synthase supramolecular complexes within their native membrane environment. The observed L-shaped arrangement of ATP synthase dimers in these studies has been later confirmed for a number of organisms including green plants (Eubel, Jänsch, & Braun, 2003), algae (van Lis, Atteia, Mendoza-Hernández, & González-Halphen, 2003), fungi (Krause et al., 2004), yeast and mammals (Cortés-Hernández, Vázquez-Memije, & García, 2007; Davies et al., 2011; Vonck & Schäfer, 2009) by native page electrophoresis and further cryo-ET studies. The picture obtained from these studies suggested that the dimerization of mitochondrial ATP synthase and the associated modulation of the membrane curvature provide the driving force for the formation of mitochondrial cristae (Davies, Anselmi, Wittig, Faraldo-Gómez, & Kühlbrandt, 2012). Studies on the yeast enzyme allowed a detailed analysis of the subunit composition of the mitochondrial ATP synthase supramolecular complex and disclosed the existence of three dimerization-specific subunits (*e*, *g* and *k*) which have not been found in the monomers. In addition, subunits *e* and *g* were identified as essential for dimer formation (Arnold, Pfeiffer, Neupert, Stuart, & Schägger, 1998; Couoh-Cardel, Uribe-Carvajal, Wilkens, & García-Trejo, 2010; Dudkina, Kouril, Peters, Braun, & Boekema, 2010). However, these dimerization-specific subunits are unique to mitochondrial ATP synthases, though the F_o part and the peripheral stalk (Fronzes, Weimann, Vaillier, Velours, & Brèthes, 2006) have been shown to further contribute to dimer formation in addition to subunits *e* and *g*. Mitochondrial ATPase dimers can adopt a variety of tilts between individual monomers (Dudkina, Sunderhaus, Braun, & Boekema, 2006; Minauro-Sanmiguel, Wilkens, & García, 2005). Tilts larger than 70° are indicative for intact mitochondria, whereas angles down to 40° are characteristic for isolated membrane fractions or inactivated complexes. In contrast to the mitochondrial enzyme, a number of studies suggested that the chloroplast ATP synthase adopts a monomeric state in the native membrane environment (Schägger, 2002; van Lis et al., 2003). Nonetheless, based on the structural similarities of the mitochondrial and the chloroplast enzyme, dimer formation of the chloroplast ATP synthase was still considered. Eventually, evidence on the existence of dimeric supracomplexes of the chloroplast ATP synthase was provided by blue-native PAGE (BN-PAGE) and colorless-native PAGE (CN-PAGE) analyses of solubilized thylakoid membranes from the green algae *Chlamydomonas reinhardtii* (Rexroth et al., 2004). Further *in vitro* studies on the enzymes from *C. reinhardtii* revealed that the chloroplast ATP synthase dimer is less stable than its mitochondrial counterpart and readily dissociated into monomers when incubated with increasing concentrations of phosphate or the transition-state analogue vanadate (Rexroth et al., 2004). The mitochondrial dimer isolated from *C. reinhardtii* was not susceptible to such treatments, suggesting a distinct dimerization mechanism of chloroplast and mitochondrial ATP synthase (Seelert & Dencher, 2011) and/or the involvement of different dimerization sites (Fronzes et al., 2006; Rexroth et al., 2004). Keeping in mind that the phosphate binding sites of the ATP synthase are located in the membrane peripheral F_1 module, dimerization of the F_oF_1 holoenzyme might be directly linked to conformational changes in the catalytic F_1 mediated by phosphate binding and/or release (Schwassmann, Rexroth, Seelert, & Dencher, 2007). Besides specific substrates such as phosphate and vanadate targeting the F_1 module, growth conditions also seem to affect the dimer to monomer equilibrium (Schwassmann et al., 2007) further supporting a metabolic control of the oligomeric state and supramolecular organization of the chloroplast ATP synthase. Dimeric complexes of *C. reinhardtii* ATP synthase were more abundant when grown under photomixotrophic compared to photoauxotrophic conditions. Metabolic control of dimerization of the chloroplast enzyme in particular by phosphate seems plausible, given that concentrations of this substrate/product of the ATP synthase are not constant within the chloroplast stroma, but vary depending on metabolic conditions and intracellular oxygen (Sharkey & Vanderveer, 1989). Further direct evidence for a dimeric supramolecular organization of the chloroplast ATP synthase comes from early cryo-ET images of thylakoid membranes (Dekker & Boekema, 2005), although it has been suggested that the observed dimers might have formed due to unspecific interaction.

5. Conclusion and future perspective

Remarkable progress has been made in ATP synthase structural characterization over the last decades largely by X-ray crystallography and cryo-EM providing detailed insights into the structure and supramolecular architecture of this exceptional multi-subunit macromolecular machine. In concert with mutagenesis and single molecule studies these high-resolution structural data provided insight into role of each subunit of the multi-subunit complex and have greatly advanced our mechanistic understanding of the membrane peripheral F_1, the membrane-embedded F_o and the ATP synthase F_oF_1 holoenzyme. Supported by computational simulations, detailed analysis of the structural and kinetic data acquired in recent years finally enabled us to elucidate the molecular origin of the driving force that leads to unidirectional stepped rotation of rotor subunits in the $\alpha_3\beta_3$ catalytic core. Further understanding of the rotary chemical function may come from atomistic simulations of the motor at μs time scales which are not possible yet due to the huge size of the molecular system. Open questions remain on biogenesis, assembly and synchronization of nuclear and plastid-encoded subunits and on the potential role of lipids or chloroplast pigments in these processes and for ATP synthase function. Answers to these questions may eventually allow efficient and stable assembly of biological motors of unsurpassed efficiency and sustainability *in vitro*. Future studies should also be directed to elucidate variations in macromolecular and supramolecular organization in response to different environmental or physiological conditions to foster our understanding on the regulatory setscrews affecting complex stability and ATP synthase catalytic activity. Future discoveries on complex assembly, stability and regulation may ultimately allow multidisciplinary translational research teams to construct efficient nanomachines driven by electrochemical gradients. However, to this end in addition to the protein biological machine further challenge will be posed by the design of stable low-conducting biomimetic or synthetic monolayer membranes.

References

Abrahams, J. P., Leslie, A. G., Lutter, R., & Walker, J. E. (1994). Structure at 2.8 A resolution of F1-ATPase from bovine heart mitochondria. *Nature, 370*(6491), 621–628. https://doi.org/10.1038/370621a0.

Akey, C. W., Crepeau, R. H., Dunn, S. D., McCarty, R. E., & Edelstein, S. J. (1983). Electron microscopy and single molecule averaging of subunit-deficient F1-ATPases from Escherichia coli and spinach chloroplasts. *The EMBO Journal, 2*(8), 1409–1415.

Allegretti, M., Klusch, N., Mills, D. J., Vonck, J., Kuhlbrandt, W., & Davies, K. M. (2015). Horizontal membrane-intrinsic alpha-helices in the stator a-subunit of an F-type ATP synthase. *Nature, 521*(7551), 237–240. https://doi.org/10.1038/nature14185.

Andersson, B., & Anderson, J. M. (1980). Lateral heterogeneity in the distribution of chlorophyll-protein complexes of the thylakoid membranes of spinach chloroplasts. *Biochimica et Biophysica Acta (BBA) Bioenergetics, 593*(2), 427–440. https://doi.org/10.1016/0005-2728(80)90078-X.

Arana, J. L., & Vallejos, R. H. (1982). Involvement of sulfhydryl groups in the activation mechanism of the ATPase activity of chloroplast coupling factor 1. *The Journal of Biological Chemistry, 257*(3), 1125–1127.

Arechaga, I., Miroux, B., Runswick, M. J., & Walker, J. E. (2003). Over-expression of Escherichia coli F 1 F o -ATPase subunit a is inhibited by instability of the uncB gene transcript. *FEBS Letters, 547*(1–3), 97–100. https://doi.org/10.1016/S0014-5793(03)00677-X.

Arnold, I., Pfeiffer, K., Neupert, W., Stuart, R. A., & Schägger, H. (1998). Yeast mitochondrial F1F0-ATP synthase exists as a dimer: Identification of three dimer-specific subunits. *The EMBO Journal, 17*(24), 7170–7178. https://doi.org/10.1093/emboj/17.24.7170.

Arvidsson, P.-O., & Sundby, C. (1999). A model for the topology of the chloroplast thylakoid membrane. *Functional Plant Biology, 26*(7), 687. https://doi.org/10.1071/PP99072.

Austin, J. R., & Staehelin, L. A. (2011). Three-dimensional architecture of grana and stroma thylakoids of higher plants as determined by electron tomography. *Plant Physiology, 155*(4), 1601–1611. https://doi.org/10.1104/pp.110.170647.

Autzen, H. E., Myasnikov, A. G., Campbell, M. G., Asarnow, D., Julius, D., & Cheng, Y. (2018). Structure of the human TRPM4 ion channel in a lipid nanodisc. *Science, 359*(6372), 228–232. https://doi.org/10.1126/science.aar4510.

Bamber, L., Harding, M., Butler, P. J. G., & Kunji, E. R. S. (2006). Yeast mitochondrial ADP/ATP carriers are monomeric in detergents. *Proceedings of the National Academy of Sciences of the United States of America, 103*(44), 16224–16229. https://doi.org/10.1073/pnas.0607640103.

Bayburt, T. H., & Sligar, S. G. (2010). Membrane protein assembly into nanodiscs. *FEBS Letters, 584*(9), 1721–1727. https://doi.org/10.1016/j.febslet.2009.10.024.

Beckers, G., Berzborn, R. J., & Strotmann, H. (1992). Zero-length crosslinking between subunits delta and I of the H(+)-translocating ATPase of chloroplasts. *Biochimica et Biophysica Acta, 1101*(1), 97–104. https://doi.org/10.1016/0167-4838(92)90473-q.

Bianchet, M. A., Hullihen, J., Pedersen, P. L., & Amzel, L. M. (1998). The 2.8-A structure of rat liver F1-ATPase: Configuration of a critical intermediate in ATP synthesis/hydrolysis. *Proceedings of the National Academy of Sciences of the United States of America, 95*(19), 11065–11070. https://doi.org/10.1073/pnas.95.19.11065.

Bill, R. M., Henderson, P. J. F., Iwata, S., Kunji, E. R. S., Michel, H., Neutze, R., et al. (2011). Overcoming barriers to membrane protein structure determination. *Nature Biotechnology*, *29*(4), 335–340. https://doi.org/10.1038/nbt.1833.

Birch, J., Axford, D., Foadi, J., Meyer, A., Eckhardt, A., Thielmann, Y., et al. (2018). The fine art of integral membrane protein crystallisation. *Methods (San Diego, Calif)*, *147*, 150–162. https://doi.org/10.1016/j.ymeth.2018.05.014.

Birkenhäger, R., Hoppert, M., Deckers-Hebestreit, G., Mayer, F., & Altendorf, K. (1995). The F0 complex of the Escherichia coli ATP synthase. Investigation by electron spectroscopic imaging and immunoelectron microscopy. *European Journal of Biochemistry*, *230*(1), 58–67.

Bockmann, R. A., & Grubmuller, H. (2002). Nanoseconds molecular dynamics simulation of primary mechanical energy transfer steps in F1-ATP synthase. *Nature Structural Biology*, *9*(3), 198–202. https://doi.org/10.1038/nsb760.

Boekema, E. J., Schmidt, G., Gräber, P., & Berden, J. A. (1988). Structure of the ATP-synthase from chloroplasts and mitochondria studied by electron microscopy. *Zeitschrift für Naturforschung C, Journal of Biosciences*, *43*(3–4), 219–225. https://doi.org/10.1515/znc-1988-3-412.

Böttcher, B., & Gräber, P. (2000). The structure of the H(+)-ATP synthase from chloroplasts and its subcomplexes as revealed by electron microscopy. *Biochimica et Biophysica Acta*, *1458*(2–3), 404–416. https://doi.org/10.1016/s0005-2728(00)00090-6.

Böttcher, B., Lücken, U., & Gräber, P. (1995). The structure of the H(+)-ATPase from chloroplasts by electron cryomicroscopy. *Biochemical Society Transactions*, *23*(4), 780–785. https://doi.org/10.1042/bst0230780.

Böttcher, B., Schwarz, L., & Gräber, P. (1998). Direct indication for the existence of a double stalk in CF0F1. *Journal of Molecular Biology*, *281*(5), 757–762. https://doi.org/10.1006/jmbi.1998.1957.

Boyer, P. D. (1975). A model for conformational coupling of membrane potential and proton translocation to ATP synthesis and to active transport. *FEBS Letters*, *58*(1), 1–6. https://doi.org/10.1016/0014-5793(75)80212-2.

Boyer, P. D. (1989). A perspective of the binding change mechanism for ATP synthesis. *The FASEB Journal*, *3*(10), 2164–2178. https://doi.org/10.1096/fasebj.3.10.2526771.

Breibeck, J., & Rompel, A. (2019). Successful amphiphiles as the key to crystallization of membrane proteins: Bridging theory and practice. *Biochimica et Biophysica Acta General Subjects*, *1863*(2), 437–455. https://doi.org/10.1016/j.bbagen.2018.11.004.

Bulygin, V. V., Duncan, T. M., & Cross, R. L. (2004). Rotor/stator interactions of the epsilon subunit in Escherichia coli ATP synthase and implications for enzyme regulation. *The Journal of Biological Chemistry*, *279*(34), 35616–35621. https://doi.org/10.1074/jbc.M405012200.

Caffrey, M. (2008). On the mechanism of membrane protein crystallization in lipidic Mesophases. *Crystal Growth & Design*, *8*(12), 4244–4254. https://doi.org/10.1021/cg800693r.

Caffrey, M. (2015). A comprehensive review of the lipid cubic phase or in meso method for crystallizing membrane and soluble proteins and complexes. *Acta Crystallographica Section F, Structural Biology Communications*, *71*(1), 3–18. https://doi.org/10.1107/S2053230X14026843.

Caffrey, M., Li, D., & Dukkipati, A. (2012). Membrane protein structure determination using crystallography and lipidic mesophases: Recent advances and successes. *Biochemistry*, *51*(32), 6266–6288. https://doi.org/10.1021/bi300010w.

Cao, E., Liao, M., Cheng, Y., & Julius, D. (2013). TRPV1 structures in distinct conformations reveal activation mechanisms. *Nature*, *504*(7478), 113–118. https://doi.org/10.1038/nature12823.

Carbajo, R. J., Silvester, J. A., Runswick, M. J., Walker, J. E., & Neuhaus, D. (2004). Solution structure of subunit F(6) from the peripheral stalk region of ATP synthase from bovine heart mitochondria. *Journal of Molecular Biology*, *342*(2), 593–603. https://doi.org/10.1016/j.jmb.2004.07.013.

Cheng, Y. (2018). Membrane protein structural biology in the era of single particle cryo-EM. *Current Opinion in Structural Biology*, *52*, 58–63. https://doi.org/10.1016/j.sbi.2018.08.008.

Cheng, R., Huang, C., Hennig, M., Nar, H., & Schnapp, G. (2019). In situ crystallography as an emerging method for structure solution of membrane proteins: The case of CCR2A. *The FEBS Journal*, *287*, 866–873. febs.15098. https://doi.org/10.1111/febs.15098.

Cherezov, V. (2011). Lipidic cubic phase technologies for membrane protein structural studies. *Current Opinion in Structural Biology*, *21*(4), 559–566. https://doi.org/10.1016/j.sbi.2011.06.007.

Chipot, C., Dehez, F., Schnell, J. R., Zitzmann, N., Pebay-Peyroula, E., Catoire, L. J., et al. (2018). Perturbations of native membrane protein structure in alkyl phosphocholine detergents: A critical assessment of NMR and biophysical studies. *Chemical Reviews*, *118*(7), 3559–3607. https://doi.org/10.1021/acs.chemrev.7b00570.

Cohen, R. D., & Pielak, G. J. (2017). A cell is more than the sum of its (dilute) parts: A brief history of quinary structure. *Protein Science: A Publication of the Protein Society*, *26*(3), 403–413. https://doi.org/10.1002/pro.3092.

Cortés-Hernández, P., Vázquez-Memije, M. E., & García, J. J. (2007). ATP6 homoplasmic mutations inhibit and destabilize the human F1F0-ATP synthase without preventing enzyme assembly and oligomerization. *The Journal of Biological Chemistry*, *282*(2), 1051–1058. https://doi.org/10.1074/jbc.M606828200.

Couoh-Cardel, S. J., Uribe-Carvajal, S., Wilkens, S., & García-Trejo, J. J. (2010). Structure of dimeric F1F0-ATP synthase. *The Journal of Biological Chemistry*, *285*(47), 36447–36455. https://doi.org/10.1074/jbc.M110.144907.

Daum, B., & Kuhlbrandt, W. (2011). Electron tomography of plant thylakoid membranes. *Journal of Experimental Botany*, *62*(7), 2393–2402. https://doi.org/10.1093/jxb/err034.

Daum, B., Nicastro, D., Austin, J., McIntosh, J. R., & Kühlbrandt, W. (2010). Arrangement of photosystem II and ATP synthase in chloroplast membranes of spinach and pea. *The Plant Cell*, *22*(4), 1299–1312. https://doi.org/10.1105/tpc.109.071431.

Davies, K. M., Anselmi, C., Wittig, I., Faraldo-Gómez, J. D., & Kühlbrandt, W. (2012). Structure of the yeast F1Fo-ATP synthase dimer and its role in shaping the mitochondrial cristae. *Proceedings of the National Academy of Sciences of the United States of America*, *109*(34), 13602–13607. https://doi.org/10.1073/pnas.1204593109.

Davies, K. M., Strauss, M., Daum, B., Kief, J. H., Osiewacz, H. D., Rycovska, A., et al. (2011). Macromolecular organization of ATP synthase and complex I in whole mitochondria. *Proceedings of the National Academy of Sciences of the United States of America*, *108*(34), 14121–14126. https://doi.org/10.1073/pnas.1103621108.

Deisenhofer, J., Epp, O., Miki, K., Huber, R., & Michel, H. (1985). Structure of the protein subunits in the photosynthetic reaction centre of Rhodopseudomonas viridis at 3Å resolution. *Nature*, *318*(6047), 618–624. https://doi.org/10.1038/318618a0.

Deisenhofer, J., Epp, O., Sinning, I., & Michel, H. (1995). Crystallographic refinement at 2.3 Å resolution and refined model of the photosynthetic reaction centre from Rhodopseudomonas viridis. *Journal of Molecular Biology*, *246*(3), 429–457. https://doi.org/10.1006/jmbi.1994.0097.

Dekker, J. P., & Boekema, E. J. (2005). Supramolecular organization of thylakoid membrane proteins in green plants. *Biochimica et Biophysica Acta (BBA) Bioenergetics*, *1706*(1–2), 12–39. https://doi.org/10.1016/j.bbabio.2004.09.009.

Del Rizzo, P. A., Bi, Y., & Dunn, S. D. (2006). ATP synthase b subunit dimerization domain: A right-handed coiled coil with offset helices. *Journal of Molecular Biology, 364*(4), 735–746. https://doi.org/10.1016/j.jmb.2006.09.028.

Del Rizzo, P. A., Bi, Y., Dunn, S. D., & Shilton, B. H. (2002). The "second stalk" of Escherichia coli ATP synthase: Structure of the isolated dimerization domain. *Biochemistry, 41*(21), 6875–6884. https://doi.org/10.1021/bi025736i.

Dilworth, M. V., Piel, M. S., Bettaney, K. E., Ma, P., Luo, J., Sharples, D., et al. (2018). Microbial expression systems for membrane proteins. *Methods (San Diego, Calif), 147*, 3–39. https://doi.org/10.1016/j.ymeth.2018.04.009.

Dmitriev, O., Altendorf, K., & Fillingame, R. H. (2004). Subunit A of the E. coli ATP synthase: Reconstitution and high resolution NMR with protein purified in a mixed polarity solvent. *FEBS Letters, 556*(1–3), 35–38. https://doi.org/10.1016/s0014-5793(03)01360-7.

Dmitriev, O. Y., Jones, P. C., Jiang, W., & Fillingame, R. H. (1999). Structure of the membrane domain of subunit b of the Escherichia coli F0F1 ATP synthase. *The Journal of Biological Chemistry, 274*(22), 15598–15604. https://doi.org/10.1074/jbc.274.22.15598.

Dobro, M. J., Melanson, L. A., Jensen, G. J., & McDowall, A. W. (2010). Plunge freezing for electron cryomicroscopy. In JensenG. J. (Ed.), *Methods in Enzymology* 481. (pp. 63–82). Elsevier. https://doi.org/10.1016/S0076-6879(10)81003-1.

Dudkina, N. V., Kouril, R., Peters, K., Braun, H.-P., & Boekema, E. J. (2010). Structure and function of mitochondrial supercomplexes. *Biochimica et Biophysica Acta, 1797*(6–7), 664–670. https://doi.org/10.1016/j.bbabio.2009.12.013.

Dudkina, N. V., Sunderhaus, S., Braun, H.-P., & Boekema, E. J. (2006). Characterization of dimeric ATP synthase and cristae membrane ultrastructure from Saccharomyces and Polytomella mitochondria. *FEBS Letters, 580*(14), 3427–3432. https://doi.org/10.1016/j.febslet.2006.04.097.

Duncan, T. M., Bulygin, V. V., Zhou, Y., Hutcheon, M. L., & Cross, R. L. (1995). Rotation of subunits during catalysis by Escherichia coli F1-ATPase. *Proceedings of the National Academy of Sciences of the United States of America, 92*(24), 10964–10968. https://doi.org/10.1073/pnas.92.24.10964.

Dürr, U. H. N., Gildenberg, M., & Ramamoorthy, A. (2012). The magic of bicelles lights up membrane protein structure. *Chemical Reviews, 112*(11), 6054–6074. https://doi.org/10.1021/cr300061w.

Duser, M. G., Zarrabi, N., Cipriano, D. J., Ernst, S., Glick, G. D., Dunn, S. D., et al. (2009). 36 degrees step size of proton-driven c-ring rotation in FoF1-ATP synthase. *The EMBO Journal, 28*(18), 2689–2696. https://doi.org/10.1038/emboj.2009.213.

Egelman, E. H. (2016). The current revolution in cryo-EM. *Biophysical Journal, 110*(5), 1008–1012. https://doi.org/10.1016/j.bpj.2016.02.001.

Engelbrecht, S., Giakas, E., Marx, O., & Lill, H. (1998). Fluorescence resonance energy transfer mapping of subunit delta in spinach chloroplast F1 ATPase. *European Journal of Biochemistry, 252*(2), 277–283. https://doi.org/10.1046/j.1432-1327.1998.2520277.x.

Engelman, D. M. (2005). Membranes are more mosaic than fluid. *Nature, 438*(7068), 578–580. https://doi.org/10.1038/nature04394.

Eubel, H., Jänsch, L., & Braun, H.-P. (2003). New insights into the respiratory chain of plant mitochondria. Supercomplexes and a unique composition of complex II. *Plant Physiology, 133*(1), 274–286. https://doi.org/10.1104/pp.103.024620.

Faham, S., & Bowie, J. U. (2002). Bicelle crystallization: A new method for crystallizing membrane proteins yields a monomeric bacteriorhodopsin structure. *Journal of Molecular Biology, 316*(1), 1–6. https://doi.org/10.1006/jmbi.2001.5295.

Fearnley, I. M., & Walker, J. E. (1986). Two overlapping genes in bovine mitochondrial DNA encode membrane components of ATP synthase. *The EMBO Journal, 5*(8), 2003–2008.

Ferguson, S. A., Cook, G. M., Montgomery, M. G., Leslie, A. G., & Walker, J. E. (2016). Regulation of the thermoalkaliphilic F1-ATPase from Caldalkalibacillus thermarum. *Proceedings of the National Academy of Sciences of the United States of America, 113*(39), 10860–10865. https://doi.org/10.1073/pnas.1612035113.

Fromme, P., Boekema, E. J., & Gräber, P. (1987). Isolation and characterization of a suprmolecular complex of subunit III of the ATP synthase from chloroplasts. *Zeitschrift für Naturforschung, 42c*, 1239–1245.

Fromme, P., & Spence, J. C. (2011). Femtosecond nanocrystallography using X-ray lasers for membrane protein structure determination. *Current Opinion in Structural Biology, 21*(4), 509–516. https://doi.org/10.1016/j.sbi.2011.06.001.

Fronzes, R., Weimann, T., Vaillier, J., Velours, J., & Brèthes, D. (2006). The peripheral stalk participates in the yeast ATP synthase dimerization independently of e and g subunits. *Biochemistry, 45*(21), 6715–6723. https://doi.org/10.1021/bi0601407.

Furuike, S., Hossain, M. D., Maki, Y., Adachi, K., Suzuki, T., Kohori, A., et al. (2008). Axle-less F1-ATPase rotates in the correct direction. *Science, 319*(5865), 955–958. https://doi.org/10.1126/science.1151343.

Garavito, R. M., & Rosenbusch, J. P. (1980). Three-dimensional crystals of an integral membrane protein: An initial x-ray analysis. *The Journal of Cell Biology, 86*(1), 327–329. https://doi.org/10.1083/jcb.86.1.327.

Gautier, A. (2014). Structure determination of α-helical membrane proteins by solution-state NMR: Emphasis on retinal proteins. *Biochimica Et Biophysica Acta, 1837*(5), 578–588. https://doi.org/10/ggm6tc.

Gautier, A., Mott, H. R., Bostock, M. J., Kirkpatrick, J. P., & Nietlispach, D. (2010). Structure determination of the seven-helix transmembrane receptor sensory rhodopsin II by solution NMR spectroscopy. *Nature Structural & Molecular Biology, 17*(6), 768–774. https://doi.org/10.1038/nsmb.1807.

Geertsma, E. R., Groeneveld, M., Slotboom, D.-J., & Poolman, B. (2008). Quality control of overexpressed membrane proteins. *Proceedings of the National Academy of Sciences of the United States of America, 105*(15), 5722–5727. https://doi.org/10.1073/pnas.0802190105.

Gibbons, C., Montgomery, M. G., Leslie, A. G., & Walker, J. E. (2000). The structure of the central stalk in bovine F(1)-ATPase at 2.4 A resolution. *Nature Structural Biology, 7*(11), 1055–1061. https://doi.org/10.1038/80981.

Girvin, M. E., Rastogi, V. K., Abildgaard, F., Markley, J. L., & Fillingame, R. H. (1998). Solution structure of the transmembrane H+-transporting subunit c of the F1F0 ATP synthase. *Biochemistry, 37*(25), 8817–8824. https://doi.org/10.1021/bi980511m.

Gohlke, H., Schlieper, D., & Groth, G. (2012). Resolving the negative potential side (n-side) water-accessible proton pathway of F-type ATP synthase by molecular dynamics simulations. *The Journal of Biological Chemistry, 287*(43), 36536–36543. https://doi.org/10.1074/jbc.M112.398396.

Gourdon, P., Andersen, J. L., Hein, K. L., Bublitz, M., Pedersen, B. P., Liu, X.-Y., et al. (2011). HiLiDe—Systematic approach to membrane protein crystallization in lipid and detergent. *Crystal Growth & Design, 11*(6), 2098–2106. https://doi.org/10.1021/cg101360d.

Gräber, P., Fromme, P., Schmidt, G., & Boekema, E. (1988). Structure of the ATP-synthase from chloroplasts as revealed from biochemical studies and electron microscopy. *Progress in Clinical and Biological Research, 273*, 67–74.

Grisshammer, R. (2017). New approaches towards the understanding of integral membrane proteins: A structural perspective on G protein-coupled receptors. *Protein Science: A Publication of the Protein Society, 26*(8), 1493–1504. https://doi.org/10.1002/pro.3200.

Groth, G. (2002). Structure of spinach chloroplast F1-ATPase complexed with the phytopathogenic inhibitor tentoxin. *Proceedings of the National Academy of Sciences of the United States of America, 99*(6), 3464–3468. https://doi.org/10.1073/pnas.052546099.

Groth, G., Hisabori, T., Lill, H., & Bald, D. (2002). Substitution of a single amino acid switches the tentoxin-resistant thermophilic F1-ATPase into a tentoxin-sensitive enzyme. *The Journal of Biological Chemistry*, *277*(23), 20117–20119. https://doi.org/10.1074/jbc.C200168200.

Groth, G., Mills, D. A., Christiansen, E., Richter, M. L., & Huchzermeyer, B. (2000). Characterization of a phosphate binding domain on the alpha-subunit of chloroplast ATP synthase using the photoaffinity phosphate analogue 4-azido-2-nitrophenyl phosphate. *Biochemistry*, *39*(45), 13781–13787. https://doi.org/10.1021/bi000991t.

Groth, G., & Pohl, E. (2001). The structure of the chloroplast F1-ATPase at 3.2 A resolution. *The Journal of Biological Chemistry*, *276*(2), 1345–1352. https://doi.org/10.1074/jbc.M008015200.

Groth, G., Tilg, Y., & Schirwitz, K. (1998). Molecular architecture of the c-subunit oligomer in the membrane domain of F-ATPases probed by tryptophan substitution mutagenesis. *Journal of Molecular Biology*, *281*(1), 49–59. https://doi.org/10.1006/jmbi.1998.1936.

Gubellini, F., Verdon, G., Karpowich, N. K., Luff, J. D., Boël, G., Gauthier, N., et al. (2011). Physiological response to membrane protein overexpression in E. coli. *Molecular & Cellular Proteomics*, *10*(10). https://doi.org/10.1074/mcp.M111.007930, M111.007930.

Guo, H., & Rubinstein, J. L. (2018). Cryo-EM of ATP synthases. *Current Opinion in Structural Biology*, *52*, 71–79. https://doi.org/10.1016/j.sbi.2018.08.005.

Guo, H., Bueler, S. A., & Rubinstein, J. L. (2017). Atomic model for the dimeric FO region of mitochondrial ATP synthase. *Science*, *358*(6365), 936–940. https://doi.org/10.1126/science.aao4815.

Hahn, A., Parey, K., Bublitz, M., Mills, D. J., Zickermann, V., Vonck, J., et al. (2016). Structure of a complete ATP synthase dimer reveals the molecular basis of inner mitochondrial membrane morphology. *Molecular Cell*, *63*(3), 445–456. https://doi.org/10.1016/j.molcel.2016.05.037.

Hahn, A., Vonck, J., Mills, D. J., Meier, T., & Kuhlbrandt, W. (2018). Structure, mechanism, and regulation of the chloroplast ATP synthase. *Science*, *360*(6389). https://doi.org/10.1126/science.aat4318.

Helms, V. (2002). Attraction within the membrane. Forces behind transmembrane protein folding and supramolecular complex assembly. *EMBO Reports*, *3*(12), 1133–1138. https://doi.org/10.1093/embo-reports/kvf245.

Hennig, J., & Herrmann, R. G. (1986). Chloroplast ATP synthase of spinach contains nine nonidentical subunit species, six of which are encoded by plastid chromosomes in two operons in a phylogenetically conserved arrangement. *Molecular & General Genetics*, *203*, 117–128.

Hisabori, T., Sunamura, E., Kim, Y., & Konno, H. (2013). The chloroplast ATP synthase features the characteristic redox regulation machinery. *Antioxidants & Redox Signaling*, *19*(15), 1846–1854. https://doi.org/10.1089/ars.2012.5044.

Huang, C.-Y., Olieric, V., Ma, P., Howe, N., Vogeley, L., Liu, X., et al. (2016). In meso in situ serial X-ray crystallography of soluble and membrane proteins at cryogenic temperatures. *Acta Crystallographica Section D, Structural Biology*, *72*(1), 93–112. https://doi.org/10.1107/S2059798315021683.

Huang, C.-Y., Olieric, V., Ma, P., Panepucci, E., Diederichs, K., Wang, M., et al. (2015). In meso in situ serial X-ray crystallography of soluble and membrane proteins. *Acta Crystallographica Section D Biological Crystallography*, *71*(6), 1238–1256. https://doi.org/10.1107/S1399004715005210.

Iino, R., Hasegawa, R., Tabata, K. V., & Noji, H. (2009). Mechanism of inhibition by C-terminal alpha-helices of the epsilon subunit of Escherichia coli FoF1-ATP synthase. *The Journal of Biological Chemistry*, *284*(26), 17457–17464. https://doi.org/10.1074/jbc.M109.003798.

Iino, R., Murakami, T., Iizuka, S., Kato-Yamada, Y., Suzuki, T., & Yoshida, M. (2005). Real-time monitoring of conformational dynamics of the epsilon subunit in F1-ATPase. *The Journal of Biological Chemistry, 280*(48), 40130–40134. https://doi.org/10.1074/jbc.M506160200.

Ishmukhametov, R., Hornung, T., Spetzler, D., & Frasch, W. D. (2010). Direct observation of stepped proteolipid ring rotation in E. coli F(0)F(1)-ATP synthase. *The EMBO Journal, 29*(23), 3911–3923. https://doi.org/10.1038/emboj.2010.259.

Jäger, H., Birkenhäger, R., Stalz, W. D., Altendorf, K., & Deckers-Hebestreit, G. (1998). Topology of subunit a of the Escherichia coli ATP synthase. *European Journal of Biochemistry, 251*(1–2), 122–132. https://doi.org/10.1046/j.1432-1327.1998.2510122.x.

Jiang, W., & Fillingame, R. H. (1998). Interacting helical faces of subunits a and c in the F1Fo ATP synthase of Escherichia coli defined by disulfide cross-linking. *Proceedings of the National Academy of Sciences of the United States of America, 95*(12), 6607–6612. https://doi.org/10.1073/pnas.95.12.6607.

Jiang, W., Hermolin, J., & Fillingame, R. H. (2001). The preferred stoichiometry of c subunits in the rotary motor sector of Escherichia coli ATP synthase is 10. *Proceedings of the National Academy of Sciences of the United States of America, 98*(9), 4966–4971. https://doi.org/10.1073/pnas.081424898.

Johnson, E. A., & McCarty, R. E. (2002). The carboxyl terminus of the epsilon subunit of the chloroplast ATP synthase is exposed during illumination. *Biochemistry, 41*(7), 2446–2451. https://doi.org/10.1021/bi011939f.

Jones, P. C., Jiang, W., & Fillingame, R. H. (1998). Arrangement of the multicopy H +-translocating subunit c in the membrane sector of the Escherichia coli F1F0 ATP synthase. *The Journal of Biological Chemistry, 273*(27), 17178–17185. https://doi.org/10.1074/jbc.273.27.17178.

Junge, W., Lill, H., & Engelbrecht, S. (1997). ATP synthase: An electrochemical transducer with rotatory mechanics. *Trends in Biochemical Sciences, 22*(11), 420–423. https://doi.org/10.1016/s0968-0004(97)01129-8.

Junge, F., Schneider, B., Reckel, S., Schwarz, D., Dötsch, V., & Bernhard, F. (2008). Large-scale production of functional membrane proteins. *Cellular and Molecular Life Sciences, 65*(11), 1729–1755. https://doi.org/10.1007/s00018-008-8067-5.

Kanonenberg, K., Royes, J., Kedrov, A., Poschmann, G., Angius, F., Solgadi, A., et al. (2019). Shaping the lipid composition of bacterial membranes for membrane protein production. *Microbial Cell Factories, 18*(1), 131. https://doi.org/10.1186/s12934-019-1182-1.

Kato, Y., Matsui, T., Tanaka, N., Muneyuki, E., Hisabori, T., & Yoshida, M. (1997). Thermophilic F1-ATPase is activated without dissociation of an endogenous inhibitor, epsilon subunit. *The Journal of Biological Chemistry, 272*(40), 24906–24912. https://doi.org/10.1074/jbc.272.40.24906.

Ketcham, S. R., Davenport, J. W., Warncke, K., & McCarty, R. E. (1984). Role of the gamma subunit of chloroplast coupling factor 1 in the light-dependent activation of photophosphorylation and ATPase activity by dithiothreitol. *The Journal of Biological Chemistry, 259*(11), 7286–7293.

Kirchhoff, H. (2013). Architectural switches in plant thylakoid membranes. *Photosynthesis Research, 116*(2–3), 481–487. https://doi.org/10/f5fpwf.

Kirchhoff, H. (2019). Chloroplast ultrastructure in plants. *The New Phytologist, 223*(2), 565–574. https://doi.org/10.1111/nph.15730.

Koga, N., & Takada, S. (2006). Folding-based molecular simulations reveal mechanisms of the rotary motor F1-ATPase. *Proceedings of the National Academy of Sciences of the United States of America, 103*(14), 5367–5372. https://doi.org/10.1073/pnas.0509642103.

Krause, F., Scheckhuber, C. Q., Werner, A., Rexroth, S., Reifschneider, N. H., Dencher, N. A., et al. (2004). Supramolecular organization of cytochrome c oxidase- and alternative oxidase-dependent respiratory chains in the filamentous fungus Podospora anserina. *The Journal of Biological Chemistry*, *279*(25), 26453–26461. https://doi.org/10.1074/jbc.M402756200.

Krogh, A., Larsson, B., von Heijne, G., & Sonnhammer, E. L. L. (2001). Predicting transmembrane protein topology with a hidden markov model: Application to complete genomes. Edited by F. Cohen *Journal of Molecular Biology*, *305*(3), 567–580. https://doi.org/10.1006/jmbi.2000.4315.

Kühlbrandt, W. (2014a). The resolution revolution. *Science*, *343*(6178), 1443–1444. https://doi.org/10.1126/science.1251652.

Kühlbrandt, W. (2014b). Cryo-EM enters a new era. *eLife*, *3*, e03678. https://doi.org/10.7554/eLife.03678.

Landau, E. M., & Rosenbusch, J. P. (1996). Lipidic cubic phases: A novel concept for the crystallization of membrane proteins. *Proceedings of the National Academy of Sciences of the United States of America*, *93*(25), 14532–14535. https://doi.org/10.1073/pnas.93.25.14532.

Long, J. C., Wang, S., & Vik, S. B. (1998). Membrane topology of subunit a of the F1F0 ATP synthase as determined by labeling of unique cysteine residues. *The Journal of Biological Chemistry*, *273*(26), 16235–16240. https://doi.org/10.1074/jbc.273.26.16235.

Martin-Garcia, J. M., Conrad, C. E., Coe, J., Roy-Chowdhury, S., & Fromme, P. (2016). Serial femtosecond crystallography: A revolution in structural biology. *Archives of Biochemistry and Biophysics*, *602*, 32–47. https://doi.org/10.1016/j.abb.2016.03.036.

McCarty, R. E., & Hammes, G. G. (1987). Molecular architecture of chloroplast coupling factor 1. *Trends in Biochemical Sciences*, *12*, 234–237.

McLachlin, D. T., Coveny, A. M., Clark, S. M., & Dunn, S. D. (2000). Site-directed cross-linking of b to the alpha, beta, and a subunits of the Escherichia coli ATP synthase. *The Journal of Biological Chemistry*, *275*(23), 17571–17577. https://doi.org/10.1074/jbc.M000375200.

Meier, T., Polzer, P., Diederichs, K., Welte, W., & Dimroth, P. (2005). Structure of the rotor ring of F-type Na+-ATPase from Ilyobacter tartaricus. *Science*, *308*(5722), 659–662. https://doi.org/10.1126/science.1111199.

Meiss, E., Konno, H., Groth, G., & Hisabori, T. (2008). Molecular processes of inhibition and stimulation of ATP synthase caused by the phytotoxin tentoxin. *The Journal of Biological Chemistry*, *283*(36), 24594–24599. https://doi.org/10.1074/jbc.M802574200.

Mellwig, C., & Bottcher, B. (2001). Dealing with particles in different conformational states by electron microscopy and image processing. *Journal of Structural Biology*, *133*(2–3), 214–220. https://doi.org/10.1006/jsbi.2001.4340.

Menz, R. I., Walker, J. E., & Leslie, A. G. (2001). Structure of bovine mitochondrial F(1)-ATPase with nucleotide bound to all three catalytic sites: Implications for the mechanism of rotary catalysis. *Cell*, *106*(3), 331–341. https://doi.org/10.1016/s0092-8674(01)00452-4.

Michel, H. (1983). Crystallization of membrane proteins. *Trends in Biochemical Sciences*, *8*(2), 56–59. https://doi.org/10.1016/0968-0004(83)90390-0.

Miki, J., Maeda, M., Mukohata, Y., & Futai, M. (1988). The gamma-subunit of ATP synthase from spinach chloroplasts. Primary structure deduced from the cloned cDNA sequence. *FEBS Letters*, *232*(1), 221–226. https://doi.org/10.1016/0014-5793(88)80421-6.

Mills, D. J., Mitchell, P., & Schürmann, P. (1980). Modulation of coupling ATPase activity in intact chloroplasts. *FEBS Letters*, *191*, 144–148.

Minauro-Sanmiguel, F., Wilkens, S., & García, J. J. (2005). Structure of dimeric mitochondrial ATP synthase: Novel F0 bridging features and the structural basis of mitochondrial

cristae biogenesis. *Proceedings of the National Academy of Sciences of the United States of America, 102*(35), 12356–12358. https://doi.org/10.1073/pnas.0503893102.
Miroux, B., & Walker, J. E. (1996). Over-production of proteins in Escherichia coli: Mutant hosts that allow synthesis of some membrane proteins and globular proteins at high levels. *Journal of Molecular Biology, 260*(3), 289–298. https://doi.org/10.1006/jmbi.1996.0399.
Mnatsakanyan, N., Krishnakumar, A. M., Suzuki, T., & Weber, J. (2009). The role of the betaDELSEED-loop of ATP synthase. *The Journal of Biological Chemistry, 284*(17), 11336–11345. https://doi.org/10.1074/jbc.M900374200.
Moraes, I., Evans, G., Sanchez-Weatherby, J., Newstead, S., & Stewart, P. D. S. (2014). Membrane protein structure determination—The next generation. *Biochimica et Biophysica Acta (BBA) Biomembranes, 1838*(1), 78–87. https://doi.org/10.1016/j.bbamem.2013.07.010.
Morales-Rios, E., Montgomery, M. G., Leslie, A. G., & Walker, J. E. (2015). Structure of ATP synthase from Paracoccus denitrificans determined by X-ray crystallography at 4.0 A resolution. *Proceedings of the National Academy of Sciences of the United States of America, 112*(43), 13231–13236. https://doi.org/10.1073/pnas.1517542112.
Murakami, S., Kondo, K., Katayama, S., Hara, S., Sunamura, E. I., Yamashita, E., et al. (2018). Structure of the gamma-epsilon complex of cyanobacterial F1-ATPase reveals a suppression mechanism of the gamma subunit on ATP hydrolysis in phototrophs. *The Biochemical Journal, 475*(18), 2925–2939. https://doi.org/10.1042/BCJ20180481.
Murphy, B. J., Klusch, N., Langer, J., Mills, D. J., Yildiz, O., & Kuhlbrandt, W. (2019). Rotary substates of mitochondrial ATP synthase reveal the basis of flexible F1-Fo coupling. *Science, 364*(6446), eaaw9128. https://doi.org/10.1126/science.aaw9128.
Mustardy, L. (2003). Granum revisited. A three-dimensional model ? Where things fall into place. *Trends in Plant Science, 8*(3), 117–122. https://doi.org/10.1016/S1360-1385(03)00015-3.
Mustárdy, L., Buttle, K., Steinbach, G., & Garab, G. (2008). The three-dimensional network of the thylakoid membranes in plants: Quasihelical model of the granum-stroma assembly. *The Plant Cell, 20*(10), 2552–2557. https://doi.org/10.1105/tpc.108.059147.
Neff, D., Tripathi, S., Middendorf, K., Stahlberg, H., Butt, H. J., Bamberg, E., et al. (1997). Chloroplast F0F1 ATP synthase imaged by atomic force microscopy. *Journal of Structural Biology, 119*(2), 139–148. https://doi.org/10.1006/jsbi.1997.3891.
Nesci, S., Trombetti, F., Ventrella, V., & Pagliarani, A. (2016). The c-ring of the F1FO-ATP synthase: Facts and perspectives. *The Journal of Membrane Biology, 249*(1–2), 11–21. https://doi.org/10.1007/s00232-015-9860-3.
Nevo, R., Charuvi, D., Tsabari, O., & Reich, Z. (2012). Composition, architecture and dynamics of the photosynthetic apparatus in higher plants: The photosynthetic apparatus in higher plants. *The Plant Journal, 70*(1), 157–176. https://doi.org/10.1111/j.1365-313X.2011.04876.x.
Nevo, R., Chuartzman, S. G., Tsabari, O., Reich, Z., Charuvi, D., & Shimoni, E. (2009). Architecture of thylakoid membrane networks. In H. Wada & N. Murata (Eds.), *Lipids in photosynthesis: Vol. 30.* (pp. 295–328). Netherlands: Springer. https://doi.org/10.1007/978-90-481-2863-1_14.
Noji, H., Yasuda, R., Yoshida, M., & Kinosita, K., Jr. (1997). Direct observation of the rotation of F1-ATPase. *Nature, 386*(6622), 299–302. https://doi.org/10.1038/386299a0.
Opella, S. J., & Marassi, F. M. (2017). Applications of NMR to membrane proteins. *Archives of Biochemistry and Biophysics, 628*, 92–101. https://doi.org/10.1016/j.abb.2017.05.011.
Owen, R. L., Juanhuix, J., & Fuchs, M. (2016). Current advances in synchrotron radiation instrumentation for MX experiments. *Archives of Biochemistry and Biophysics, 602*, 21–31. https://doi.org/10.1016/j.abb.2016.03.021.
Pandey, A., Shin, K., Patterson, R. E., Liu, X.-Q., & Rainey, J. K. (2016). Current strategies for protein production and purification enabling membrane protein structural biology. *Biochemistry and Cell Biology, 94*(6), 507–527. https://doi.org/10.1139/bcb-2015-0143.

Paolillo, D. J. (1970). The three-dimensional arrangement of intergranal lamellae in chloroplasts. *Journal of Cell Science*, *6*(1), 243–255.

Parker, J. L., & Newstead, S. (2012). Current trends in α-helical membrane protein crystallization: An update: α-Helical MP crystallization. *Protein Science*, *21*(9), 1358–1365. https://doi.org/10.1002/pro.2122.

Pogoryelov, D., Yu, J., Meier, T., Vonck, J., Dimroth, P., & Muller, D. J. (2005). The c15 ring of the Spirulina platensis F-ATP synthase: F1/F0 symmetry mismatch is not obligatory. *EMBO Reports*, *6*(11), 1040–1044. https://doi.org/10.1038/sj.embor.7400517.

Poulos, S., Morgan, J. L. W., Zimmer, J., & Faham, S. (2015). Bicelles coming of age. In ShuklaA. K. (Ed.), *Methods in enzymology*. Elsevier. https://doi.org/10.1016/bs.mie.2014.12.024, Bd. 557, S. 393–416.

Preiss, L., Langer, J. D., Yildiz, O., Eckhardt-Strelau, L., Guillemont, J. E., Koul, A., et al. (2015). Structure of the mycobacterial ATP synthase Fo rotor ring in complex with the anti-TB drug bedaquiline. *Science Advances*, *1*(4), e1500106. https://doi.org/10.1126/sciadv.1500106.

Pribil, M., Labs, M., & Leister, D. (2014). Structure and dynamics of thylakoids in land plants. *Journal of Experimental Botany*, *65*(8), 1955–1972. https://doi.org/10.1093/jxb/eru090.

Pu, J., & Karplus, M. (2008). How subunit coupling produces the gamma-subunit rotary motion in F1-ATPase. *Proceedings of the National Academy of Sciences of the United States of America*, *105*(4), 1192–1197. https://doi.org/10.1073/pnas.0708746105.

Puthenveetil, R., & Vinogradova, O. (2019). Solution NMR: A powerful tool for structural and functional studies of membrane proteins in reconstituted environments. *The Journal of Biological Chemistry*, *294*(44), 15914–15931. https://doi.org/10.1074/jbc.REV119.009178.

Rasmussen, S. G. F., Choi, H.-J., Rosenbaum, D. M., Kobilka, T. S., Thian, F. S., Edwards, P. C., et al. (2007). Crystal structure of the human β2 adrenergic G-protein-coupled receptor. *Nature*, *450*(7168), 383–387. https://doi.org/10.1038/nature06325.

Rastogi, V. K., & Girvin, M. E. (1999a). 1H, 13C, and 15N assignments and secondary structure of the high pH form of subunit c of the F1F0 ATP synthase. *Journal of Biomolecular NMR*, *13*(1), 91–92. https://doi.org/10.1023/a:1008379624478.

Rastogi, V. K., & Girvin, M. E. (1999b). Structural changes linked to proton translocation by subunit c of the ATP synthase. *Nature*, *402*(6759), 263–268. https://doi.org/10.1038/46224.

RCSB PDB. (2020). *PDB statistics*. Received from https://www.rcsb.org/stats.

Rexroth, S., Meyer Zu Tittingdorf, J. M. W., Schwassmann, H. J., Krause, F., Seelert, H., & Dencher, N. A. (2004). Dimeric H+-ATP synthase in the chloroplast of Chlamydomonas reinhardtii. *Biochimica et Biophysica Acta*, *1658*(3), 202–211. https://doi.org/10.1016/j.bbabio.2004.05.014.

Richter, M. L. (2004). Gamma-epsilon interactions regulate the chloroplast ATP synthase. *Photosynthesis Research*, *79*(3), 319–329. https://doi.org/10.1023/B:PRES.0000017157.08098.36.

Richter, M. L., & McCarty, R. E. (1987). Energy-dependent changes in the conformation of the epsilon subunit of the chloroplast ATP synthase. *The Journal of Biological Chemistry*, *262*(31), 15037–15040.

Richter, M. L., Snyder, B., McCarty, R. E., & Hammes, G. G. (1985). Binding stoichiometry and structural mapping of the epsilon polypeptide of chloroplast coupling factor 1. *Biochemistry*, *24*(21), 5755–5763. https://doi.org/10.1021/bi00342a011.

Sabbert, D., Engelbrecht, S., & Junge, W. (1996). Intersubunit rotation in active F-ATPase. *Nature*, *381*(6583), 623–625. https://doi.org/10.1038/381623a0.

Saita, E., Iino, R., Suzuki, T., Feniouk, B. A., Kinosita, K., Jr., & Yoshida, M. (2010). Activation and stiffness of the inhibited states of F1-ATPase probed by single-molecule

manipulation. *The Journal of Biological Chemistry, 285*(15), 11411–11417. https://doi.org/10.1074/jbc.M109.099143.
Sanders, C. R., & Prestegard, J. H. (1990). Magnetically orientable phospholipid bilayers containing small amounts of a bile salt analogue, CHAPSO. *Biophysical Journal, 58*(2), 447–460. https://doi.org/10.1016/S0006-3495(90)82390-0.
Saroussi, S., Schushan, M., Ben-Tal, N., Junge, W., & Nelson, N. (2012). Structure and flexibility of the C-ring in the electromotor of rotary F(0)F(1)-ATPase of pea chloroplasts. *PLoS One*, 7(9), e43045. https://doi.org/10.1371/journal.pone.0043045.
Schägger, H. (2002). Respiratory chain supercomplexes of mitochondria and bacteria. *Biochimica et Biophysica Acta, 1555*(1–3), 154–159. https://doi.org/10.1016/s0005-2728(02)00271-2.
Schlegel, S., Löfblom, J., Lee, C., Hjelm, A., Klepsch, M., Strous, M., et al. (2012). Optimizing membrane protein overexpression in the Escherichia coli strain Lemo21 (DE3). *Journal of Molecular Biology, 423*(4), 648–659. https://doi.org/10.1016/j.jmb.2012.07.019.
Schnick, C., Kortgen, N., & Groth, G. (2002). Complete inhibition of the tentoxin-resistant F1-ATPase from Escherichia coli by the phytopathogenic inhibitor tentoxin after substitution of critical residues in the alpha- and beta -subunit. *The Journal of Biological Chemistry*, 277(52), 51003–51007. https://doi.org/10.1074/jbc.M206095200.
Schulz, S., Wilkes, M., Mills, D. J., Kuhlbrandt, W., & Meier, T. (2017). Molecular architecture of the N-type ATPase rotor ring from Burkholderia pseudomallei. *EMBO Reports, 18*(4), 526–535. https://doi.org/10.15252/embr.201643374.
Schumann, J., Richter, M. L., & McCarty, R. E. (1985). Partial proteolysis as a probe of the conformation of the gamma subunit in activated soluble and membrane-bound chloroplast coupling factor 1. *The Journal of Biological Chemistry, 260*(21), 11817–11823.
Schwarz, O., Schürmann, P., & Strotmann, H. (1997). Kinetics and thioredoxin specificity of thiol modulation of the chloroplast H+-ATPase. *The Journal of Biological Chemistry, 272*(27), 16924–16927. https://doi.org/10.1074/jbc.272.27.16924.
Schwassmann, H. J., Rexroth, S., Seelert, H., & Dencher, N. A. (2007). Metabolism controls dimerization of the chloroplast FoF1 ATP synthase in Chlamydomonas reinhardtii. *FEBS Letters, 581*(7), 1391–1396. https://doi.org/10.1016/j.febslet.2007.02.057.
Seelert, H., & Dencher, N. A. (2011). ATP synthase superassemblies in animals and plants: Two or more are better. *Biochimica et Biophysica Acta (BBA) - Bioenergetics, 1807*(9), 1185–1197. https://doi.org/10.1016/j.bbabio.2011.05.023.
Seelert, H., Poetsch, A., Dencher, N. A., Engel, A., Stahlberg, H., & Muller, D. J. (2000). Structural biology. Proton-powered turbine of a plant motor. *Nature, 405*(6785), 418–419. https://doi.org/10.1038/35013148.
Sharkey, T. D., & Vanderveer, P. J. (1989). Stromal phosphate concentration is low during feedback limited photosynthesis. *Plant Physiology, 91*(2), 679–684. https://doi.org/10.1104/pp.91.2.679.
Shimizu, K., Cao, W., Saad, G., Shoji, M., & Terada, T. (2018). Comparative analysis of membrane protein structure databases. *Biochimica et Biophysica Acta (BBA) Biomembranes, 1860*(5), 1077–1091. https://doi.org/10.1016/j.bbamem.2018.01.005.
Shimoni, E., Rav-Hon, O., Ohad, I., Brumfeld, V., & Reich, Z. (2005). Three-dimensional organization of higher-plant chloroplast thylakoid membranes revealed by electron tomography. *The Plant Cell, 17*(9), 2580–2586. https://doi.org/10.1105/tpc.105.035030.
Shirakihara, Y., Leslie, A. G., Abrahams, J. P., Walker, J. E., Ueda, T., Sekimoto, Y., et al. (1997). The crystal structure of the nucleotide-free alpha 3 beta 3 subcomplex of F1-ATPase from the thermophilic Bacillus PS3 is a symmetric trimer. *Structure, 5*(6), 825–836. https://doi.org/10.1016/s0969-2126(97)00236-0.

Sim, D.-W., Lu, Z., Won, H.-S., Lee, S.-N., Seo, M.-D., Lee, B.-J., et al. (2017). Application of solution NMR to structural studies on α-helical integral membrane proteins. *Molecules (Basel, Switzerland)*, *22*(8), 1347. https://doi.org/10.3390/molecules22081347.
Singh, S., Turina, P., Bustamante, C. J., Keller, D. J., & Capaldi, R. (1996). Topographical structure of membrane-bound Escherichia coli F1F0 ATP synthase in aqueous buffer. *FEBS Letters*, *397*(1), 30–34. https://doi.org/10.1016/s0014-5793(96)01127-1.
Sjöstrand, F. S. (1963). A new repeat structural element of mitochondrial and certain cytoplasmic membranes. *Nature*, *199*(4900), 1262–1264. https://doi.org/10.1038/1991262a0.
Snyder, B., & Hammes, G. G. (1984). Structural mapping of chloroplast coupling factor. *Biochemistry*, *23*(24), 5787–5795. https://doi.org/10.1021/bi00319a018.
Sobti, M., Smits, C., Wong, A. S., Ishmukhametov, R., Stock, D., Sandin, S., et al. (2016). Cryo-EM structures of the autoinhibited E. coli ATP synthase in three rotational states. *eLife*, *5*, e21598. https://doi.org/10.7554/eLife.21598.
Sorgen, P. L., Bubb, M. R., & Cain, B. D. (1999). Lengthening the second stalk of F(1)F(0) ATP synthase in Escherichia coli. *The Journal of Biological Chemistry*, *274*(51), 36261–36266. https://doi.org/10.1074/jbc.274.51.36261.
Sorgen, P. L., Caviston, T. L., Perry, R. C., & Cain, B. D. (1998). Deletions in the second stalk of F1F0-ATP synthase in Escherichia coli. *The Journal of Biological Chemistry*, *273*(43), 27873–27878. https://doi.org/10.1074/jbc.273.43.27873.
Staehelin, L. A. (1986). Chloroplast structure and supramolecular organization of photosynthetic membranes. In L. Andrew Staehelin & C. J. Arntzen (Eds.), *Photosynthesis III* (pp. 1–84). Berlin Heidelberg: Springer. https://doi.org/10.1007/978-3-642-70936-4_1.
Sternweis, P. C., & Smith, J. B. (1980). Characterization of the inhibitory (epsilon) subunit of the proton-translocating adenosine triphosphatase from Escherichia coli. *Biochemistry*, *19*(3), 526–531. https://doi.org/10.1021/bi00544a021.
Stetsenko, A., & Guskov, A. (2017). An overview of the top ten detergents used for membrane protein crystallization. *Crystals*, 7(7), 197. https://doi.org/10.3390/cryst7070197.
Stock, D., Leslie, A. G., & Walker, J. E. (1999). Molecular architecture of the rotary motor in ATP synthase. *Science*, *286*(5445), 1700–1705. https://doi.org/10.1126/science.286.5445.1700.
Suzuki, T., Murakami, T., Iino, R., Suzuki, J., Ono, S., Shirakihara, Y., et al. (2003). F0F1-ATPase/synthase is geared to the synthesis mode by conformational rearrangement of epsilon subunit in response to proton motive force and ADP/ATP balance. *The Journal of Biological Chemistry*, *278*(47), 46840–46846. https://doi.org/10.1074/jbc.M307165200.
Takeyasu, K., Omote, H., Nettikadan, S., Tokumasu, F., Iwamoto-Kihara, A., & Futai, M. (1996). Molecular imaging of Escherichia coli F0F1-ATPase in reconstituted membranes using atomic force microscopy. *FEBS Letters*, *392*(2), 110–113. https://doi.org/10.1016/0014-5793(96)00796-x.
Tsunoda, S. P., Rodgers, A. J., Aggeler, R., Wilce, M. C., Yoshida, M., & Capaldi, R. A. (2001). Large conformational changes of the epsilon subunit in the bacterial F1F0 ATP synthase provide a ratchet action to regulate this rotary motor enzyme. *Proceedings of the National Academy of Sciences of the United States of America*, *98*(12), 6560–6564. https://doi.org/10.1073/pnas.111128098.
Uhlin, U., Cox, G. B., & Guss, J. M. (1997). Crystal structure of the epsilon subunit of the proton-translocating ATP synthase from Escherichia coli. *Structure*, *5*(9), 1219–1230. https://doi.org/10.1016/s0969-2126(97)00272-4.
Valiyaveetil, F. I., & Fillingame, R. H. (1998). Transmembrane topography of subunit a in the Escherichia coli F1F0 ATP synthase. *The Journal of Biological Chemistry*, *273*(26), 16241–16247. https://doi.org/10.1074/jbc.273.26.16241.
van Lis, R., Atteia, A., Mendoza-Hernández, G., & González-Halphen, D. (2003). Identification of novel mitochondrial protein components of Chlamydomonas

reinhardtii. A proteomic approach. *Plant Physiology, 132*(1), 318–330. https://doi.org/10.1104/pp.102.018325.

van Raaij, M. J., Abrahams, J. P., Leslie, A. G., & Walker, J. E (1996). The structure of bovine F1-ATPase complexed with the antibiotic inhibitor aurovertin B. *Proceedings of the National Academy of Sciences of the United States of America, 93*(14), 6913–6917. https://doi.org/10/cbqsx6.

Varco-Merth, B., Fromme, R., Wang, M., & Fromme, P. (2008). Crystallization of the c14-rotor of the chloroplast ATP synthase reveals that it contains pigments. *Biochimica et Biophysica Acta, 1777*(7–8), 605–612. https://doi.org/10.1016/j.bbabio.2008.05.009.

Vik, S. B., & Antonio, B. J. (1994). A mechanism of proton translocation by F1F0 ATP synthases suggested by double mutants of the a subunit. *The Journal of Biological Chemistry, 269*(48), 30364–30369.

Vik, S. B., Long, J. C., Wada, T., & Zhang, D. (2000). A model for the structure of subunit a of the Escherichia coli ATP synthase and its role in proton translocation. *Biochimica et Biophysica Acta, 1458*(2–3), 457–466. https://doi.org/10.1016/s0005-2728(00)00094-3.

Vlasov, A. V., Kovalev, K. V., Marx, S. H., Round, E. S., Gushchin, I. Y., Polovinkin, V. A., et al. (2019). Unusual features of the c-ring of F1FO ATP synthases. *Scientific Reports, 9*(1), 18547. https://doi.org/10.1038/s41598-019-55092-z.

Vollmar, M., Schlieper, D., Winn, M., Buchner, C., & Groth, G. (2009). Structure of the c14 rotor ring of the proton translocating chloroplast ATP synthase. *The Journal of Biological Chemistry, 284*(27), 18228–18235. https://doi.org/10.1074/jbc.M109.006916.

Vonck, J., & Schäfer, E. (2009). Supramolecular organization of protein complexes in the mitochondrial inner membrane. *Biochimica et Biophysica Acta, 1793*(1), 117–124. https://doi.org/10.1016/j.bbamcr.2008.05.019.

Wachter, A., Bi, Y., Dunn, S. D., Cain, B. D., Sielaff, H., Wintermann, F., et al. (2011). Two rotary motors in F-ATP synthase are elastically coupled by a flexible rotor and a stiff stator stalk. *Proceedings of the National Academy of Sciences of the United States of America, 108*(10), 3924–3929. https://doi.org/10.1073/pnas.1011581108.

Wada, T., Long, J. C., Zhang, D., & Vik, S. B. (1999). A novel labeling approach supports the five-transmembrane model of subunit a of the Escherichia coli ATP synthase. *The Journal of Biological Chemistry, 274*(24), 17353–17357. https://doi.org/10.1074/jbc.274.24.17353.

Wadsten, P., Wöhri, A. B., Snijder, A., Katona, G., Gardiner, A. T., Cogdell, R. J., et al. (2006). Lipidic sponge phase crystallization of membrane proteins. *Journal of Molecular Biology, 364*(1), 44–53. https://doi.org/10.1016/j.jmb.2006.06.043.

Wagner, S., Baars, L., Ytterberg, A. J., Klussmeier, A., Wagner, C. S., Nord, O., et al. (2007). Consequences of membrane protein overexpression in Escherichia coli. *Molecular & Cellular Proteomics, 6*(9), 1527–1550. https://doi.org/10.1074/mcp.M600431-MCP200.

Wagner, S., Bader, M. L., Drew, D., & de Gier, J.-W. (2006). Rationalizing membrane protein overexpression. *Trends in Biotechnology, 24*(8), 364–371. https://doi.org/10.1016/j.tibtech.2006.06.008.

Wagner, S., Klepsch, M. M., Schlegel, S., Appel, A., Draheim, R., Tarry, M., et al. (2008). Tuning Escherichia coli for membrane protein overexpression. *Proceedings of the National Academy of Sciences of the United States of America, 105*(38), 14371–14376. https://doi.org/10.1073/pnas.0804090105.

Walker, J. E., Lutter, R., Dupuis, A., & Runswick, M. J. (1991). Identification of the subunits of F1F0-ATPase from bovine heart mitochondria. *Biochemistry, 30*(22), 5369–5378. https://doi.org/10.1021/bi00236a007.

Wallin, E., & Heijne, G. V. (2008). Genome-wide analysis of integral membrane proteins from eubacterial, archaean, and eukaryotic organisms. *Protein Science*, 7(4), 1029–1038. https://doi.org/10.1002/pro.5560070420.

Watt, I. N., Montgomery, M. G., Runswick, M. J., Leslie, A. G. W., & Walker, J. E. (2010). Crystal structure of bovine F1-c8 sub-complex of ATP synthase. *Proceedings of the National Academy of Sciences of the United States of America*, *107*, 16823–16827. https://doi.org/10.1073/pnas.1011099107.

Wehrmeyer, W. (1963). Über Membranbildungsprozesse im Chloroplasten: I. Mitteilung Zur Morphogenese der Granamembranen. *Planta*, *59*(3), 280–295. https://doi.org/10.1007/BF01872299.

Wehrmeyer, W. (1964). Zur Klärung der strukturellen Variabilität der Chloroplastengrana des Spinats in Profil und Aufsicht. *Planta*, *62*(3), 272–293. https://doi.org/10.1007/BF01954049.

Weierstall, U., James, D., Wang, C., White, T. A., Wang, D., Liu, W., et al. (2014). Lipidic cubic phase injector facilitates membrane protein serial femtosecond crystallography. *Nature Communications*, *5*(1), 3309. https://doi.org/10.1038/ncomms4309.

White, S. H. (2020). *Membrane proteins of known structure [database]*. Retrieved from https://blanco.biomol.uci.edu/mpstruc/.

Wilkens, S., Borchardt, D., Weber, J., & Senior, A. E. (2005). Structural characterization of the interaction of the delta and alpha subunits of the Escherichia coli F1F0-ATP synthase by NMR spectroscopy. *Biochemistry*, *44*(35), 11786–11794. https://doi.org/10.1021/bi0510678.

Wilkens, S., & Capaldi, R. A. (1998). Solution structure of the epsilon subunit of the F1-ATPase from Escherichia coli and interactions of this subunit with beta subunits in the complex. *The Journal of Biological Chemistry*, *273*(41), 26645–26651. https://doi.org/10.1074/jbc.273.41.26645.

Wilkens, S., Dahlquist, F. W., McIntosh, L. P., Donaldson, L. W., & Capaldi, R. A. (1995). Structural features of the epsilon subunit of the Escherichia coli ATP synthase determined by NMR spectroscopy. *Nature Structural Biology*, *2*(11), 961–967. https://doi.org/10.1038/nsb1195-961.

Wilkens, S., Dunn, S. D., Chandler, J., Dahlquist, F. W., & Capaldi, R. A. (1997). Solution structure of the N-terminal domain of the delta subunit of the E. coli ATPsynthase. *Nature Structural Biology*, *4*(3), 198–201. https://doi.org/10.1038/nsb0397-198.

Wirth, A. J., & Gruebele, M. (2013). Quinary protein structure and the consequences of crowding in living cells: Leaving the test-tube behind. *BioEssays: News and Reviews in Molecular, Cellular and Developmental Biology*, *35*(11), 984–993. https://doi.org/10.1002/bies.201300080.

Yagi, H., Kajiwara, N., Tanaka, H., Tsukihara, T., Kato-Yamada, Y., Yoshida, M., et al. (2007). Structures of the thermophilic F1-ATPase epsilon subunit suggesting ATP-regulated arm motion of its C-terminal domain in F1. *Proceedings of the National Academy of Sciences of the United States of America*, *104*(27), 11233–11238. https://doi.org/10.1073/pnas.0701045104.

Yagi, H., Konno, H., Murakami-Fuse, T., Isu, A., Oroguchi, T., Akutsu, H., et al. (2010). Structural and functional analysis of the intrinsic inhibitor subunit epsilon of F1-ATPase from photosynthetic organisms. *The Biochemical Journal*, *425*, 85–94.

Yamada, H., Moriyama, Y., Maeda, M., & Futai, M. (1996). Transmembrane topology of Escherichia coli H(+)-ATPase (ATP synthase) subunit a. *FEBS Letters*, *390*(1), 34–38. https://doi.org/10.1016/0014-5793(96)00621-7.

Zhou, A., Rohou, A., Schep, D. G., Bason, J. V., Montgomery, M. G., Walker, J. E., et al. (2015). Structure and conformational states of the bovine mitochondrial ATP synthase by cryo-EM. *eLife*, *4*, e10180. https://doi.org/10.7554/eLife.10180.

Zorman, S., Botte, M., Jiang, Q., Collinson, I., & Schaffitzel, C. (2015). Advances and challenges of membrane–protein complex production. *Current Opinion in Structural Biology*, *32*, 123–130. https://doi.org/10.1016/j.sbi.2015.03.010.

CHAPTER THREE

Chloroplast ATP synthase from green microalgae

Felix E. Buchert*
Institute of Plant Biology and Biotechnology, University of Münster, Münster, Germany
*Corresponding author: f.buchert@uni-muenster.de

Contents

Abstract

The chloroplast ATP synthase (CF_OCF_1) is crucial for photosynthesis since it controls the transmembrane electrochemical proton potential gradient which is generated by light-driven electron and proton transfer processes. Thereby, CF_OCF_1 produces ATP and, together with the provided reducing equivalents, it powers CO_2 fixation. Since CF_OCF_1 represents the major proton-conducting element in photosynthesis, it is important to understand its tight integration in a feedback-regulated cellular energetics framework.

Advances in Botanical Research, Volume 96
ISSN 0065-2296
https://doi.org/10.1016/bs.abr.2020.07.001
© 2020 Elsevier Ltd
All rights reserved.

CF_oCF_1 from green microalgae resembles its counterpart from vascular plants by consisting of a membrane-embedded CF_o proton channel and a catalytic CF_1 portion. The algal enzyme has been subject of numerous functional studies owing to the feasible generation and maintenance of point mutants and knockout strains. Thereby, a wealth of information has been obtained *in vitro* and *in vivo* concerning the genetic control of complex assembly and regulatory mechanisms of the enzyme on various levels. While many findings are transferable to vascular plants, some may concern specific adaptations of the microalgal enzyme. This chapter will present a selection of previous outcomes and summarize our current understanding of algal CF_oCF_1 regulation.

1. Introduction

The chloroplast ATP synthase (CF_oCF_1) is a member of the F-type ATP synthase family and has a core structure very similar to bacterial and mitochondrial homologs. The oxidation of water and the reduction $NADP^+$ is facilitated by a light-driven coupling of vectorial movements of electrons and protons across the thylakoid membrane. Thereby, a membrane potential ($\Delta\Psi$) is generated by primary photochemistry of both photosystems and secondary electron bifurcation in the cytochrome b_6f complex. The latter is also engaged in proton uptake from the stroma and proton release into the lumen which, together with water splitting at photosystem II, generates a ΔpH. Since the membrane conductivities for protons and ions are limited, the fast light-driven reactions result in the formation of a transmembrane electrochemical proton potential difference, or proton motive force (pmf), that consists of $\Delta\Psi$ and ΔpH. The pmf is eventually consumed by CF_oCF_1, in a process referred to as photophosphorylation, to produce ATP from ADP and inorganic phosphate (Pi). Light-driven electron transfer also produces NADPH which, besides ATP, is required to fix CO_2 in the Calvin-Benson-Bassham cycle. However, linear electron flow from water to NADPH yields an ATP/NADPH ratio slightly below 1.5, which is required to fix CO_2. To cope with this ATP shortage, photosynthetic organisms evolved sophisticated electron transfer rerouting strategies that fuel CF_oCF_1 with extra pmf. Like CF_oCF_1 biogenesis and assembly, the activity of CF_oCF_1 is regulated on various levels *in vivo*. Over the years, the algal CF_oCF_1 has become a useful model to unravel the mechanism of ATP synthesis as well as the significance of CF_oCF_1 as a photoprotective and metabolic key player in everchanging environments. This chapter will focus on microalgal CF_oCF_1 properties such as structure and *in membrana* organization, as well as the genetic regulation. We will revisit CF_oCF_1 finetuning mechanisms that have been described in algal strains and, in some cases, differ from vascular plant CF_oCF_1. Finally, we will focus on the substantial pmf that is maintained in dark-adapted algal cells and elaborate the role of CF_oCF_1 in this phenomenon as well as possible consequences on various cellular processes.

2. Structural composition

2.1 CF_oCF_1

CF_oCF_1 is a typical F-type ATP synthase enzyme which has a molecular weight of ~540 kDa and the catalytic CF_1 head is anchored to the membrane-spanning CF_o motor (Fig. 1). Coupling of the ion-conducting CF_o motor, consisting of subunits $a_1b_1b'_1c_n$, to the sites of nucleotide catalysis in CF_1 follows the same principles as in bacterial and mitochondrial enzymes. Free energy, stored as the pmf, is used to synthesize ATP from ADP and Pi during rotary catalysis (Abrahams, Leslie, Lutter, & Walker, 1994; Boyer, Cross, & Momsen, 1973; Mitchell, 1961). The following outline, which is based on the spinach CF_oCF_1 cryo-electron microscopy structure (Hahn, Vonck, Mills, Meier, & Kuhlbrandt, 2018), will begin from the membrane plane and end at the crown of the CF_1 head. The spinach structure confirmed earlier findings that the c-ring is composed of 14 subunits in vascular plant chloroplasts (Saroussi, Schushan, Ben-Tal, Junge, & Nelson, 2012; Seelert et al., 2000). However, the c-ring in the green alga *Chlamydomonas reinhardtii* (hereafter, Chlamydomonas) is believed to be formed by 13 subunits (Meyer Zu Tittingdorf et al., 2004) although the high resolution structure from Chlamydomonas is still missing. Taking into account the cylinder of 15 c-monomers in *Spirulina platensis* (Pogoryelov et al., 2005), a certain c-ring flexibility in the green lineage of photosynthetic organisms can be expected. As explained in the following paragraphs, this inevitably leads to alterations of the (H^+) ion/ATP ratio which is a function of (de-)protonation events per 360° rotation. Based on initial experiments in *E. coli*, the c-ring stoichiometry was proposed to depend on the metabolic state of the cell (Schemidt, Qu, Williams, & Brusilow, 1998). This was not the case in Chlamydomonas (Meyer Zu Tittingdorf et al., 2004) and further studies revealed that, in fact, the c-ring stoichiometry depends on a membrane protein-packing motif (GxGxGxG, or AxAxAxA) which is located in the N-terminal α helix of the c-subunits (Pogoryelov et al., 2012; Preiss et al., 2013). The corresponding sequence in the spinach

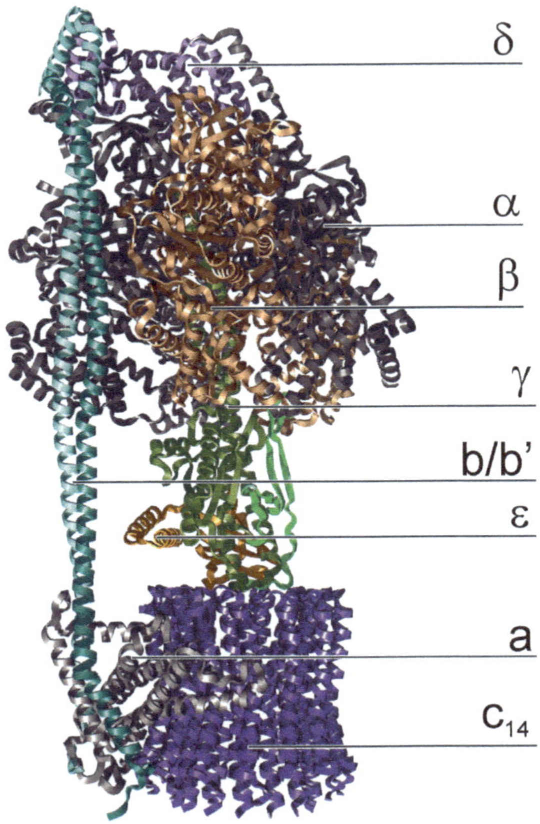

Fig. 1 Cryo-electron microscopy structure of chloroplast ATP synthase from spinach (PDB ID 6fkh). The differently colored subunits are labeled in Greek (CF_1) and Latin letters (CF_o). The redox domain in the γ-subunit, exclusively found in chloroplasts of the green lineage, is colored in light green. For more structural details, refer to Section 2.1.

c-subunit (CF_o numbering ^{23}GPGVGQG^{29}) shows slight packing motif differences compared to Chlamydomonas (^{23}GPGMGQG^{29}). Whether this results in a smaller algal c-ring needs to be tested.

Proton translocation via the channel involves (de-)protonation of the conserved E61 in the c-subunit that is part of an interface between c- and a-subunits (Hahn et al., 2018). The a-subunit, consisting of α helices H1 to H6, forms the proton translocation pathway. Two long, membrane-intrinsic hairpins are formed by H3/H4 and H5/H6. The loop that connects H3/H4 has access to the lumen surface and protons are believed to be funneled to E61 in the c-subunit via a narrow gap between H5/H6 which is decorated with charged and polar a-subunit residues. Upon protonation of E61 in the c-subunit, the residue can dwell on the lipid phase and after an almost complete rotation of the c-ring, the glutamate is deprotonated due to the high lumen pH. The proton exit cavity in the enzyme contains charged and polar a-subunit residues.

The rotation of the c-ring motor in the CF_o sector also drives rotation of the CF_1 subunits γ and ε. The CF_1 subunits consist of $\alpha_3\beta_3\gamma_1\delta_1\varepsilon_1$ and early reports, using isotopic labeling, predicted such a CF_1 stoichiometry in Chlamydomonas (Merchant, Shaner, & Selman, 1983). Both N- and C-terminus of the γ-subunit are α helices that form the rotor shaft, composed of a left-handed coiled coil. This cranked shaft will be discussed further below in the context of nucleotide binding properties. Furthermore, the central globular domain of the γ-subunit, which harbors the redox loop, will be discussed in Section 4.3. The loops adjacent to the terminal α helices in γ form close contacts with the c-ring, mainly via electrostatic interactions with the conserved c-subunit residues $^{41}RQP^{43}$. The abundance of interacting γ-residues allows contacts with the majority of c-monomers within a ring, independently of c-ring stoichiometry (Dautant, Velours, & Giraud, 2010; Hahn et al., 2018).

The third rotating subunit is ε, which consists of an antiparallel eight-stranded β sandwich on the N-terminus and an α helical domain in the C-terminus. Like γ, the ε-subunit strengthens the interaction between the rotary cylinder and the rotor shaft by forming contacts with $^{41}RQ^{42}$ in the c-subunit, via conserved εH37. The importance of the ε-subunit in coupling proton translocation to ATP synthesis has been reported in spinach thylakoids (Richter, Patrie, & McCarty, 1984). CF_1-depleted membranes were able to rebind ε-deficient CF_1 but photophosphorylation was only restored upon additional titrations with isolated ε-subunit preparations. As expected from the structure (Hahn et al., 2018), removal of H37 in the ε-subunit increases the proton slip (Cruz, Harfe, Radkowski, Dann, & McCarty, 1995). In terms of membrane reconstitution, CF_1 differs from the bacterial F_1 since only the ε-deficient chloroplast complex binds to washed membranes (Kagawa, 1978; Sternweis, 1978). The C-terminus in the CF_1 ε-subunit has been subjected to a series of pre-structure studies (Johnson & McCarty, 2002; Nowak & McCarty, 2004; Nowak, Tabidze, & McCarty, 2002), which will be discussed in Section 4.4.

ATP is produced and hydrolyzed in the catalytic sites which reside in three asymmetric αβ heterodimers, more precisely in the β-subunit. The heterodimers form an $\alpha_3\beta_3$ heterohexamer and each heterodimer also has a noncatalytic ATP binding site, sitting in the α-subunits. Both amino acid sequences of the α- and β-subunit are highly conserved among chloroplast enzymes, even when compared with mitochondrial ATP synthases. The nucleotide binding sites of the chloroplast enzyme are identical to the mitochondrial counterparts (Groth & Pohl, 2001). The so-called Walker-A motif (CF_1 numbering ^{161}GXXXXGKT^{168}) lines the nucleotide binding pocket in the β-subunit (Walker, Saraste, Runswick, & Gay, 1982) and the essential arginine finger in the adjacent α-subunit, formed by CF_1 R366, coordinates the γ phosphate and turns away in absence of ATP (Abrahams et al., 1994; Hahn et al., 2018). The three catalytic nucleotide binding sites exist in different conformations which alternate their state cooperatively (Abrahams et al., 1994; Gresser, Myers, & Boyer, 1982) and contain either Mg-ATP, Mg-ADP or are empty without nucleotides. The rotating convex shaft, which is formed by the α helical termini of the γ-subunit, pushes a helix-turn-helix in the C-terminus of the β-subunit aside and thus changes the vicinal binding site conformation. The helix-turn-helix element harbors the βDELSEED motif which is a conserved cluster of charged residues. Loop rigidity of the helix-turn-helix element is important for torque transmission (Tanigawara et al., 2012) and the charged DELSEED motif is believed to be a guiding element during γ-subunit rotation by transiently forming ionic interactions (Ma et al., 2002).

F-type ATP synthase enzymes have a conserved structural element that connects the soluble and membrane-spanning portions. This peripheral stalk is composed of the CF_o subunits b and b′, and of the CF_1 δ-subunit. The latter is composed of two domains and binds at a central position of the CF_1 crown via its N-terminus (Hahn et al., 2018). The N-terminus of the δ-subunit interacts with N-terminal helices of two α-subunits. The N-terminal helix of the third α-subunit is a long loop that interacts with the C-terminus of subunit b′, thus facilitating the fixation of CF_1 to CF_o. However, there are reports concluding that, in contrast to its necessity to restore functional coupling, the δ-subunit is not strictly required for binding of CF_1 to CF_o (Andreo, Patrie, & McCarty, 1982). The C-terminus of the δ-subunit is the binding site for the kinked C-terminus of subunit b. Both α helical subunits b and b′ form a loose right-handed coiled coil until the helices separate near the membrane surface to clamp the CF_o a-subunit to the rotating c-ring.

2.2 *In membrana* organization of CF_oCF_1 in *C. reinhardtii*

Here, a selection of distinct algal features will be highlighted with an emphasis on the thylakoid ultrastructure. The discourse aims to draw on findings that stimulated discussions about the possible oligomerization of Chlamydomonas CF_oCF_1.

In the family of F-type ATP synthase, the mitochondrial enzyme shows the most sophisticated assembly of subunits. Compared to bacteria and chloroplasts, the additional mitochondrial subunits form an interface that enables the enzyme to dimerize and shape the cristae structure (recently reviewed in Kuhlbrandt, 2019). Like bacterial F-type ATP synthases, CF_oCF_1 has a simpler structural blueprint. Owing to the lack of interface-forming subunits and inability to participate in thylakoid shaping, it is believed that CF_oCF_1 does not form dimers. Recently, this view has gained strong support by electron cryotomography of vitreous chloroplast sections from spinach and pea (Daum, Nicastro, Austin 2nd, McIntosh, & Kuhlbrandt, 2010). Although the technique is demanding and samples are fragile (Daum & Kuhlbrandt, 2011), the vitreous sections also confirmed an early suggestion that localization of vascular plant CF_oCF_1 is confined to non-appressed membranes (Miller & Staehelin, 1976).

The spatio-temporal organization of the photosynthetic machinery undergoes dynamic changes since the thylakoid ultrastructure changes upon illumination. For instance, it was shown that the expansion of thylakoid lumen increases upon light adaptation (Daum et al., 2010; Engel et al., 2015; Kirchhoff et al., 2011). Here, Chlamydomonas displays some unique features that influence *in membrana* organization of the photosynthetic machinery, including CF_oCF_1 (possible functional consequences will be presented in Section 5.2). It is well-known that the thylakoid ultrastructure in Chlamydomonas differs from vascular plants by lacking a clear division into appressed and non-appressed membranes (Engel et al., 2015). Appressed membranes are enriched in photosystem II (PSII) and the cytochrome b_6f complex (b_6f), whereas non-appressed membranes contain mostly photosystem I (PSI) and CF_oCF_1. According to the idea that all thylakoids form one continuous compartment (Mustardy & Garab, 2003; Staehelin, 2003), a lower degree of defined thylakoid stacking regions in Chlamydomonas would coincide with a more homogeneous distribution of photosynthetic complexes. Furthermore, appressed membranes have fewer plies in Chlamydomonas, compared to grana stacks from vascular plants. Thus, the abundant non-appressed membranes are more likely to share their lumenal interfaces with a vicinal appressed membrane (Fig. 2A). In vascular plants, the corresponding lumenal interfaces would be found between flat grana end membranes and the adjacent membranes of stacked grana disks (Daum et al., 2010). A variety of lumenal interfaces as well as the corresponding stromal spaces have been analyzed in a recent study of Chlamydomonas thylakoid membrane tomograms (Wietrzynski et al., 2020). The authors could show that CF_oCF_1 distributes evenly across

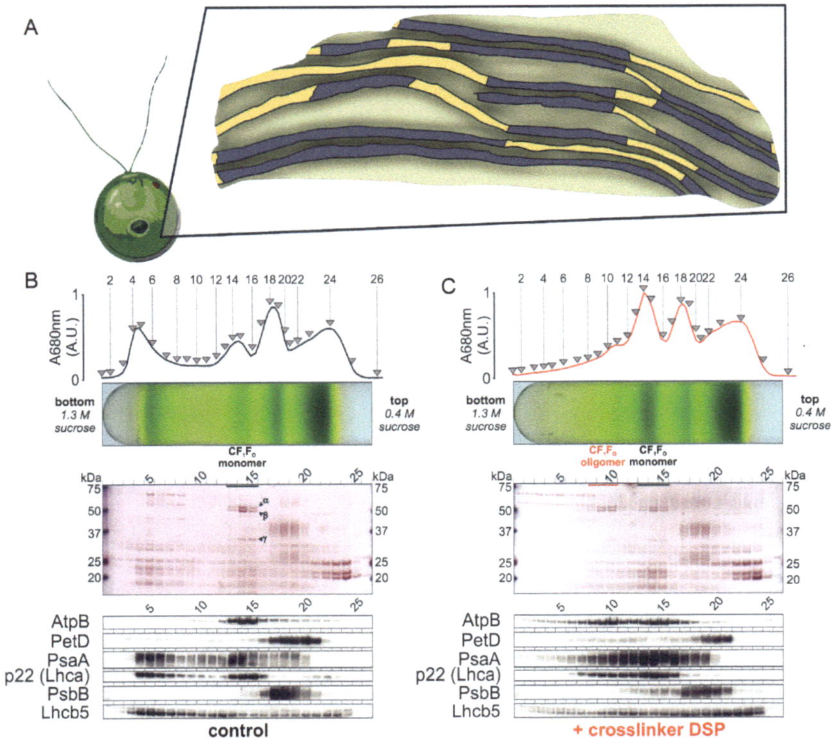

Fig. 2 Thylakoid ultrastructure and oligomerization of Chlamydomonas CF_oCF_1 *in membrana*. (A) A schematic drawing is shown that highlights the loosely defined appressed and non-appressed membrane regions in Chlamydomonas. Unlike in vascular plant chloroplasts, grana stacks do not exist, and the appressed algal membranes have a smaller number of plies. The dark-shaded sections indicate lumenal spaces that are formed between the two types of appressed and non-appressed membranes. The former harbors PSII and most of the b_6f, whereas PSI and CF_oCF_1 localize in non-appressed membranes and a possible functional role of the interfaces in Chlamydomonas will be presented in Section 5.2. In vascular plants, these shared lumenal interfaces will be found between flat grana end membranes and the adjacent membranes of stacked grana disks. Panels (B) and (C) show fractions of solubilized thylakoid membranes from Chlamydomonas which were obtained by sucrose density gradient centrifugation (Buchert et al., 2018). The A_{680} sample profile for reducing SDS-PAGE, Ponceau staining of the PvDF membrane for Western blot transfer efficiency, and various immunodetection signals are shown (AtpB, CF_oCF_1; PetD, b_6f; PsaA, PSI; p22, light-harvesting complex I; PsbB, PSII; Lhcb5, light-harvesting complex II). At variance with panel (B), the reversible lysine crosslinker dithiobis(succinimidyl propionate), DSP, was used before solubilization in (C) and stabilized labile interactions during the centrifugation step. Before SDS-PAGE, crosslinks were cleaved. DSP treatment produced, besides other effects, a second CF_oCF_1 population of a higher density. Further investigations are required if DSP-stabilized CF_oCF_1 dimers exist, in support of earlier observations (Rexroth et al., 2004).

the portion of analyzed membranes with a mean distance of 16.3 nm, from the CF_1 center-to-center. Although some depicted CF_1 particles were very close (Wietrzynski et al., 2020), no tendency to form dimers was observed. These close particles could result from stochastic co-localization in the analyzed data set. It will be interesting to determine the mean distance between CF_oCF_1 when the lumenal interface is formed by both appressed and non-appressed membranes vs. the mean distance in non-appressed lamellae.

Apparently, Chlamydomonas CF_oCF_1 localizes as a monomer *in membrana*, which is in line with the electron cryotomography study of spinach and pea chloroplasts but disagrees with earlier biochemical experiments using 2D Blue-Native/Tricine SDS-PAGE. Therein, it was suggested that Chlamydomonas CF_oCF_1 dimerizes under certain conditions. After establishing mild thylakoid isolation conditions, the authors could show that the majority of solubilized CF_oCF_1 still migrated as a monomer, but a CF_oCF_1 subpopulation migrated as a dimer (Rexroth et al., 2004). The latter was destabilized by addition of phosphate and the phosphatase inhibitor vanadate, which presumably affected CF_oCF_1 activity as well. Accordingly, dimerization was proposed to represent inactive CF_oCF_1 state. In a later study, a metabolic control of dimerization was reinforced, since dimers were less abundant when cells were grown photoautotrophically compared to cultures grown in the presence of acetate (Schwassmann, Rexroth, Seelert, & Dencher, 2007). Since acetate is metabolized in the mitochondria and provides ATP to the whole cell (reviewed in Johnson & Alric, 2013), the authors suggested that chloroplast ATP demand was met and the ~30% dimeric CF_oCF_1 were interpreted as a transiently inactive state (Schwassmann et al., 2007). It remains to be seen if the above-mentioned CF_oCF_1 mean distance of 16.3 nm, obtained with acetate-supplemented cells (Wietrzynski et al., 2020), is increased when cells are grown photoautotrophically.

In an independent approach, heavy CF_oCF_1 assemblies were obtained when separating solubilized thylakoid membrane proteins via rate-zonal centrifugation (using sucrose density gradients). Despite using mild detergents and low ionic strength conditions, CF_oCF_1 fractionated as a single population (Fig. 2B). However, when using a homobifunctional crosslinker before solubilization, a second CF_oCF_1 population was observed which corresponded to roughly 50% of the total CF_oCF_1 signal (Fig. 2C). The membrane-soluble crosslinker contained functional groups that were 1.2 nm apart and apparently stabilized the high-molecular weight assembly which contained CF_oCF_1. Certainly, this preliminary result from Chlamydomonas demonstrates crosslinker-assisted stabilization of a labile association between a significant CF_oCF_1 population with another heavy protein complex, which remains to be identified. Assuming the ~50% signal represents a CF_oCF_1 dimer (different in form and function from the mitochondrial one), the relatively short distance between reactive crosslinker groups argues against stabilization on the level of the extrinsic CF_1 domain. The latter should be, on average, in ~16 nm distance to its neighboring CF_oCF_1. Still, dimer stabilization by the lipid-soluble crosslinker may happen via yet unresolvable, small transmembrane proteins that are loosely attached to the CF_oCF_1 core structure. It is known that photosynthetic complexes in Chlamydomonas have specific labile subunits that can undergo post-translational modifications and are not present in vascular plants, for instance on the level of the b_6f with its single-transmembrane subunit PETO (Buchert et al., 2018; Hamel, Olive, Pierre, Wollman, & de Vitry, 2000). Visualization of the latter in membrane tomograms might prove difficult. By increasing the membrane tomography dataset and following spatio-temporal resolution approaches, further details might be obtained on the transient localization behavior of CF_oCF_1, as has been observed for the mitochondrial counterpart (Salewskij et al., 2020).

Taken together, Chlamydomonas CF_oCF_1 co-localization as dimer cannot be unambiguously confirmed or excluded by biochemical and imaging experiments yet. While the former results may simply represent stochastic artifacts, the latter comprise reliable statistics from a limited portion of total thylakoid membranes only. Certainly, if it exists, CF_oCF_1 dimerization will differ from mitochondria and the algal thylakoid membrane holds many yet undiscovered secrets which might be revealed thanks to current breakthroughs in visualizing the native thylakoid architecture by cryo-electron tomography.

2.3 Chlamydomonas CF_oCF_1 purification and complex integrity: Distinct features of the algal enzyme

The first isolation of a coupling factor involved in photophosphorylation was reported nearly 6 decades ago (Avron, 1963). This paved the way for biochemical analysis of the complex and the discovery of thiol modulation that is exclusively found in the chloroplast enzyme of green algae and plants. Preceding substantial structural insights, *in vitro* studies of the wild type and site-directed mutants have revealed some remarkable functional features, of which a selection will be presented in Section 4. Contributions of numerous laboratories improved the initial isolation procedures, which have been reviewed elsewhere (Jagendorf, 1982; Merchant & Selman, 1985; Strotmann, Shavit, & Leu, 1998). Still, the isolation protocol is a laborious and time-consuming procedure, and pure CF_oCF_1 can only be obtained within at least a day's work.

When intending to release CF_1, many protocols based on Avron's discovery use EDTA, and thus aim at chelating divalent cations in the medium which are crucial for CF_1 attachment. This treatment appears more efficient when thylakoids are diluted but CF_1 is often contaminated with ribulose bisphosphate carboxylase. Alternatively, a pyrophosphate washing step and resuspension of thylakoid pellets in sucrose-containing buffers yields pure CF_1 (Strotmann, Hesse, & Edelmann, 1973), which can be further purified by anion exchange chromatography (Lien & Racker, 1971). Another method involves short exposure of thylakoids to chloroform, thereby releasing CF_1 at a relatively high concentration compared to EDTA and pyrophosphate procedures (Younis, Winget, & Racker, 1977). Pioneer studies on isolated CF_1 were performed in vascular plants, and only later in Chlamydomonas using the EDTA/pyrophosphate procedure (Selman-Reimer, Merchant, & Selman, 1981) and chloroform extraction (Piccioni, Bennoun, & Chua, 1981). In the following years, improved genetic manipulation techniques of the nuclear and chloroplast genome made it possible to characterize site-directed algal CF_oCF_1 mutants (Ross, Zhang, & Selman, 1995) and introduce affinity tags for facilitated purification (Johnson, Rosenberg, & McCarty, 2007).

In vitro studies of the algal enzyme revealed several differences compared to the vascular plant counterparts in similar experimental conditions. This applies to both complex stability and, as elaborated in Section 4, catalytic features. Isolated spinach CF_1 is cold labile (Lien, Berzborn, & Racker, 1972), unless nucleotides are present (Posorske & Jagendorf, 1976). On the other hand, the complex from Chlamydomonas is stable at 0–4°C (Merchant & Selman, 1985). Whether this feature is an adaptation to freshwater habitat requires further investigation. At least in another unicellular freshwater alga, *Euglena gracilis*, CF_1 cold stability is also observed (Chang & Kahn, 1966). The chloroform extraction method yields Chlamydomonas CF_1 which lacks the δ-subunit completely and thus fails to restore photophosphorylation in CF_1-depleted thylakoids (Lemaire & Wollman, 1989a; Piccioni et al., 1981; Selman-Reimer et al., 1981). Vascular plant CF_1 retains the δ-subunit after chloroform treatment, although it dissociates when the complex is further purified via anion exchange chromatography (Younis et al., 1977). Since the δ-subunit is not essential for CF_1 catalysis, ATP hydrolysis studies were performed with EDTA- and chloroform-extracted complexes from Chlamydomonas. When using the chloroform extraction method, isolation buffers which contain dithiothreitol (DTT) were reported to promote the reversible dissociation of the ε-subunit in Chlamydomonas and spinach (Duhe & Selman, 1989, 1990). Taken together, the algal enzyme can be obtained with commonly used isolation protocols, although modifications are required to preserve integrity of the oligomeric structure.

3. Regulation of ATP gene expression and complex assembly

3.1 Gene arrangements and evolution

As mentioned in Section 2.1, the subunit composition of CF_oCF_1 is very similar to the bacterial F-ATP synthase. Nevertheless, the CF_oCF_1 genes were separated during evolution into nuclear- (*ATP*) and chloroplast-encoded genes (*atp*). Two *atp* clusters, *atpIHFA* and *atpBE*, exist in vascular plants and three genes are nuclear-encoded. It is believed that *atpA* (α-subunit) and *atpB* (β-subunit) resulted from gene duplication, considering their sequence identity (Strotmann et al., 1998). Chlamydomonas evolved a remarkably spread *atp* arrangement from which the only exception is a tetracistronic transcription unit (Levy, Kindle, & Stern, 1997), containing *atpA* and *atpH* (c-subunit). Probably because of gene rearrangements and as outlined in Section 3.2, the translation control of the CF_oCF_1 genes and the intertwined assembly of the gene products differs from vascular plants. Certainly, more sophisticated interplays on transcriptional and (post-)translational levels were established by transferring *ATPC*, *ATPD* and *ATPG* into the nuclear genome, since cyanobacterial ancestors encode all subunits in one operon (with two clusters, *atpA* and *atpB*). After endosymbiosis, *ATPC* (γ-subunit) was likely first integrated into the nucleus since it is separately encoded or attached to the *atpA* cluster in cyanobacteria (Strotmann et al., 1998), and it is the only nuclear *ATP* gene in certain diatoms (Pancic, Strotmann, & Kowallik, 1992) and red algae (Reith & Munholland, 1995). *ATPD* (δ-subunit) and *ATPG* (b′-subunit, shares ancestor with *atpF*) were transferred to the nucleus exclusively in the green lineage. These photosynthetic organisms also experienced sequence modifications within *ATPC*, compared to cyanobacterial ancestors, which probably involved an intra-gene duplication (Ponomarenko, 2007). The resulting mechanistic consequences will be discussed in Section 4.3.

3.2 Genetic regulation of CF_oCF_1 assembly and mutants identified in *C. reinhardtii*

Regulation of chloroplast gene expression has been studied in detail in Chlamydomonas and the chloroplast *atp* genes underly a similar feedback mechanism as other chloroplast-encoded photosynthetic complexes. Nuclear-encoded factors are involved, like in vascular plants, but massive rearrangements in the chloroplast genome of Chlamydomonas (Maul et al., 2002) might have contributed to a slightly different genetic regulation mechanism of the algal CF_oCF_1 genes. These nuclear-encoded factors, which mostly belong to helical repeat protein families (reviewed in Barkan & Small, 2014), have specific chloroplast mRNA targets. Their function is to control either the stability and maturation (M factors) and/or translation activation (T factors) of their targets. T factors are not well understood yet and it is speculated that they facilitate ribosome recruitment (Scharff et al., 2017). The majority of identified M factors in Chlamydomonas bind their targets on the very 5′-end, thereby preventing target degradation by 5′→3′ exonuclease activity (Cavaiuolo, Kuras, Wollman, Choquet, & Vallon, 2017).

Concerning Chlamydomonas *atpA*, the T factor TDA1 activates the translation but does not stabilize the mRNA (Drapier, Girard-Bascou, & Wollman, 1992; Eberhard et al., 2011). The M factor for *atpA* is MDA1 (Drapier, Girard-Bascou, Stern, & Wollman, 2002). As an exceptional M factor, MDA1 does not bind on the very 5′-end of *atpA* but in the middle of the 5′-untranslated region (UTR) near the TDA1 binding region (Viola et al., 2019). It was suggested that MDA1, TDA1 and *atpA* transcripts form a complex that dissociates after translation, thus initiating *atpA* mRNA degradation. As mentioned in Section 3.1, *atpA* is the first gene of a tetracistronic transcription unit (*atpA*, *psbI*, *cemA*, *atpH*) and translation of mono- and polycistronic *atpA* transcripts has been reported (Drapier et al., 1998; Levy et al., 1997).

On the contrary, *atpH* translation can only be achieved from monocistronic mRNAs using its own promoter. A mutant lacking MDH1, the corresponding M factor, fails to accumulate monocistronic *atpH* mRNA (Majeran, Olive, Drapier, Vallon, & Wollman, 2001). In fact, MDH1 is also an exceptional M factor since it was recently reported to represent a dual targeted trans-acting factor by binding upstream of the *atpI* initiation codon (Ozawa et al., 2020). Unexpectedly, it controls expression of both rotor and stator subunits and was therefore renamed to MTHI1 (maturation/stability and translation of the *atpH* and *atpI* mRNA).

In the case of the *atpB* transcript stabilization, *MDB1* has been described (Drapier et al., 1992). The 3′-UTR of the Chlamydomonas *atpB* gene drew considerable attention to 3′-end maturation studies since it is essential for transcript association with polysomes (Rott, Levy, Drager, Stern, & Schuster, 1998). The *atpB* 3′-UTR contains a stem/loop which is an RNA stability determinant, impeding 3′→5′ exonuclease digestion (Stern, Radwanski, & Kindle, 1991). Modulation of *atpB* mRNA stability also involves polyadenylation of the 3′-UTR, including the stem/loop (Komine, Kikis, Schuster, & Stern, 2002).

For the nuclear-encoded *ATP* genes (*ATPC*, *ATPD* and *ATPG*) it was shown in spinach that they are expressed synchronously and, besides light and organ-specific factors, chloroplast-derived signal pathways are involved (Bolle, Kusnetsov, Herrmann, & Oelmuller, 1996). The situation might differ in unicellular algae but early experiments, using the *y-1* mutant which does not green in the dark, demonstrated that CF_oCF_1 activity preceded chlorophyll accumulation (Wallach, Bar-Nun, & Ohad, 1972). The expression of *atpA* and *atpB* occurred before synthesis of chlorophylls and photosystems in pulse labeling experiments of photoperiodically grown Chlamydomonas (Herrin & Michaels, 1984). More recently, a diurnal transcriptome analysis backed up earlier findings and revealed that all *atp* genes were expressed after the first 30-min of light (Strenkert et al., 2019). Expression peaks for the nuclear *ATP* genes were measured at ~3-h after the onset of light, before mRNAs of both photosystems accumulated.

This suggests that chloroplast-encoded CF_oCF_1 subunits accumulate earlier than the nuclear-encoded ones and that the assembly process of CF_oCF_1 is coordinated once the latter are imported into the chloroplast. The assembly sequence has not been clarified in detail yet but it was observed that synthesis of interface-forming subunits, such as the c-subunit and a-subunit, require the presence of the interaction partner (Lemaire & Wollman, 1989b). Furthermore, the γ-subunit seems to mediate the cross-talk between the two genetic compartments in algae (Drapier, Rimbault, Vallon, Wollman, & Choquet, 2007). Although the stoichiometric excess of $\alpha_3\beta_3$ subunits might be facilitated by relatively early expression at the onset of light (Strenkert et al., 2019), subunit accumulation is only possible up to a critical concentration (Drapier et al., 2007). Accordingly, the β-subunit transactivates the translation initiation of the α-subunit and the spontaneously formed heterodimers of αβ assemble to a competent $\alpha_3\beta_3$ heterohexamer which can bind the γ-subunit. If the γ-subunit is not present or the $\alpha_3\beta_3$ is in excess, the latter exerts a negative feedback loop on translation initiation of the β-subunit and thus on the α-subunit translation as well. A recent report suggests that the ClpP protease is involved in degradation of non-assembled α-subunits (Majeran, Wostrikoff, Wollman, & Vallon, 2019) and further assembly studies are necessary to identify other key players in the feedback-controlled degradation pathways. Furthermore, CF_o does not accumulate without assembly of CF_1 and vice versa (Lemaire & Wollman, 1989b). Despite a different translation regulation mechanism of the CF_oCF_1 genes, nuclear-imposed feedback loops on CF_oCF_1 accumulation are also present in Arabidopsis, at least on the levels of γ- and δ-subunits (Bosco et al., 2004; Maiwald et al., 2003). The consequences of a missing CF_oCF_1 in those mutant plants are profound defects in growth and viability. On the other hand, maintenance of CF_oCF_1-lacking Chlamydomonas is feasible in acetate-supplemented growth medium and, except for *ATPG*, a variety of knockout mutants are available.

The available CF_oCF_1-deficient mutants will stand out when following a screening protocol that measures chlorophyll fluorescence yields after exposure to moderate light (Majeran et al., 2001). Owing to a strongly impaired H^+ conductivity in such CF_oCF_1 mutants, light-induced lumen acidification was proposed to change the fluorescence pattern upon altering the stromal acceptor side of PSII, eventually promoting cessation of linear electron transfer. This protocol, however, is only suitable for severely affected mutants and will not serve as a straight-forward tool to identify strains with mild to moderate CF_oCF_1 accumulation defects. There are mutants with reduced CF_oCF_1 content, for instance the *ncc1* assembly mutant in Chlamydomonas (Drapier et al., 1992) and a weak antisense line in tobacco (Rott et al., 2011). These photoautotrophic mutants show similar chlorophyll fluorescence parameters as the wild type. This underpins the excessive capacity of (diminished) CF_oCF_1 to govern the pmf within the continuous thylakoid compartment. Likewise, for future research on genetic regulation of CF_oCF_1, it corroborates the need for alternative (mass spectrometric and biochemical) screening strategies.

4. Selected regulatory mechanisms and catalytic features of the algal (sub-) complex *in vitro* and *in vivo*

4.1 Miscellaneous factors that influence catalytic properties

Numerous investigations of CF_oCF_1 and its catalytic portion have been conducted in the pre-structure era, and further progress was made after the structure of the mitochondrial complex was resolved (Abrahams et al., 1994). The studies led to a profound understanding of how the rotary machine is functioning, while also documenting subtle CF_oCF_1 differences within the green lineage. Although some features may only have limited validity for CF_oCF_1 function *in vivo*, findings from the Chlamydomonas enzyme that were mainly gathered *in vitro*, and in part varied from the vascular plant counterparts, will be reviewed first.

Since vascular plant CF_1 is isolated as a latent enzyme, ATPase activity can be elicited by various treatments such as heat or partial proteolytic digestion. Both treatments do not activate Chlamydomonas CF_1 (Selman-Reimer et al., 1981). Since they affect enzyme integrity, this might be related to stability features of the algal enzyme outlined in Section 2.3.

A variety of CF_1-binding inhibitors have been employed to gain insights in catalytic sub-steps. An example is tentoxin, which is a cyclic tetrapeptide from the phytopathogenic fungus *Alternaria alternata*. Interestingly, its phytopathogenic efficiency and affinity to CF_1 depends on the species (Steele, Durbin, Uchytil, & Rich, 1978; Steele, Uchytil, Durbin, Bhatnagar, & Rich, 1976). Chlamydomonas CF_oCF_1 is relatively weakly affected by low tentoxin concentrations and replacing βE82D converted the algal enzyme to a tentoxin-sensitive variant (Hu et al., 1997; Pick, Conrad, Conrad, Durbin, & Selman, 1982).

Although Mg-ADP and Mg-ATP are the nucleotide substrates for CF_oCF_1 *in vivo* (Zhou & Boyer, 1992), nucleotide complexes with other divalent cations were extensively studied to describe enzymatic features *in vitro*. Like in land plants, free metal ions showed an inhibitory effect whereas, on the contrary, Chlamydomonas CF_1 showed higher Mg-ATP and Mn-ATP hydrolysis activity, and Ca-ATP was a poor substrate (Kneusel, Merchant, & Selman, 1982; Selman-Reimer et al., 1981). On the other hand, in contrast to vascular plants (Pick & Bassilian, 1982), the detergent octyl glucoside stimulates the algal Ca-ATPase activity (Hu, Strotmann, Shavit, & Leu, 1998). However, octyl glucoside activates CF_1 Mg-ATPase activity both in vascular plants and Chlamydomonas, although the stimulation effect by the detergent was lower in Chlamydomonas than in spinach CF_1 (Pick et al., 1982).

The octyl glucoside effect is linked to a lower ADP inhibition (reviewed in Lapashina & Feniouk, 2018) which is an intrinsic feature of all F-type ATP synthase enzymes and was first observed in CF_oCF_1 (Avron, 1962). This mode of regulation involves transient inhibition upon binding of Mg-ADP in a catalytic site (Zhou, Xue, Du, Melese, & Boyer, 1988) and generation of a pmf reactivates CF_oCF_1 (Strotmann & Bickel-Sandkotter, 1977).

Furthermore, ADP inhibition, which can be locked by azide binding to CF_1 (Minkov & Strotmann, 1989), requires Mg-ADP to bind to the catalytic site in absence of Pi (Dunham & Selman, 1981). Interestingly, the affinity of Chlamydomonas CF_OCF_1 for ADP is substantially lower, but that for Pi is comparable to vascular plants (Fiedler, Schlesinger, Strotmann, Shavit, & Leu, 1997; Strotmann et al., 1998). Whether algal ADP affinity is related to the above-mentioned differences in Mg-ATPase stimulation by octyl glucoside, and whether it contributes to algal ADP inhibition tuning *in vivo* remains to be investigated. Attempts to modulate ADP inhibition have been made by modifying the Walker-A motif of Chlamydomonas CF_OCF_1 via βT168S substitution (Hu et al., 1998), of which various homologous mutants have been studied before (Jault et al., 1996; Mueller, 1989; Omote, Maeda, & Futai, 1992). Displaying a lower ADP inhibition, the βT168S mutant showed a 10-fold increase of ATPase activity and was less efficiently locked in the ADP-inhibited state by azide (Hu et al., 1998). When supplemented with acetate but kept in the dark, the βT168S mutant cells grew faster. As will be explored in Section 5.1, tight metabolic coupling ensures the supply of mitochondrial ATP to the chloroplast (Lemaire, Wollman, & Bennoun, 1988). It remains to be tested whether enhanced growth in the ATP-hydrolyzing βT168S cells can be linked to a stimulation of mitochondrial respiration.

4.2 Activation by the proton motive force

Early studies with isolated chloroplasts showed that CF_OCF_1 transitioned to a highly active state after a critical number of laser flashes was fired in a short sequence to generate, altogether, a sufficient membrane potential that served as the driving force for photophosphorylation (Junge, Rumberg, & Schroder, 1970). It was postulated that an energetic barrier, the threshold pmf, had to be exceeded to trigger the conversion to the pmf-activated CF_OCF_1 state (Junge, 1970). Surpassing the threshold pmf overpowers other regulatory levels, such as mitigating ADP inhibition (Strotmann & Bickel-Sandkotter, 1977) or, as outlined in Section 4.3, cancelling restricted turnover due to thiol modulation. It is important to note that both pmf components, $\Delta\Psi$ and ΔpH, are thermodynamically equivalent in driving ATP synthesis (Jagendorf & Uribe, 1966; Witt, Schlodder, & Graber, 1976) and it was observed that a minimum $\Delta\Psi$ is required to activate the enzyme (Kaim & Dimroth, 1999). Soon after the findings by Junge and colleagues, first indications of light-dependent structural changes in spinach

CF_oCF_1 were reported by exchange of tritiated water from the medium to CF_1 (Ryrie & Jagendorf, 1971). Several biochemical approaches followed to examine the changes in CF_oCF_1. The partial tryptic digestion pattern differed in illuminated thylakoid samples which suggested energization-dependent exposure of CF_oCF_1 cleavage sites, in particular within the γ-subunit (Schumann, Richter, & McCarty, 1985). As a side effect of illumination in the presence of trypsin, ATPase activity was elicited after purifying CF_1 from those membranes (Moroney & McCarty, 1982a, 1982b). Digestion experiments have been continued with isolated CF_1 and the ε-deficient CF_1, showing that removal of the ε-subunit exposed the trypsin cleavage sites in solution that are exposed in membrane-bound CF_1 upon illumination (Hightower & McCarty, 1996). Alternatively to tryptic digestion, various chemical labeling experiments were conducted which revealed light-dependent modification of the highly conserved γC89 (Moroney, Andreo, Vallejos, & McCarty, 1980) and γK24/γK30 (Komatsu-Takaki, 1996). Similarly, there have also been reports on energization-dependent structural changes in the β-subunit hinge region of the DELSEED-loop (βK399) and a surface βK447 (Komatsu-Takaki, 1995). From today's perspective, with the resolved spinach particles available (Hahn et al., 2018), these reactive residues are located in a torque-generating γ-/β-subunit interface (Pu & Karplus, 2008). The labeling studies also revealed γK222/γK231 (Komatsu-Takaki, 1996) which are located in a structure that is relevant for thiol modulation (see Section 4.3). Other light-dependent reactivity changes were found in εK109 (Komatsu-Takaki, 1989), which is in the hairpin of the C-terminal α helix (see Section 4.4). The light-dependent labeling of εK109 is supported by antibody-induced uncoupling of photophosphorylation which indicates structural rearrangements in the ε-subunit C-terminus when membranes are energized (Johnson & McCarty, 2002; Richter & McCarty, 1987). It has been proposed that pmf-induced changes involve transient formations/cleavages of salt bridges between acidic and basic residues (Strotmann et al., 1998) and some of the labeled sites above are indeed in vicinity of salt bridge-forming residues (Hahn et al., 2018). To visualize the light-dependent labeling and trypsin cleavage sites in the resolved CF_oCF_1 structure, a map is presented in Fig. 3 which suggests the potential pmf impact on the CF_oCF_1 structure at critical interfaces.

4.3 Thiol modulation

Unlike bacterial or mitochondrial F_1, CF_1 is usually isolated as a latent ATPase enzyme and several ATPase elicitation methods were explored in

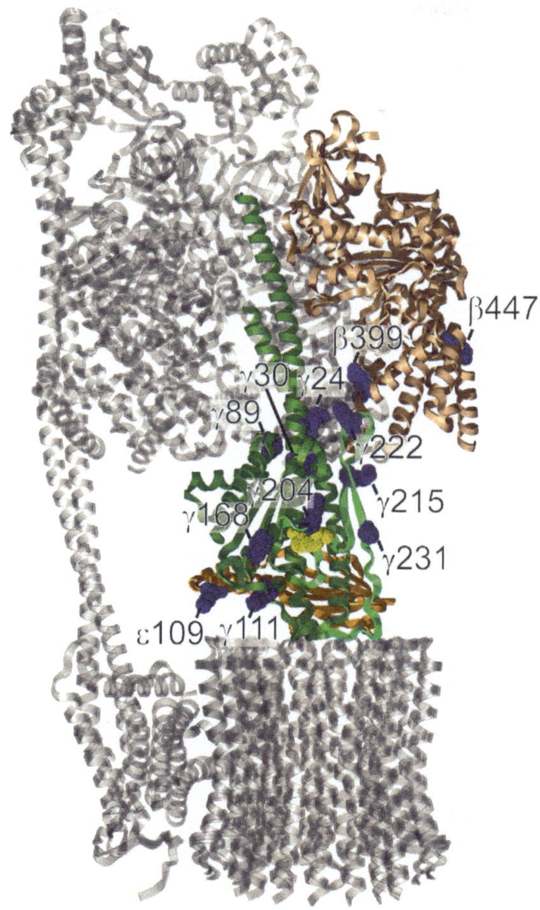

Fig. 3 Pmf-dependent reactivity map of various amino acids in spinach CF_oCF_1. The representation is an adaptation from Fig. 1. Except for subunits β (empty), γ and ε, all other subunits are semi-transparent in gray. One αβ heterodimer was omitted for clarity. Pmf-dependent reactive sites are represented as blue dots and the redox cysteine couple is marked in yellow. Lysines in the γ-subunit (numbering 24, 30, 168, 222, 231), the β-subunit (399, 447) and ε109 reacted with pyridoxal 5′-phosphate (Komatsu-Takaki, 1989, 1995, 1996). The solvent exposure of a γ-cysteine at position 89 depends on the pmf (Moroney et al., 1980), as well as the disulfide-forming couple (Ketcham, Davenport, Warncke, & McCarty, 1984; Konno et al., 2012). The γ-residues at position 204, 215 and 231 become hypersensitive to trypsin digestion (Moroney & McCarty, 1982b; Schumann et al., 1985). Reactivity of γ-residues in subunit-deficient CF_1 have been reviewed elsewhere (Richter, Hein, & Huchzermeyer, 2000).

Section 4.1. Here, the brief focus will be on thiol modulation which involves an intrinsic redox-active disulfide couple in the CF_oCF_1 γ-subunits of the green lineage. The two γ-cysteines are located in a ~40-amino acid domain before the C-terminal α helix (Hahn et al., 2018). This domain contains two hairpins that form an L shape. The first hairpin harbors the disulfide couple and the second hairpin stretches along the γ-subunit N-terminus, getting in close contact with the DELSEED motif of the empty β-subunit. Thus, the second hairpin is believed to block rotation upon ATP hydrolysis in the presence of the disulfide. In fact, a thiol modulation-dependent interplay between the γ-subunit and the βDELSEED motif has been reported previously (Buchert, Konno, & Hisabori, 2015), which points to structural rearrangements of the L-shaped redox domain upon disulfide cleavage. Treatment of the enzyme with DTT elicits ATPase activity *in vitro* and *in situ* (Arana & Vallejos, 1982; McCarty & Racker, 1968; Nalin & McCarty, 1984). *In vivo*, disulfide cleavage relies on photosynthetic electron transfer and is very efficiently realized either by the NADPH thioredoxin reductase C in dim light (Carrillo, Froehlich, Cruz, Savage, & Kramer, 2016; Naranjo et al., 2016) or, at higher light intensities, by the ferredoxin-thioredoxin regulatory system (Ketcham et al., 1984; Mills, Mitchell, & Schurmann, 1980; Schwarz, Schurmann, & Strotmann, 1997; Yoshida, Matsuoka, Hara, Konno, & Hisabori, 2014). Oxidation of the γ-dithiol is not spontaneous and involves a redox cascade, as shown in a recent work in Arabidopsis where H_2O_2 acts as ultimate electron sink (Yoshida, Hara, Sugiura, Fukaya, & Hisabori, 2018).

As outlined in Section 4.2, CF_oCF_1 is regulated on other levels such as the pmf itself (Junge, 1970). In vascular plants, the threshold pmf for activation was shown to be elevated in the γ-disulfide conformation (Junesch & Graber, 1987). On the other hand, once photosynthetic pmf generation exceeds the corresponding threshold, both γ-redox states yield similar maximal ATP synthesis rates (Junesch & Graber, 1987; Konno et al., 2012). Thus, the γ-disulfide cleavage enables higher ATP synthesis rates at low to moderate pmf levels. On grounds of thermodynamics, this would be true for the reverse reaction as well.

The impact of the pmf on the CF_oCF_1 structure and thiol modulation is tightly intertwined. Chemical and thioredoxin-mediated reduction of the γ-disulfide are more efficient in the presence of a pmf in energized membranes (Ketcham et al., 1984; Konno et al., 2012), suggesting that the γ-redox domain experiences pmf-dependent structural rearrangements as well. As shown in Fig. 3, this is backed up by light-dependent labeling of γK222/γK231 in the second hairpin that blocks ATP hydrolysis (Hahn et al., 2018; Komatsu-Takaki, 1996). Thiol modulation is also intertwined with binding and release of ADP, since it appears to weaken the extent of

ADP inhibition *in vitro*. It was shown that F_1 particles show a lower probability to lapse into the ADP-inhibited state when the γ-dithiol is present (Kim, Konno, Sugano, & Hisabori, 2011).

Several observations have been published regarding thiol modulation in Chlamydomonas. DTT treatment yielded a two-fold activated Mg-ATPase activity (Hu et al., 1998; Selman-Reimer et al., 1981), whereas higher activation ratios were observed in spinach CF_1 (McCarty & Racker, 1968). This might be related to the fact that isolated Chlamydomonas CF_oCF_1 displayed slightly elevated basal ATPase activity compared to the spinach enzyme. Some authors attributed this observation to protease activity during isolation (Strotmann et al., 1998). Alternatively, it could be that ADP inhibition efficiency in the presence of the γ-disulfide is weaker in the alga. This would favor ATP hydrolysis in Chlamydomonas and, if mitochondrial ATP supply is functional, maintain an elevated pmf in the dark (see Section 5). Still, it remains to be tested in Chlamydomonas whether there is a causal relationship between the relatively weak ATP hydrolysis stimulation by DTT on the one hand, and the mitigated octyl glucoside effect as well as the lower affinity for ADP on the other (Fiedler et al., 1997; Pick et al., 1982; Strotmann et al., 1998).

The physiological role of establishing the γ-redox domain in chloroplasts of the green lineage is still not fully understood. It was suggested that thiol modulation facilitates ATP conservation when the pmf in darkness is low (Ort & Oxborough, 1992), and the chock-like hairpin of the disulfide-containing γ-redox domain supports this suggestion (Hahn et al., 2018). However, *Arabidopsis thaliana* (Wu & Ort, 2008) and *Chlamydomonas reinhardtii* (Ross et al., 1995) are not severely affected under photoperiodic growth conditions when the redox-active γ-cysteines were mutated. This suggests a more subtle physiological role of thiol modulation than merely ATP conservation in the dark. In fact, as elaborated in Section 5.2, thiol modulation in Arabidopsis could be involved in pmf-dependent protein import (Kohzuma et al., 2017). An additional regulatory mechanism that might be intertwined with thiol modulation will be presented in Section 4.5.

4.4 The ε-subunit: An intrinsic ATP hydrolysis inhibitor

Based on the early observation that the CF_1 ε-subunit inhibits ATP hydrolysis (Nelson, Nelson, & Racker, 1972), a profound body of data demonstrates that this small subunit is an intrinsic inhibitor. Until recently, it was assumed that the situation in the chloroplast enzyme resembles bacterial F_1, in which the extended ε-subunit C-terminus reaches into the F_1 head (Cingolani & Duncan, 2011) and links this structural feature to a well-documented block of bacterial F_1 ATPase activity (Kato-Yamada et al., 1999). Thereby, the CF_1 ε-subunit could be part of an ATP conservation mechanism. However, in the recent high resolution structure of spinach CF_oCF_1, the ε-subunit C-terminus was in a retracted conformation and unable to be engaged in such an inhibition mechanism (Hahn et al., 2018). It remains enigmatic how the ε-subunit regulates the catalytic activity of the chloroplast enzyme. Further structural investigations are necessary to resolve this issue, for instance, by incorporating pmf-induced ε-subunit adjustments in the complex (Johnson & McCarty, 2002; Komatsu-Takaki, 1989).

In the pre-structure era, various groups have focused on the C-terminus of the Chlamydomonas CF_1 ε-subunit by investigating to which extent it is involved in functional CF_oCF_1 integrity. Harboring a Chlamydomonadales-specific segment extension that links the fifth and sixth strand of the β sandwich, the Chlamydomonas CF_1 ε-subunit has a rather low identity of 42% with the spinach homolog (Fig. 4A).

The *FUD17* mutant was rescued by chloroplast transformation to test whether the C-terminal amino acid sequence affects Chlamydomonas CF_oCF_1 *in vivo*. *FUD17* harbors a frameshift in *atpE* and produces a non-functional ε-subunit (Johnson & Anastasios, 2004; Robertson et al., 1990). When the C-terminal ε-subunit segment was shortened by 10 residues in a previous study (Johnson, 2008), changed ATP hydrolysis properties were observed *in vitro* in addition to a dramatic decrease of CF_oCF_1 accumulation (see Section 3.2). Truncation of the ε-subunit beyond the last 10 residues did not allow for phototrophic growth in Chlamydomonas (Johnson & Anastasios, 2004). Fig. 4A also shows the ε-subunit sequence of *Arabidopsis thaliana* and *Pisum sativum* (pea). There are similarities in the C-terminus from Chlamydomonas and pea, such as vicinal positively charged residues. On the other hand, the amino acid sequence from Arabidopsis and spinach are similar. Compared to isolated spinach CF_1, which requires activation treatment by DTT, pea CF_1 constitutively produces higher ATPase activities *in vitro* (Buckley, Cruz, Cohen, & McCarty, 1999). This resembles isolated Chlamydomonas CF_1 which displays higher basal ATPase activity as well (McCarty & Racker, 1968; Selman-Reimer et al., 1981). It has been suggested that, among other possibilities, variation of the C-terminal ε-subunit sequence between pea and spinach could produce the different activities (Buckley et al., 1999). Given the above-mentioned resemblance

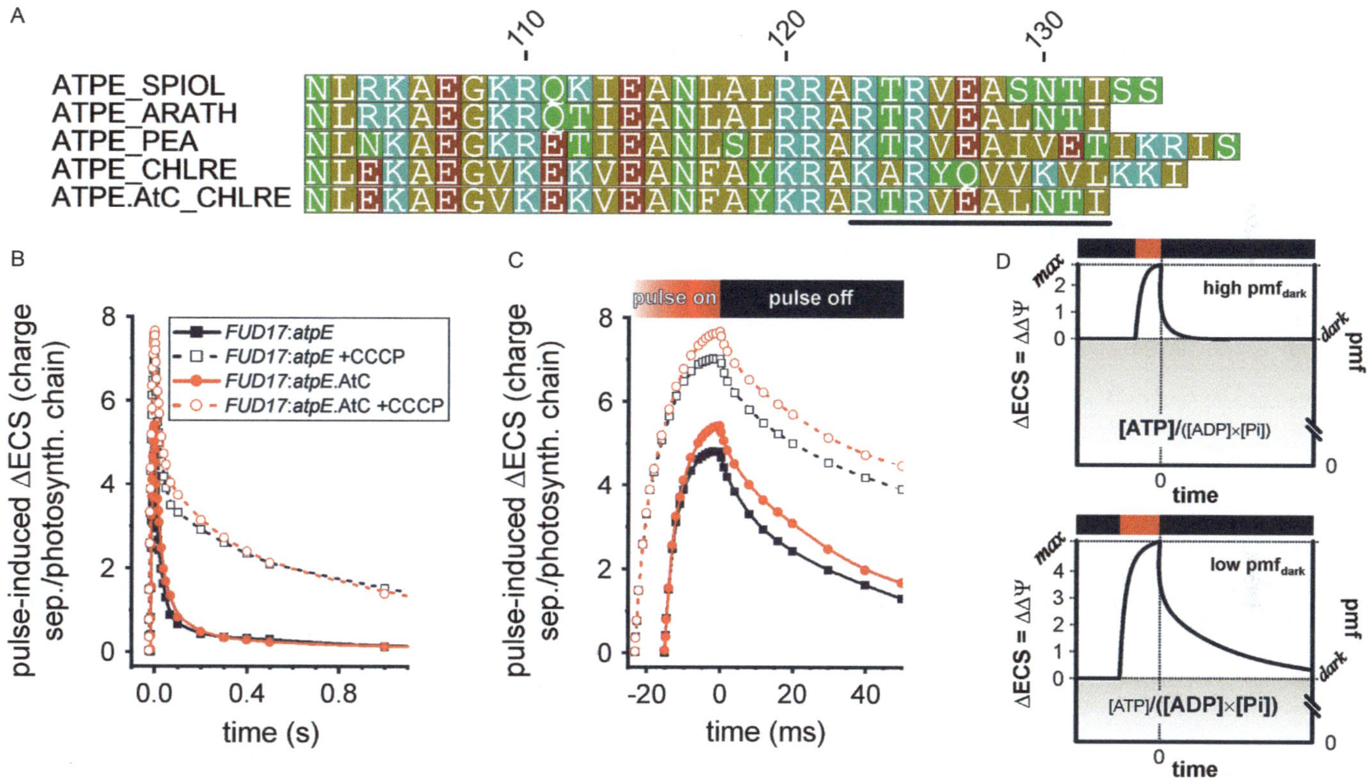

Fig. 4 See figure legend on next page.

with pea and a potential role in ATP conservation (via ATPase inhibition), vicinal positively charged residues were removed and the Chlamydomonas ^{128}KARYQVVKVLKKI^{140} ε-segment was replaced with the Arabidopsis sequence ^{123}RTRVEALNTI^{132} (black bar in Fig. 4A). The mutation also shortened the C-terminus by three amino acids, but still allowed for photoautotrophic growth. Furthermore, as shown in Fig. 4B and C, the swapped segment did not change the ATP-conserving CF_oCF_1 performance. This can be demonstrated by the light pulse-induced electrochromic shift (ECS) amplitudes (Joliot & Joliot, 2008) which were comparable upon extended dark-adaptation. The same was observed when using uncouplers to collapse the significant pmf established in the dark (explored further in Section 5). This suggests that the pmf in the dark was comparable in both strains (Fig. 4D). Furthermore, unlike for truncation (Cruz et al., 1995; Johnson & Anastasios, 2004; Johnson, 2008), there was a certain tolerance for C-terminal amino acid variations in the Chlamydomonas ε-subunit that maintained CF_oCF_1 performance in the presence or absence of light-driven pmf generation. It remains to be tested whether other C-terminal residues change

Fig. 4 C-terminal variations of the ε-subunit and CF_oCF_1 function in an algal mutant. (A) The C-terminal amino acid alignment of spinach, Arabidopsis, pea and Chlamydomonas reveals slight alterations in the very terminal region. Accordingly, pea and Chlamydomonas harbor positively charged residues (blue) whereas spinach and Arabidopsis contain hydrophobic and neutral amino acids (olive and green, respectively). UniProt IDs with spinach numbering are shown. The black bar indicates a swapped segment from the Arabidopsis ε-subunit, yielding Chlamydomonas ATPE. AtC in panels (B) and (C). Therein, the *FUD17* mutant (Robertson, Boynton, & Gillham, 1990) was rescued with the wild type and AtC version of *atpE*, restoring photosynthesis in both strains. Dark-adapted cells developed an electrochromic shift (ECS) during a short saturating illumination since pigments in the photosynthetic membranes respond to the membrane potential $\Delta\Psi$ by changing their absorption behavior, functioning as spectroscopic voltmeters (reviewed in Bailleul, Cardol, Breyton, & Finazzi, 2010). The ECS signal during the pulse was comparable in both strains. The same applied to the CF_oCF_1 activity in the dark, which determines the ECS decay rate (B) during several 100-ms, (C) as well as immediately after the pulse. (D) The used pulse method visualizes the pedestal pmf which is present before the measurement, thus producing variable ECS amplitudes during the pulse until the maximal $\Delta\Psi$ is obtained that a thylakoid membrane can sustain (Joliot & Joliot, 2008). According to Eq. (1) and Section 5, the pmf in the dark equilibrates as a function of the [ATP]/([ADP] $\times$ [Pi]) ratio with the contribution of the mitochondrial ATP pool. By inhibiting respiration, the latter pool is depleted in the presence of 0.5 μM carbonyl cyanide m-chlorophenyl hydrazine, CCCP, in panels (B) and (C). The inhibitor collapsed the dark-pmf similarly in both strains to a low level, compared to the control. Taken together, the AtC variant of the ε-subunit yielded a wild type-like CF_oCF_1.

the latent CF_1 ATP hydrolysis activity in various species (Fig. 4A), for instance in the hairpin that harbors the pmf-responsive εK109 (Komatsu-Takaki, 1989).

4.5 Post-transcriptional regulation by reactive oxygen species

The photosynthetic machinery is confronted with reactive oxygen species (ROS), which are side products of electron transfer reactions. ROS are potentially harmful signaling molecules for which the cell possesses an array of detoxification mechanisms (reviewed in Foyer, 2018). Electron flow regulation prevents ROS formation and several key mechanisms (extensively reviewed in Eberhard, Finazzi, & Wollman, 2008) are contributing to the protection of the photosynthetic apparatus. (i) The ΔpH across the membrane has several regulatory functions, one of which is termed photosynthetic control (reviewed in Finazzi, Minagawa, & Johnson, 2016). At low lumen pH, the electron transfer through the b_6f is slowed down to protect the PSI acceptor side from overreduction, thus limiting ROS formation. (ii) Thermal dissipation of excitation energy in response to low lumen pH protects both photosystems and is observed as nonphotochemical chlorophyll fluorescence quenching, termed NPQ (reviewed in Niyogi & Truong, 2013).

It was shown that (i) and (ii) can be finetuned by a modified H^+ permeability of CF_oCF_1, and it was proposed that, among other possibilities, ROS-induced amino acid oxidations within the γ-subunit are involved. A modified H^+ permeability has been observed in a number of studies during CO_2-limited conditions (Kanazawa & Kramer, 2002) or drought (Kohzuma et al., 2009), where CF_oCF_1 activity is downregulated and facilitates (i) and (ii). The effect of low CO_2 on H^+ permeability has been termed metabolic control of CF_oCF_1 (Kanazawa & Kramer, 2002). There are indications that post-translational modifications are causing the downregulation of CF_oCF_1 activity under CO_2-limited conditions, apart from the redox-active cysteine couple (Kohzuma, Dal Bosco, Meurer, & Kramer, 2013). Alternatively, stromal Pi availability could be restricted under these conditions, thus modulating the H^+ conductivity via CF_oCF_1 (Kramer, Avenson, & Edwards, 2004; Kramer, Cruz, & Kanazawa, 2003). However, due to substrate limitation of the Calvin Benson cycle, low CO_2 concentrations in combination with light could promote ROS formation upon strong reduction of the stroma. Therefore, downregulation of CF_oCF_1 activity under ROS-favoring conditions is crucial and recent studies of Arabidopsis γ-subunit point mutations documented the adverse effects of elevated H^+ conductance via CF_oCF_1 on the PSI redox state (Kanazawa et al., 2017; Takagi et al., 2017).

It was observed that Mg-ATPase activity of isolated spinach thylakoids and CF_1 variants was inhibited by ROS-mediated oxidations in the γ-subunit (Buchert & Forreiter, 2010; Buchert, Schober, Rompp, Richter, & Forreiter, 2012). Therein, it was ruled out that the redox-active cysteine couple was involved. Instead, the inhibition was the result of oxidations within a conserved methionine/cysteine cluster in the γ-subunit, located in a torque-generating γ-/β-subunit interface (Pu & Karplus, 2008). As seen in Fig. 3, solvent exposure of the ROS targets γM23/γC89/γM279/γM282 depends on the pmf (Komatsu-Takaki, 1996; Moroney et al., 1980), which could link this type of modification to a light-dependent process. When bypassing the impact of ROS target oxidations in Chlamydomonas, a highly active CF_oCF_1 was obtained after high light treatment (Buchert, Bailleul, & Hisabori, 2017). The ROS-resistant Chlamydomonas mutant γR279Q might tolerate oxidations in the energy coupling cluster consisting of γM23/γC89 that interacts with the βDELSEED motif (Li, Ma, & Weber, 2019). Interestingly, a similar mechanism might be effective in the above-mentioned Arabidopsis studies, which involved random γ-subunit point mutants either in proximity to γC89 (Takagi et al., 2017) or which were related to altered thiol modulation (Kanazawa et al., 2017). The γ-/β-subunit interface, which harbors the ROS targets, is subject to structural changes within CF_1 that are associated with thiol modulation (Buchert et al., 2015). This could explain why, depending on the γ-redox state, ROS have a different impact on Mg-ATPase activity in isolated spinach thylakoids. The relative loss of activity was more pronounced in the presence of the γ-disulfide (Buchert & Forreiter, 2010). It remains to be tested whether ROS regulation of CF_oCF_1 is interconnected with the newly discovered photosynthetic γ-disulfide formation, depending on thioredoxin-like2 and H_2O_2 as electron sink (Yoshida et al., 2018). Besides thioredoxin-like2, which has a significantly higher midpoint redox potential than other thioredoxins, 2-Cys peroxiredoxins were involved. This type of disulfide formation could either coincide with or even promote a more efficient CF_oCF_1 downregulation by ROS, and thus would extend the list of CF_oCF_1 regulation levels that are intertwined with thiol modulation. In Chlamydomonas, however, Thioredoxin-like2 is missing and it remains to be tested if other (algae-specific) thioredoxins with a high midpoint redox potential could mediate this reaction.

5. Maintenance of the chloroplast proton motive force in dark-adapted algae via ATP hydrolysis

5.1 Origin, amplitude and partitioning of the pmf in the dark

Not Chlamydomonas but *Chlorella* spp. were first used as a microalgal system to explore the pmf that is present in the dark. These studies led to the conclusion that dark-adapted algae maintain a significant pmf in the presence and absence of oxygen by ATP hydrolysis via CF_OCF_1 (Diner & Joliot, 1976; Joliot & Joliot, 1980). The required ATP is either of mitochondrial origin and produced by respiration or, if mitochondria are inactive under anaerobiosis, the ATP is produced via glycolysis (reviewed in Johnson & Alric, 2013). An efficient metabolic coupling between chloroplasts and mitochondria is also present in diatoms (Bailleul et al., 2015). Although profound insights are still missing, it could be that metabolic organellar interlacing in photosynthetic unicellular organisms culminated in a highly flexible exchange network between reducing equivalents from chloroplasts and ATP from mitochondria. An example is found in a suppressor screen of Chlamydomonas, which yielded the *FUD50*su strain (Lemaire et al., 1988). The authors showed that the CF_OCF_1-deficient *FUD50* mutant, which contains an *atpB* gene deletion (Woessner et al., 1984), regained the ability to fix CO_2 photosynthetically in *FUD50*su, which harbored multiple uncharacterized suppressor mutations (Lemaire et al., 1988). CO_2 fixation was strictly dependent on mitochondrial respiration since, unlike wild type, *FUD50*su failed to grow phototrophically in the presence of mitochondrial inhibitors. The efficient ATP import from the cytosol into the chloroplast, which was initially observed in isolated spinach chloroplasts (Heldt, 1969), is accomplished by different types of nucleotide transporters. One type, the chloroplast nucleotide transporters NTT1 and NTT2 (Reiser, Linka, Lemke, Jeblick, & Neuhaus, 2004), are H^+/ATP symporters and ATP/dNTP exchangers, respectively. NTT1 and NTT2 are ancient transporters and they have also been identified in glaucophytes and red algae (Weber & Linka, 2011), as well as in diatoms where they might function slightly differently (Ast et al., 2009). In Arabidopsis, NTT1 and NTT2 are most important for cytosolic ATP import into the chloroplast in young tissues, whereas they become less important in older leaves (Voon et al., 2018). It remains to be tested how and to which extent NTT1/NTT2 are involved in ATP import in Chlamydomonas and other green microalgae.

The other chloroplast nucleotide transporter type is the thylakoid ATP/ADP carrier TAAC, which is highly conserved in the green lineage and located in the thylakoid membrane (Thuswaldner et al., 2007). The ATP that is provided via TAAC was shown to be necessary for PSII repair in Arabidopsis (Yin et al., 2010).

This shows that certain chloroplast functions rely on metabolic coupling. Furthermore, in the dark when photophosphorylation is inactive, a fraction of the imported ATP will contribute to the concentration equilibrium of the [ATP]/([ADP]×[Pi]) ratio, i.e.,? the dark-pmf that is governed by CF_oCF_1. The pmf, $\Delta\tilde{\mu}_{H^+}$, is defined at equilibrium by Eq. (1):

$$n \times \Delta\tilde{\mu}_{H^+} = \Delta G_{ATP} = \Delta G^0_{ATP} + RT \times \ln\left(\frac{[ATP]}{[ADP] \times [Pi]}\right) \quad (1)$$

Therein, the standard Gibbs free energy of ATP synthesis, ΔG^0_{ATP}, the gas constant, R, and the temperature in Kelvin, T, are denoted. As outlined in Section 2.1, n is the H^+/ATP ratio that is defined by the stoichiometry in the c-subunit ring. Compared to vascular plants with $n = 14/3$, a smaller c-subunit ring in Chlamydomonas, yielding $n = 13/3$ (Meyer Zu Tittingdorf et al., 2004), could be an operational adaptation to slightly elevated pmf in algae. The pmf that exists in the dark-adapted chloroplasts has been first quantified in *Chlorella sorokiniana* under anaerobic conditions (Joliot & Joliot, 1989), yielding an estimate of 20–50 mV. The osmotic pmf component has been later determined to be around 1.5 ΔpH units (Finazzi & Rappaport, 1998), which would correspond to a total pmf in the dark of ~110–140 mV.

The partitioning of the dark-pmf into ΔΨ and ΔpH depends on the presence of CF_oCF_1. Dark-adapted CF_oCF_1-deficient *Chlorella sorokiniana* mutants displayed no ΔpH and the pmf was dominated by the electric component (Rappaport, Finazzi, Pierre, & Bennoun, 1999), which suggested the existence of (an) enigmatic ATP-dependent ion pump(s). The authors propose that the pump contributes only marginally to the pmf in physiological conditions since it is outcompeted by CF_oCF_1 activity. Although further characterization is required, the enigmatic pump could be engaged in pmf partitioning in concert with other ion channels (reviewed in Marchand, Heydarizadeh, Schoefs, & Spetea, 2018) or when chloroplast ATP metabolism is restricted, such as in ribulose-1,5-bisphosphate carboxylase/oxygenase mutants (Finazzi et al., 2006). Additionally, this ATP-dependent ion pump would be beneficial in case CF_oCF_1 transitioned into an inactivate state during severe ATP depletion in the dark. Once respiration resumes, the pump could generate the minimal $\Delta\Psi$ that is required for CF_oCF_1 activity (Kaim & Dimroth, 1999). Provided that mitochondrial respiration and nucleotide import activities are high enough, and that ATP hydrolysis via such a $\Delta\Psi$-activated CF_oCF_1 is governed by the [ATP]/([ADP] × [Pi]) ratio *in vivo*, the pmf in the dark will then be re-established by CF_oCF_1.

5.2 Possible consequences of the pmf in the dark

Since CF_oCF_1 maintains the elevated pmf in the dark when mitochondrial respiration is active, protein import processes could be facilitated and, as outlined further below, processes at the onset of photosynthesis could be modified.

Translocon complexes within the outer and inner chloroplast membrane facilitate the import of nuclear-encoded photosynthetic proteins (reviewed in Cline & Theg, 2007; Paila, Richardson, & Schnell, 2015). Once the peptides entered the stoma, four pathways (Tat, Sec, SRP, and spontaneous) are engaged in thylakoid membrane targeting of subunits from both photosystems and the b_6f. While substrates for the Sec and SRP pathways are unfolded proteins, thus requiring chaperones, the Tat pathway operates with folded proteins as well. The Sec and SRP pathways require ATP, whereas the Tat pathway does not. All three pathways depend on a ΔpH, based on *in vitro* results. The requirement for a ΔpH-dependent Tat pathway *in vivo* was recently challenged by using the CF_oCF_1-deficient Chlamydomonas mutant *FUD50* (Finazzi, Chasen, Wollman, & de Vitry, 2003). The mutant showed Tat-dependent thylakoid targeting and it was proposed that the presence of unknown factors facilitates the *in vivo* targeting in the absence of a ΔpH. Alternatively, Tat-dependent targeting might be less discriminatory in favor of the osmotic pmf component since it is also accomplished by $\Delta\Psi$ (Braun, Davis, & Theg, 2007), which dominates the pmf in CF_oCF_1-deficient algal mutants under certain experimental conditions (Rappaport et al., 1999).

As mentioned in Section 4.3, thiol modulation could be involved in pmf-dependent protein import in Arabidopsis (Kohzuma et al., 2017). It was shown that a mutant which fails to form the γ-disulfide developed a stay green phenotype after several days in the dark, coinciding with higher levels of PSII and b_6f subunits. The authors suggested that, among other possible processes, the elevated pmf in the ATP-hydrolyzing mutant impacted protein import and integrity of the photosynthetic machinery in the dark. However, a clear picture of the various processes in the dark-adapted Arabidopsis mutant is still missing.

As mentioned in Section 4.5, adjusting H^+ conductivity via CF_oCF_1 could modulate the energy-dependent NPQ efficiency in the light, which relies on an acidification of the lumen (reviewed in Niyogi & Truong, 2013). Restricting H^+ conductivity of CF_oCF_1 at low CO_2 levels has been observed and resulted in an increase of the pH-dependent qE component of NPQ (Kanazawa & Kramer, 2002). Maintenance of a significant osmotic pmf component could ensure photoprotection in dynamic light environments. Accordingly, violaxanthin de-epoxidase could remain active in the presence of a ΔpH in the dark, synthesizing zeaxanthin, which can bind to certain light-harvesting antennae proteins. In addition, the latter switch from a light harvesting to an energy-dissipating form upon protonation of critical residues in the lumen. A recent work in Chlamydomonas showed that, when grown under field conditions with a sinusoidal light amplitude, the NPQ-inducing light-harvesting proteins are expressed and showed a slow degradation in the dark (Nawrocki, Liu, & Croce, 2020). Thus, the photosynthetic apparatus could be more protected during an abrupt onset of light since NPQ capacity is sustained in the dark.

An elevated dark-pmf might be even more crucial in Chlamydomonas than in vascular plants. Such an equilibrium in the dark could lower the energetic barrier that needs to be overcome, by light-induced pmf generation, to obtain the pmf-activated CF_oCF_1. The energetic barrier might be kept small by efficient mitochondrial ATP import and mitigated ADP inhibition in Chlamydomonas CF_oCF_1, thus facilitating ATP hydrolysis in the dark. As elaborated in Sections 2.2 and 4.2, a rapid generation of the pmf-activated CF_oCF_1 might be required due to features of the algal photosynthetic machinery and ultrastructure of the thylakoid membrane. Shared lumenal interfaces are more likely occurring in Chlamydomonas than in vascular plants, owing to a less strict division into appressed and non-appressed algal membranes. Those lumenal interfaces contain the light-dependent ΔpH generators PSII and b_6f on the vicinal appressed membrane (Olive, Vallon, Wollman, Recouvreur, & Bennoun, 1986) which face CF_oCF_1 in the non-appressed membrane (Fig. 2A). At the onset of light, proximity of pmf-generating b_6f (and PSII) to CF_oCF_1 could prevent local ΔpH maxima if CF_oCF_1 is readily pmf-activated and yields high ATP synthesis rates. The latter was shown previously in a reconstituted *E. coli* system where the authors found that, due to limited H^+ diffusion, ATP synthesis rates increased with decreasing distance to the liposome-acidifying cytochrome bo3 complex (Sjöholm et al., 2017). Other studies on lateral pH profiles suggest that the local pH is more alkaline in proximity of ATP synthase (Rieger, Junge, & Busch, 2014). In the shared lumenal interfaces, limitation of local ΔpH maxima by efficient ATP synthesis would also affect the b_6f at the onset of light by transiently weakening photosynthetic control. The latter is defined as b_6f activity downregulation due to hampered PQ formation and H^+ release into the acidified lumen, preventing PSI overreduction in the light. Quantitative mass spectrometry determined the plastocyanin to PSI ratio as ~1 in Chlamydomonas (Hammel, Zimmer, Sommer, Muhlhaus, & Schroda, 2018), compared to estimates of ~2–3 in vascular plants (Graan & Ort, 1984; Haehnel, Ratajczak, & Robenek, 1989). If the estimates are correct and the algal plastocyanin pool is smaller, a relatively weak photosynthetic control will adversely influence total electron flow by exhausting primary PSI electron donors, given that PSI is not acceptor side-limited. Especially in soil-dwelling Chlamydomonas, electron flow limitations in a light-limited environment are deleterious to meeting the requirements for CO_2 fixation and it is generally accepted that alternative electron sinks are essential for optimizing CO_2 fixation. Accordingly, oxidation of the algal plastocyanin pool is facilitated by flavodiiron proteins which are absent in vascular plants and drive electron flow during the induction phase of photosynthesis by accepting electrons from NADPH (Chaux et al., 2017). They also catalyze photoreduction of O_2 at the PSI acceptor side and thereby prevent PSI photodamage in fluctuating light conditions. Taken together, the algal photosynthetic machinery and thylakoid ultrastructure might rely on the elevated pmf in the dark which yields, compared to vascular plants, a modified H^+ conductivity tuning via CF_oCF_1 at the onset of light. Thus, while maximizing ATP synthesis efficiency, flavodiiron proteins could lower the potential photooxidative risk of a softened photosynthetic control that may exist in the absence of a strict division into appressed and non-appressed thylakoid membranes. Interestingly, similar features are observed in mosses in terms of the low degree of appressed membranes in the presence of flavodiiron proteins (Gerotto et al., 2016, 2019), although the dark-pmf has not been extensively studied. It will be interesting to see in the future if algal CF_oCF_1 mutants can be designed which fail to maintain the elevated dark-pmf, owing to hampered ATP hydrolysis. Thereby, it can be tested if readily pmf-activated WT CF_oCF_1 is necessary to sustain algal photosynthesis in a dynamic low-light environment, in the presence or absence of flavodiiron proteins. At the same time, it can be tested if facilitated ATPase activity in WT is involved in establishing photosynthetic control at the onset of light.

6. Conclusion and future perspective

In conclusion, CF_OCF_1 from green microalgae has several distinct features that are different from the enzyme in vascular plants, such as a smaller c-ring stoichiometry and the elevated dark-pmf by ATP hydrolysis. These features might underlie an evolutionary strategy in unicellular organisms, which often contain a single chloroplast only and perform very efficient metabolic coupling between the organelle and its cellular surroundings. Owing to its genetic accessibility, Chlamydomonas is the perfect organism for future investigations to clarify whether bioenergetics are affected upon alteration of the algal-specific CF_OCF_1 features. Moreover, it needs to be elucidated to which extent the algal structure blueprint deviates from vascular plants, and how spatio-temporal CF_OCF_1 organization is realized under different environments. By disentangling the flexible CF_OCF_1 embedment in algal bioenergetics, we could improve our understanding of how green microalgae optimize photosynthetic work in a shared biotope with other respiring microbes.

References

Abrahams, J. P., Leslie, A. G., Lutter, R., & Walker, J. E. (1994). Structure at 2.8 Å resolution of F_1-ATPase from bovine heart mitochondria. *Nature, 370*(6491), 621–628.

Andreo, C. S., Patrie, W. J., & McCarty, R. E. (1982). Effect of ATPase activation and the δ subunit of coupling factor-1 on reconstitution of photophosphorylation. *Journal of Biological Chemistry, 257*(17), 9968–9975.

Arana, J. L., & Vallejos, R. H. (1982). Involvement of sulfhydryl groups in the activation mechanism of the ATPase activity of chloroplast coupling factor-1. *Journal of Biological Chemistry, 257*(3), 1125–1127.

Ast, M., Gruber, A., Schmitz-Esser, S., Neuhaus, H. E., Kroth, P. G., Horn, M., et al. (2009). Diatom plastids depend on nucleotide import from the cytosol. *Proceedings of the National Academy of Sciences of the United States of America, 106*(9), 3621–3626. https://doi.org/10.1073/pnas.0808862106.

Avron, M. (1962). Light-dependent adenosine triphosphatase in chloroplasts. *Journal of Biological Chemistry, 237*, 2011–2017.

Avron, M. (1963). A coupling factor in photophosphorylation. *Biochimica et Biophysica Acta (BBA): Bioenergetics, 77*, 699–702.

Bailleul, B., Berne, N., Murik, O., Petroutsos, D., Prihoda, J., Tanaka, A., et al. (2015). Energetic coupling between plastids and mitochondria drives CO_2 assimilation in diatoms. *Nature, 524*(7565), 366–369. https://doi.org/10.1038/nature14599.

Bailleul, B., Cardol, P., Breyton, C., & Finazzi, G. (2010). Electrochromism: A useful probe to study algal photosynthesis. *Photosynthesis Research, 106*(1-2), 179–189. https://doi.org/10.1007/s11120-010-9579-z.

Barkan, A., & Small, I. (2014). Pentatricopeptide repeat proteins in plants. *Annual Review of Plant Biology, 65*(1), 415–442. https://doi.org/10.1146/annurev-arplant-050213-040159.

Bolle, C., Kusnetsov, V. V., Herrmann, R. G., & Oelmuller, R. (1996). The spinach *AtpC* and *AtpD* genes contain elements for light-regulated, plastid-dependent and organ-specific expression in the vicinity of the transcription start sites. *Plant Journal*, *9*(1), 21–30. https://doi.org/10.1046/j.1365-313x.1996.09010021.x.

Bosco, C. D., Lezhneva, L., Biehl, A., Leister, D., Strotmann, H., Wanner, G., et al. (2004). Inactivation of the chloroplast ATP synthase gamma subunit results in high non-photochemical fluorescence quenching and altered nuclear gene expression in *Arabidopsis thaliana*. *Journal of Biological Chemistry*, *279*(2), 1060–1069.

Boyer, P. D., Cross, R. L., & Momsen, W. (1973). A new concept for energy coupling in oxidative phosphorylation based on a molecular explanation of the oxygen exchange reactions. *Proceedings of the National Academy of Sciences of the United States of America*, *70*(10), 2837–2839.

Braun, N. A., Davis, A. W., & Theg, S. M. (2007). The chloroplast Tat pathway utilizes the transmembrane electric potential as an energy source. *Biophysical Journal*, *93*(6), 1993–1998. https://doi.org/10.1529/biophysj.106.098731.

Buchert, F., Bailleul, B., & Hisabori, T. (2017). A γ-subunit point mutation in *Chlamydomonas reinhardtii* chloroplast F_1F_o-ATP synthase confers tolerance to reactive oxygen species. *Biochimica et Biophysica Acta (BBA): Bioenergetics*, *1858*(12), 966–974. https://doi.org/10.1016/j.bbabio.2017.09.001.

Buchert, F., & Forreiter, C. (2010). Singlet oxygen inhibits ATPase and proton translocation activity of the thylakoid ATP synthase CF_1CF_o. *FEBS Letters*, *584*(1), 147–152. https://doi.org/10.1016/j.febslet.2009.11.040.

Buchert, F., Hamon, M., Gabelein, P., Scholz, M., Hippler, M., & Wollman, F. A. (2018). The labile interactions of cyclic electron flow effector proteins. *Journal of Biological Chemistry*, *293*(45), 17559–17573. https://doi.org/10.1074/jbc.RA118.004475.

Buchert, F., Konno, H., & Hisabori, T. (2015). Redox regulation of CF_1-ATPase involves interplay between the γ-subunit neck region and the turn region of the βDELSEED-loop. *Biochimica et Biophysica Acta (BBA): Bioenergetics*, *1847*(4-5), 441–450. https://doi.org/10.1016/j.bbabio.2015.01.013.

Buchert, F., Schober, Y., Rompp, A., Richter, M. L., & Forreiter, C. (2012). Reactive oxygen species affect ATP hydrolysis by targeting a highly conserved amino acid cluster in the thylakoid ATP synthase γ subunit. *Biochimica et Biophysica Acta (BBA): Bioenergetics*, *1817*(11), 2038–2048. https://doi.org/10.1016/j.bbabio.2012.06.007.

Buckley, N. M., Cruz, J. A., Cohen, W. S., & McCarty, R. E. (1999). The ε subunit of the chloroplast ATP synthase of *Pisum sativum*. *Photosynthesis Research*, *59*(2–3), 137–145.

Carrillo, L. R., Froehlich, J. E., Cruz, J. A., Savage, L. J., & Kramer, D. M. (2016). Multi-level regulation of the chloroplast ATP synthase: The chloroplast NADPH thioredoxin reductase C (NTRC) is required for redox modulation specifically under low irradiance. *Plant Journal*, *87*(6), 654–663. https://doi.org/10.1111/tpj.13226.

Cavaiuolo, M., Kuras, R., Wollman, F. A., Choquet, Y., & Vallon, O. (2017). Small RNA profiling in Chlamydomonas: Insights into chloroplast RNA metabolism. *Nucleic Acids Research*, *45*(18), 10783–10799. https://doi.org/10.1093/nar/gkx668.

Chang, I. C., & Kahn, J. S. (1966). Isolation of a possible coupling factor for photophosphorylation from chloroplasts of *Euglena gracilis*. *Archives of Biochemistry and Biophysics*, *117*(2), 282–288. https://doi.org/10.1016/0003-9861(66)90414-0.

Chaux, F., Burlacot, A., Mekhalfi, M., Auroy, P., Blangy, S., Richaud, P., et al. (2017). Flavodiiron proteins promote fast and transient O_2 photoreduction in Chlamydomonas. *Plant Physiology*, *174*(3), 1825–1836. https://doi.org/10.1104/pp.17.00421.

Cingolani, G., & Duncan, T. M. (2011). Structure of the ATP synthase catalytic complex F_1 from *Escherichia coli* in an autoinhibited conformation. *Nature Structural & Molecular Biology*, *18*(6), 701–707. https://doi.org/10.1038/nsmb.2058.

Cline, K., & Theg, S. M. (2007). The Sec and Tat protein translocation pathways in chloroplasts. In R. E. Dalbey, C. M. Koehler, & F. Tamanoi (Eds.), *Vol. 25. The enzymes* (pp. 463–492). Academic Press.

Cruz, J. A., Harfe, B., Radkowski, C. A., Dann, M. S., & McCarty, R. E. (1995). Molecular dissection of the ε subunit of the chloroplast ATP synthase of spinach. *Plant Physiology*, *109*(4), 1379–1388. https://doi.org/10.1104/pp.109.4.1379.

Daum, B., & Kuhlbrandt, W. (2011). Electron tomography of plant thylakoid membranes. *Journal of Experimental Botany*, *62*(7), 2393–2402. https://doi.org/10.1093/jxb/err034.

Daum, B., Nicastro, D., Austin, J., 2nd, McIntosh, J. R., & Kuhlbrandt, W. (2010). Arrangement of photosystem II and ATP synthase in chloroplast membranes of spinach and pea. *Plant Cell*, *22*(4), 1299–1312. https://doi.org/10.1105/tpc.109.071431.

Dautant, A., Velours, J., & Giraud, M. F. (2010). Crystal structure of the Mg-ADP-inhibited state of the yeast F_1c_{10}-ATP synthase. *Journal of Biological Chemistry*, *285*(38), 29502–29510. https://doi.org/10.1074/jbc.M110.124529.

Diner, B., & Joliot, P. (1976). Effect of the transmembrane electric field on the photochemical and quenching properties of photosystem II *in vivo*. *Biochimica et Biophysica Acta (BBA): Bioenergetics*, *423*(3), 479–498.

Drapier, D., Girard-Bascou, J., Stern, D. B., & Wollman, F. A. (2002). A dominant nuclear mutation in Chlamydomonas identifies a factor controlling chloroplast mRNA stability by acting on the coding region of the *atpA* transcript. *Plant Journal*, *31*(6), 687–697. https://doi.org/10.1046/j.1365-313x.2002.01387.x.

Drapier, D., Girard-Bascou, J., & Wollman, F. A. (1992). Evidence for nuclear control of the expression of the *atpA* and *atpB* chloroplast genes in Chlamydomonas. *Plant Cell*, *4*(3), 283–295. https://doi.org/10.1105/tpc.4.3.283.

Drapier, D., Rimbault, B., Vallon, O., Wollman, F. A., & Choquet, Y. (2007). Intertwined translational regulations set uneven stoichiometry of chloroplast ATP synthase subunits. *EMBO Journal*, *26*(15), 3581–3591. https://doi.org/10.1038/sj.emboj.7601802.

Drapier, D., Suzuki, H., Levy, H., Rimbault, B., Kindle, K. L., Stern, D. B., et al. (1998). The chloroplast *atpA* gene cluster in *Chlamydomonas reinhardtii*. Functional analysis of a polycistronic transcription unit. *Plant Physiology*, *117*(2), 629–641. https://doi.org/10.1104/pp.117.2.629.

Duhe, R. J., & Selman, B. R. (1989). Studies on the heterogeneity of the soluble chloroplast coupling factor 1: The formation of ε-deficient isozymes. *Biochimica et Biophysica Acta (BBA): Bioenergetics*, *974*(3), 294–302. https://doi.org/10.1016/S0005-2728(89)80247-6.

Duhe, R. J., & Selman, B. R. (1990). The dithiothreitol-stimulated dissociation of the chloroplast coupling factor-1 ε subunit is reversible. *Biochimica et Biophysica Acta (BBA): Bioenergetics*, *1017*(1), 70–78.

Dunham, K. R., & Selman, B. R. (1981). Interactions of inorganic phosphate with spinach coupling factor 1. Effects on ATPase and ADP binding activities. *Journal of Biological Chemistry*, *256*(19), 10044–10049.

Eberhard, S., Finazzi, G., & Wollman, F. A. (2008). The dynamics of photosynthesis. *Annual Review of Genetics*, *42*, 463–515. https://doi.org/10.1146/annurev.genet.42.110807.091452.

Eberhard, S., Loiselay, C., Drapier, D., Bujaldon, S., Girard-Bascou, J., Kuras, R., et al. (2011). Dual functions of the nucleus-encoded factor TDA1 in trapping and translation activation of *atpA* transcripts in *Chlamydomonas reinhardtii* chloroplasts. *Plant Journal*, *67*(6), 1055–1066. https://doi.org/10.1111/j.1365-313X.2011.04657.x.

Engel, B. D., Schaffer, M., Kuhn Cuellar, L., Villa, E., Plitzko, J. M., & Baumeister, W. (2015). Native architecture of the Chlamydomonas chloroplast revealed by *in situ* cryo-electron tomography. *eLife*, *4*, 1–16. https://doi.org/10.7554/eLife.04889.

Fiedler, H. R., Schlesinger, J., Strotmann, H., Shavit, N., & Leu, S. (1997). Characterization of *atpA* and *atpB* deletion mutants produced in *Chlamydomonas reinhardtii* cw15: Electron transport and photophosphorylation activities of isolated thylakoids. *Biochimica et Biophysica Acta (BBA): Bioenergetics, 1319*(1), 109–118. https://doi.org/10.1016/S0005-2728(96)00123-5.

Finazzi, G., Chasen, C., Wollman, F. A., & de Vitry, C. (2003). Thylakoid targeting of Tat passenger proteins shows no delta pH dependence *in vivo*. *EMBO Journal, 22*(4), 807–815. https://doi.org/10.1093/emboj/cdg081.

Finazzi, G., Johnson, G. N., Dall'Osto, L., Zito, F., Bonente, G., Bassi, R., et al. (2006). Nonphotochemical quenching of chlorophyll fluorescence in *Chlamydomonas reinhardtii*. *Biochemistry, 45*(5), 1490–1498. https://doi.org/10.1021/bi0521588.

Finazzi, G., Minagawa, J., & Johnson, G. N. (2016). The cytochrome b_6f complex: A regulatory hub controlling electron flow and the dynamics of photosynthesis? In W. A. Cramer & T. Kallas (Eds.), *Cytochrome complexes: Evolution, structures, energy transduction, and signaling* (pp. 437–452). Dordrecht: Springer Netherlands.

Finazzi, G., & Rappaport, F. (1998). *In vivo* characterization of the electrochemical proton gradient generated in darkness in green algae and its kinetic effects on cytochrome b_6f turnover. *Biochemistry, 37*(28), 9999–10005. https://doi.org/10.1021/bi980320j.

Foyer, C. H. (2018). Reactive oxygen species, oxidative signaling and the regulation of photosynthesis. *Environmental and Experimental Botany, 154*, 134–142. https://doi.org/10.1016/j.envexpbot.2018.05.003.

Gerotto, C., Alboresi, A., Meneghesso, A., Jokel, M., Suorsa, M., Aro, E. M., et al. (2016). Flavodiiron proteins act as safety valve for electrons in *Physcomitrella patens*. *Proceedings of the National Academy of Sciences of the United States of America, 113*(43), 12322–12327. https://doi.org/10.1073/pnas.1606685113.

Gerotto, C., Trotta, A., Bajwa, A. A., Mancini, I., Morosinotto, T., & Aro, E. M. (2019). Thylakoid protein phosphorylation dynamics in a moss mutant lacking serine/threonine protein kinase STN8. *Plant Physiology, 180*(3), 1582–1597. https://doi.org/10.1104/pp.19.00117.

Graan, T., & Ort, D. R. (1984). Quantitation of the rapid electron donors to P700, the functional plastoquinone pool, and the ratio of the photosystems in spinach chloroplasts. *Journal of Biological Chemistry, 259*(22), 14003–14010.

Gresser, M. J., Myers, J. A., & Boyer, P. D. (1982). Catalytic site cooperativity of beef heart mitochondrial F_1 adenosine triphosphatase. Correlations of initial velocity, bound intermediate, and oxygen exchange measurements with an alternating three-site model. *Journal of Biological Chemistry, 257*(20), 2030–2038.

Groth, G., & Pohl, E. (2001). The structure of the chloroplast F_1-ATPase at 3.2 angstrom resolution. *Journal of Biological Chemistry, 276*(2), 1345–1352.

Haehnel, W., Ratajczak, R., & Robenek, H. (1989). Lateral distribution and diffusion of plastocyanin in chloroplast thylakoids. *Journal of Cell Biology, 108*(4), 1397–1405. https://doi.org/10.1083/jcb.108.4.1397.

Hahn, A., Vonck, J., Mills, D. J., Meier, T., & Kuhlbrandt, W. (2018). Structure, mechanism, and regulation of the chloroplast ATP synthase. *Science, 360*(6389), eaat4318. https://doi.org/10.1126/science.aat4318.

Hamel, P., Olive, J., Pierre, Y., Wollman, F. A., & de Vitry, C. (2000). A new subunit of cytochrome b_6f complex undergoes reversible phosphorylation upon state transition. *Journal of Biological Chemistry, 275*(22), 17072–17079. https://doi.org/10.1074/jbc.M001468200.

Hammel, A., Zimmer, D., Sommer, F., Muhlhaus, T., & Schroda, M. (2018). Absolute quantification of major photosynthetic protein complexes in *Chlamydomonas reinhardtii* using quantification concatamers (QconCATs). *Frontiers in Plant Science, 9*, 1265. https://doi.org/10.3389/fpls.2018.01265.

Heldt, H. W. (1969). Adenine nucleotide translocation in spinach chloroplasts. *FEBS Letters*, *5*(1), 11–14. https://doi.org/10.1016/0014-5793(69)80280-2.
Herrin, D., & Michaels, A. (1984). Gene expression during the cell cycle of *Chlamydomonas reinhardtii*. In G. S. Stein & J. L. Stein (Eds.), *Recombinant DNA and cell proliferation* (pp. 87–106). Academic Press.
Hightower, K. E., & McCarty, R. E. (1996). Proteolytic cleavage within a regulatory region of the γ subunit of chloroplast coupling factor-1. *Biochemistry*, *35*(15), 4846–4851.
Hu, D., Fiedler, H. R., Golan, T., Edelman, M., Strotmann, H., Shavit, N., et al. (1997). Catalytic properties and sensitivity to tentoxin of *Chlamydomonas reinhardtii* ATP synthases changed in codon 83 of *atpB* by site-directed mutagenesis. *Journal of Biological Chemistry*, *272*(9), 5457–5463. https://doi.org/10.1074/jbc.272.9.5457.
Hu, D., Strotmann, H., Shavit, N., & Leu, S. (1998). The *C. reinhardtii* CF_1 with the mutation βT168S has high ATPase activity. *FEBS Letters*, *421*(1), 65–68. https://doi.org/10.1016/s0014-5793(97)01531-7.
Jagendorf, A. (1982). Isolation of chloroplast coupling factor (CF_1) and of its subunits. In M. Edelman, R. Hallick, & N.-H. Chua (Eds.), *Methods of chloroplast molecular biology*. Amsterdam: Elsevier/North Holland Press.
Jagendorf, A. T., & Uribe, E. (1966). ATP formation caused by acid-base transition of spinach chloroplasts. *Proceedings of the National Academy of Sciences of the United States of America*, *55*(1), 170–177. https://doi.org/10.1073/pnas.55.1.170.
Jault, J. M., Dou, C., Grodsky, N. B., Matsui, T., Yoshida, M., & Allison, W. S. (1996). The $\alpha_3\beta_3\gamma$ subcomplex of the F_1-ATPase from the thermophilic *Bacillus* PS3 with the βT165S substitution does not entrap inhibitory MgADP in a catalytic site during turnover. *Journal of Biological Chemistry*, *271*(46), 28818–28824.
Johnson, E., & Anastasios, M. (2004). Functional characterization of *Chlamydomonas reinhardtii* with alterations in the *atpE* gene. *Photosynthesis Research*, *82*(2), 131–140. https://doi.org/10.1007/s11120-004-6567-1.
Johnson, E. A. (2008). Altered expression of the chloroplast ATP synthase through site-directed mutagenesis in *Chlamydomonas reinhardtii*. *Photosynthesis Research*, *96*(2), 153–162. https://doi.org/10.1007/s11120-008-9296-z.
Johnson, E. A., & McCarty, R. E. (2002). The carboxyl terminus of the ε subunit of the chloroplast ATP synthase is exposed during illumination. *Biochemistry*, *41*(7), 2446–2451. https://doi.org/10.1021/bi011939f.
Johnson, E. A., Rosenberg, J., & McCarty, R. E. (2007). Expression by *Chlamydomonas reinhardtii* of a chloroplast ATP synthase with polyhistidine-tagged β subunits. *Biochimica et Biophysica Acta (BBA): Bioenergetics*, *1767*(5), 374–380. https://doi.org/10.1016/j.bbabio.2007.03.003.
Johnson, X., & Alric, J. (2013). Central carbon metabolism and electron transport in *Chlamydomonas reinhardtii*: Metabolic constraints for carbon partitioning between oil and starch. *Eukaryotic Cell*, *12*(6), 776–793. https://doi.org/10.1128/EC.00318-12.
Joliot, P., & Joliot, A. (1980). Dependence of delayed luminescence upon adenosine triphosphatase activity in *Chlorella*. *Plant Physiology*, *65*(4), 691–696.
Joliot, P., & Joliot, A. (1989). Characterization of linear and quadratic electrochromic probes in *Chlorella sorokiniana* and *Chlamydomonas reinhardtii*. *Biochimica et Biophysica Acta (BBA): Bioenergetics*, *975*(3), 355–360. https://doi.org/10.1016/S0005-2728(89)80343-3.
Joliot, P., & Joliot, A. (2008). Quantification of the electrochemical proton gradient and activation of ATP synthase in leaves. *Biochimica et Biophysica Acta (BBA): Bioenergetics*, *1777*(7-8), 676–683. https://doi.org/10.1016/j.bbabio.2008.04.010.
Junesch, U., & Graber, P. (1987). Influence of the redox state and the activation of the chloroplast ATP synthase on proton-transport-coupled ATP synthesis hydrolysis. *Biochimica et Biophysica Acta (BBA): Bioenergetics*, *893*(2), 275–288.

Junge, W. (1970). The critical electric potential difference for photophosphorylation. Its relation to the chemiosmotic hypothesis and to the triggering requirements of the ATPase system. *European Journal of Biochemistry, 14*(3), 582–592.

Junge, W., Rumberg, B., & Schroder, H. (1970). The necessity of an electric potential difference and its use for photophosphorylation in short flash groups. *European Journal of Biochemistry, 14*(3), 575–581.

Kagawa, Y. (1978). Reconstitution of the energy transformer, gate and channel subunit reassembly, crystalline ATPase and ATP synthesis. *Biochimica et Biophysica Acta (BBA): Bioenergetics, 505*(1), 45–93. https://doi.org/10.1016/0304-4173(78)90008-3.

Kaim, G., & Dimroth, P. (1999). ATP synthesis by F-type ATP synthase is obligatorily dependent on the transmembrane voltage. *EMBO Journal, 18*(15), 4118–4127. https://doi.org/10.1093/emboj/18.15.4118.

Kanazawa, A., & Kramer, D. M. (2002). *In vivo* modulation of nonphotochemical exciton quenching (NPQ) by regulation of the chloroplast ATP synthase. *Proceedings of the National Academy of Sciences of the United States of America, 99*(20), 12789–12794.

Kanazawa, A., Ostendorf, E., Kohzuma, K., Hoh, D., Strand, D. D., Sato-Cruz, M., et al. (2017). Chloroplast ATP synthase modulation of the thylakoid proton motive force: Implications for photosystem I and photosystem II photoprotection. *Frontiers in Plant Science, 8*, 719. https://doi.org/10.3389/fpls.2017.00719.

Kato-Yamada, Y., Bald, D., Koike, M., Motohashi, K., Hisabori, T., & Yoshida, M. (1999). ε-subunit, an endogenous inhibitor of bacterial F_1-ATPase, also inhibits F_oF_1-ATPase. *Journal of Biological Chemistry, 274*(48), 33991–33994. https://doi.org/10.1074/jbc.274.48.33991.

Ketcham, S. R., Davenport, J. W., Warncke, K., & McCarty, R. E. (1984). Role of the γ subunit of chloroplast coupling factor-1 in the light-dependent activation of photophosphorylation and ATPase activity by dithiothreitol. *Journal of Biological Chemistry, 259*(11), 7286–7293.

Kim, Y., Konno, H., Sugano, Y., & Hisabori, T. (2011). Redox regulation of rotation of the cyanobacterial F_1-ATPase containing thiol regulation switch. *Journal of Biological Chemistry, 286*(11), 9071–9078. https://doi.org/10.1074/jbc.M110.200584.

Kirchhoff, H., Hall, C., Wood, M., Herbstová, M., Tsabari, O., Nevo, R., et al. (2011). Dynamic control of protein diffusion within the granal thylakoid lumen. *Proceedings of the National Academy of Sciences of the United States of America, 108*(50), 20248–20253. https://doi.org/10.1073/pnas.1104141109.

Kneusel, R. E., Merchant, S., & Selman, B. R. (1982). Properties of the solvent-stimulated ATPase activity of chloroplast coupling factor 1 from *Chlamydomonas reinhardii*. *Biochimica et Biophysica Acta (BBA): Bioenergetics, 681*(3), 337–344.

Kohzuma, K., Cruz, J. A., Akashi, K., Hoshiyasu, S., Munekage, Y. N., Yokota, A., et al. (2009). The long-term responses of the photosynthetic proton circuit to drought. *Plant, Cell & Environment, 32*(3), 209–219. https://doi.org/10.1111/j.1365-3040.2008.01912.x.

Kohzuma, K., Dal Bosco, C., Meurer, J., & Kramer, D. M. (2013). Light- and metabolism-related regulation of the chloroplast ATP synthase has distinct mechanisms and functions. *Journal of Biological Chemistry, 288*(18), 13156–13163. https://doi.org/10.1074/jbc.M113.453225.

Kohzuma, K., Froehlich, J. E., Davis, G. A., Temple, J. A., Minhas, D., Dhingra, A., et al. (2017). The role of light-dark regulation of the chloroplast ATP synthase. *Frontiers in Plant Science, 8*, 1248. https://doi.org/10.3389/fpls.2017.01248.

Komatsu-Takaki, M. (1989). Energy-dependent conformational changes in the ε subunit of the chloroplast ATP synthase (CF_oCF_1). *Journal of Biological Chemistry, 264*(30), 17750–17753.

Komatsu-Takaki, M. (1995). Effects of energization and substrates on the reactivities of lysine residues of the chloroplast ATP synthase β subunit. *European Journal of Biochemistry, 228*(2), 265–270. https://doi.org/10.1111/j.1432-1033.1995.0265n.x.

Komatsu-Takaki, M. (1996). Energizing effects of illumination on the reactivities of lysine residues of the γ-subunit of chloroplast ATP synthase. *European Journal of Biochemistry, 236*(2), 470–475.

Komine, Y., Kikis, E., Schuster, G., & Stern, D. (2002). Evidence for *in vivo* modulation of chloroplast RNA stability by 3'-UTR homopolymeric tails in *Chlamydomonas reinhardtii*. *Proceedings of the National Academy of Sciences of the United States of America, 99*(6), 4085–4090. https://doi.org/10.1073/pnas.052327599.

Konno, H., Nakane, T., Yoshida, M., Ueoka-Nakanishi, H., Hara, S., & Hisabori, T. (2012). Thiol modulation of the chloroplast ATP synthase is dependent on the energization of thylakoid membranes. *Plant & Cell Physiology, 53*(4), 626–634. https://doi.org/10.1093/Pcp/Pcs018.

Kramer, D. M., Avenson, T. J., & Edwards, G. E. (2004). Dynamic flexibility in the light reactions of photosynthesis governed by both electron and proton reactions. *Trends in Plant Science, 9*(7), 349–357. https://doi.org/10.1016/j.tplants.2004.05.001.

Kramer, D. M., Cruz, J. A., & Kanazawa, A. (2003). Balancing the central roles of the thylakoid proton gradient. *Trends in Plant Science, 8*(1), 27–32. https://doi.org/10.1016/s1360-1385(02)00010-9.

Kuhlbrandt, W. (2019). Structure and mechanisms of F-type ATP synthases. *Annual Review of Biochemistry, 88*, 515–549. https://doi.org/10.1146/annurev-biochem-013118-110903.

Lapashina, A. S., & Feniouk, B. A. (2018). ADP-inhibition of H^+-F_OF_1-ATP synthase. *Biochemistry (Mosc), 83*(10), 1141–1160. https://doi.org/10.1134/S0006297918100012.

Lemaire, C., & Wollman, F. A. (1989a). The chloroplast ATP synthase in *Chlamydomonas reinhardtii*. I. Characterization of its nine constitutive subunits. *Journal of Biological Chemistry, 264*(17), 10228–10234.

Lemaire, C., & Wollman, F. A. (1989b). The chloroplast ATP synthase in *Chlamydomonas reinhardtii*. II. Biochemical studies on its biogenesis using mutants defective in photophosphorylation. *Journal of Biological Chemistry, 264*(17), 10235–10242.

Lemaire, C., Wollman, F. A., & Bennoun, P. (1988). Restoration of phototrophic growth in a mutant of *Chlamydomonas reinhardtii* in which the chloroplast *atpB* gene of the ATP synthase has a deletion: An example of mitochondria-dependent photosynthesis. *Proceedings of the National Academy of Sciences of the United States of America, 85*(5), 1344–1348.

Levy, H., Kindle, K. L., & Stern, D. B. (1997). A nuclear mutation that affects the 3' processing of several mRNAs in Chlamydomonas chloroplasts. *Plant Cell, 9*(5), 825–836. https://doi.org/10.1105/tpc.9.5.825.

Li, Y., Ma, X., & Weber, J. (2019). Interaction between γC87 and γR242 residues participates in energy coupling between catalysis and proton translocation in *Escherichia coli* ATP synthase. *Biochimica et Biophysica Acta (BBA): Bioenergetics, 1860*(8), 679–687. https://doi.org/10.1016/j.bbabio.2019.06.016.

Lien, S., Berzborn, R. J., & Racker, E. (1972). Partial resolution of the enzymes catalyzing photophosphorylation. IX. Studies on the subunit structure of coupling factor 1 from chloroplasts. *Journal of Biological Chemistry, 247*(11), 3520–3524.

Lien, S., & Racker, E. (1971). Preparation and assay of chloroplast coupling factor CF_1. In *Vol. 23. Methods in enzymology* (pp. 547–555). Academic Press.

Ma, J. P., Flynn, T. C., Cui, Q., Leslie, A. G. W., Walker, J. E., & Karplus, M. (2002). A dynamic analysis of the rotation mechanism for conformational change in F_1-ATPase. *Structure, 10*(7), 921–931.

Maiwald, D., Dietzmann, A., Jahns, P., Pesaresi, P., Joliot, P., Joliot, A., et al. (2003). Knock-out of the genes coding for the Rieske protein and the ATP-synthase δ-subunit of

Arabidopsis. Effects on photosynthesis, thylakoid protein composition, and nuclear chloroplast gene expression. *Plant Physiology, 133*(1), 191–202. https://doi.org/10.1104/pp.103.024190.
Majeran, W., Olive, J., Drapier, D., Vallon, O., & Wollman, F. A. (2001). The light sensitivity of ATP synthase mutants of *Chlamydomonas reinhardtii*. *Plant Physiology, 126*(1), 421–433. https://doi.org/10.1104/pp.126.1.421.
Majeran, W., Wostrikoff, K., Wollman, F. A., & Vallon, O. (2019). Role of ClpP in the biogenesis and degradation of RuBisCO and ATP synthase in *Chlamydomonas reinhardtii*. *Plants (Basel), 8*(7). https://doi.org/10.3390/plants8070191.
Marchand, J., Heydarizadeh, P., Schoefs, B., & Spetea, C. (2018). Ion and metabolite transport in the chloroplast of algae: Lessons from land plants. *Cellular and Molecular Life Sciences, 75*(12), 2153–2176. https://doi.org/10.1007/s00018-018-2793-0.
Maul, J. E., Lilly, J. W., Cui, L., dePamphilis, C. W., Miller, W., Harris, E. H., et al. (2002). The *Chlamydomonas reinhardtii* plastid chromosome: Islands of genes in a sea of repeats. *Plant Cell, 14*(11), 2659–2679. https://doi.org/10.1105/tpc.006155.
McCarty, R. E., & Racker, E. (1968). Partial resolution of the enzymes catalyzing photophosphorylation. 3. Activation of adenosine triphosphatase and ^{32}P-labeled orthophosphate-adenosine triphosphate exchange in chloroplasts. *Journal of Biological Chemistry, 243*(1), 129–137.
Merchant, S., & Selman, B. R. (1985). Photosynthetic ATPases: Purification, properties, subunit isolation and function. *Photosynthesis Research, 6*(1), 3–31. https://doi.org/10.1007/BF00029044.
Merchant, S., Shaner, S. L., & Selman, B. R. (1983). Molecular weight and subunit stoichiometry of the chloroplast coupling factor 1 from *Chlamydomonas reinhardi*. *Journal of Biological Chemistry, 258*(2), 1026–1031.
Meyer Zu Tittingdorf, J. M., Rexroth, S., Schafer, E., Schlichting, R., Giersch, C., Dencher, N. A., et al. (2004). The stoichiometry of the chloroplast ATP synthase oligomer III in *Chlamydomonas reinhardtii* is not affected by the metabolic state. *Biochimica et Biophysica Acta (BBA): Bioenergetics, 1659*(1), 92–99. https://doi.org/10.1016/j.bbabio.2004.08.008.
Miller, K. R., & Staehelin, L. A. (1976). Analysis of the thylakoid outer surface. Coupling factor is limited to unstacked membrane regions. *Journal of Cell Biology, 68*(1), 30–47. https://doi.org/10.1083/jcb.68.1.30.
Mills, J. D., Mitchell, P., & Schurmann, P. (1980). Modulation of coupling factor ATPase activity in intact chloroplasts—The role of the thioredoxin system. *FEBS Letters, 112*(2), 173–177. https://doi.org/10.1016/0014-5793(80)80173-6.
Minkov, I. B., & Strotmann, H. (1989). The effect of azide on regulation of the chloroplast H^+-ATPase by ADP and phosphate. *Biochimica et Biophysica Acta (BBA): Bioenergetics, 973*(1), 7–12.
Mitchell, P. (1961). Coupling of phosphorylation to electron and hydrogen transfer by a chemi-osmotic type of mechanism. *Nature, 191*, 144–148.
Moroney, J. V., Andreo, C. S., Vallejos, R. H., & McCarty, R. E. (1980). Uncoupling and energy transfer inhibition of photophosphorylation by sulfhydryl reagents. *Journal of Biological Chemistry, 255*(14), 6670–6674.
Moroney, J. V., & McCarty, R. E. (1982a). Effect of proteolytic digestion on the Ca^{2+}-ATPase activity and subunits of latent and thiol-activated chloroplast coupling factor-1. *Journal of Biological Chemistry, 257*(10), 5910–5914.
Moroney, J. V., & McCarty, R. E. (1982b). Light-dependent cleavage of the γ subunit of coupling factor-1 by trypsin causes activation of Mg^{2+}-ATPase activity and uncoupling of photophosphorylation in spinach chloroplasts. *Journal of Biological Chemistry, 257*(10), 5915–5920.
Mueller, D. M. (1989). A mutation altering the kinetic responses of the yeast mitochondrial F_1-ATPase. *Journal of Biological Chemistry, 264*(28), 16552–16556.

Mustardy, L., & Garab, G. (2003). Granum revisited. A three-dimensional model—Where things fall into place. *Trends in Plant Science*, *8*(3), 117–122. https://doi.org/10.1016/S1360-1385(03)00015-3.

Nalin, C. M., & McCarty, R. E. (1984). Role of a disulfide bond in the γ subunit in activation of the ATPase of chloroplast coupling factor-1. *Journal of Biological Chemistry*, *259*(11), 7275–7280.

Naranjo, B., Mignee, C., Krieger-Liszkay, A., Hornero-Mendez, D., Gallardo-Guerrero, L., Cejudo, F. J., et al. (2016). The chloroplast NADPH thioredoxin reductase C, NTRC, controls non-photochemical quenching of light energy and photosynthetic electron transport in Arabidopsis. *Plant, Cell & Environment*, *39*(4), 804–822. https://doi.org/10.1111/pce.12652.

Nawrocki, W. J., Liu, X., & Croce, R. (2020). *Chlamydomonas reinhardtii* exhibits de facto constitutive NPQ capacity in physiologically relevant conditions. *Plant Physiology*, *182*(1), 472–479. https://doi.org/10.1104/pp.19.00658.

Nelson, N., Nelson, H., & Racker, E. (1972). Partial resolution of the enzymes catalyzing photophosphorylation. XII. Purification and properties of an inhibitor isolated from chloroplast coupling factor 1. *Journal of Biological Chemistry*, *247*(23), 7657–7662.

Niyogi, K. K., & Truong, T. B. (2013). Evolution of flexible non-photochemical quenching mechanisms that regulate light harvesting in oxygenic photosynthesis. *Current Opinion in Plant Biology*, *16*(3), 307–314. https://doi.org/10.1016/j.pbi.2013.03.011.

Nowak, K. F., & McCarty, R. E. (2004). Regulatory role of the C-terminus of the ε-subunit from the chloroplast ATP synthase. *Biochemistry*, *43*(11), 3273–3279. https://doi.org/10.1021/bi035820d.

Nowak, K. F., Tabidze, V., & McCarty, R. E. (2002). The C-terminal domain of the ε-subunit of the chloroplast ATP synthase is not required for ATP synthesis. *Biochemistry*, *41*(51), 15130–15134. https://doi.org/10.1021/bi026594v.

Olive, J., Vallon, O., Wollman, F. A., Recouvreur, M., & Bennoun, P. (1986). Studies on the cytochrome b_6f complex. II. Localization of the complex in the thylakoid membranes from spinach and *Chlamydomonas reinhardtii* by immunocytochemistry and freeze-fracture analysis of b_6f mutants. *Biochimica et Biophysica Acta (BBA): Bioenergetics*, *851*(2), 239–248. https://doi.org/10.1016/0005-2728(86)90131-3.

Omote, H., Maeda, M., & Futai, M. (1992). Effects of mutations of conserved Lys-155 and Thr-156 residues in the phosphate-binding glycine-rich sequence of the F_1-ATPase β-subunit of *Escherichia coli*. *Journal of Biological Chemistry*, *267*(29), 20571–20576.

Ort, D. R., & Oxborough, K. (1992). *In situ* regulation of chloroplast coupling factor activity. *Annual Review of Plant Physiology and Plant Molecular Biology*, *43*, 269–291.

Ozawa, S.-I., Cavaiuolo, M., Jarrige, D., Kuras, R., Rutgers, M., Eberhard, S., et al. (2020). The OPR protein MTHI1 controls the expression of two different subunits of ATP synthase CF_o in *Chlamydomonas reinhardtii*. *Plant Cell*, *32*(4), 1179–1203. https://doi.org/10.1105/tpc.19.00770.

Paila, Y. D., Richardson, L. G. L., & Schnell, D. J. (2015). New insights into the mechanism of chloroplast protein import and its integration with protein quality control, organelle biogenesis and development. *Journal of Molecular Biology*, *427*(5), 1038–1060. https://doi.org/10.1016/j.jmb.2014.08.016.

Pancic, P. G., Strotmann, H., & Kowallik, K. V. (1992). Chloroplast ATPase genes in the diatom *Odontella sinensis* reflect cyanobacterial characters in structure and arrangement. *Journal of Molecular Biology*, *224*(2), 529–536. https://doi.org/10.1016/0022-2836(92)91017-j.

Piccioni, R. G., Bennoun, P., & Chua, N. H. (1981). A nuclear mutant of *Chlamydomonas reinhardtii* defective in photosynthetic photophosphorylation. Characterization of the algal coupling factor ATPase. *European Journal of Biochemistry*, *117*(1), 93–102. https://doi.org/10.1111/j.1432-1033.1981.tb06307.x.

Pick, U., & Bassilian, S. (1982). Activation of magnesium ion specific adenosinetriphosphatase in chloroplast coupling factor 1 by octyl glucoside. *Biochemistry*, *21*(24), 6144–6152. https://doi.org/10.1021/bi00267a019.

Pick, U., Conrad, P. L., Conrad, J. M., Durbin, R. D., & Selman, B. R. (1982). Synergistic activation of an Mg-specific ATPase activity in chloroplast coupling factor by octylglucoside and tentoxin. *Biochimica et Biophysica Acta (BBA): Bioenergetics*, *682*(1), 55–58. https://doi.org/10.1016/0005-2728(82)90118-9.

Pogoryelov, D., Klyszejko, A. L., Krasnoselska, G. O., Heller, E. M., Leone, V., Langer, J. D., et al. (2012). Engineering rotor ring stoichiometries in the ATP synthase. *Proceedings of the National Academy of Sciences of the United States of America*, *109*(25), E1599–E1608. https://doi.org/10.1073/pnas.1120027109.

Pogoryelov, D., Yu, J., Meier, T., Vonck, J., Dimroth, P., & Muller, D. J. (2005). The c_{15} ring of the *Spirulina platensis* F-ATP synthase: F_1/F_o symmetry mismatch is not obligatory. *EMBO Reports*, *6*(11), 1040–1044. https://doi.org/10.1038/sj.embor.7400517.

Ponomarenko, S. V. (2007). Evolutionary modifications of molecular structure of ATP-synthase γ-subunit. *Journal of Evolutionary Biochemistry and Physiology*, *43*(5), 467–475. https://doi.org/10.1134/S0022093007050027.

Posorske, L., & Jagendorf, A. T. (1976). Nucleotide and metal interactions affecting inactivation of spinach chloroplast coupling factor by NaCl in the cold. *Archives of Biochemistry and Biophysics*, *177*(1), 276–283. https://doi.org/10.1016/0003-9861(76)90437-9.

Preiss, L., Klyszejko, A. L., Hicks, D. B., Liu, J., Fackelmayer, O. J., Yildiz, O., et al. (2013). The c-ring stoichiometry of ATP synthase is adapted to cell physiological requirements of alkaliphilic *Bacillus pseudofirmus* OF4. *Proceedings of the National Academy of Sciences of the United States of America*, *110*(19), 7874–7879. https://doi.org/10.1073/pnas.1303333110.

Pu, J. Z., & Karplus, M. (2008). How subunit coupling produces the γ subunit rotary motion in F_1-ATPase. *Proceedings of the National Academy of Sciences of the United States of America*, *105*(4), 1192–1197. https://doi.org/10.1073/pnas.0708746105.

Rappaport, F., Finazzi, G., Pierre, Y., & Bennoun, P. (1999). A new electrochemical gradient generator in thylakoid membranes of green algae. *Biochemistry*, *38*(7), 2040–2047. https://doi.org/10.1021/bi982351k.

Reiser, J., Linka, N., Lemke, L., Jeblick, W., & Neuhaus, H. E. (2004). Molecular physiological analysis of the two plastidic ATP/ADP transporters from Arabidopsis. *Plant Physiology*, *136*(3), 3524–3536. https://doi.org/10.1104/pp.104.049502.

Reith, M., & Munholland, J. (1995). Complete nucleotide sequence of the *Porphyra purpurea* chloroplast genome. *Plant Molecular Biology Reporter*, *13*(4), 333–335.

Rexroth, S., Meyer Zu Tittingdorf, J. M., Schwassmann, H. J., Krause, F., Seelert, H., & Dencher, N. A. (2004). Dimeric H^+-ATP synthase in the chloroplast of *Chlamydomonas reinhardtii*. *Biochimica et Biophysica Acta (BBA): Bioenergetics*, *1658*(3), 202–211. https://doi.org/10.1016/j.bbabio.2004.05.014.

Richter, M. L., Hein, R., & Huchzermeyer, B. (2000). Important subunit interactions in the chloroplast ATP synthase. *Biochimica et Biophysica Acta (BBA): Bioenergetics*, *1458*(2–3), 326–342. pii: S0005-2728(00)00084-0.

Richter, M. L., & McCarty, R. E. (1987). Energy-dependent changes in the conformation of the ε subunit of the chloroplast ATP synthase. *Journal of Biological Chemistry*, *262*(31), 15037–15040.

Richter, M. L., Patrie, W. J., & McCarty, R. E. (1984). Preparation of the ε-subunit and ε-subunit-deficient chloroplast coupling factor-1 in reconstitutively active forms. *Journal of Biological Chemistry*, *259*(12), 7371–7373.

Rieger, B., Junge, W., & Busch, K. B. (2014). Lateral pH gradient between OXPHOS complex IV and F_oF_1 ATP-synthase in folded mitochondrial membranes. *Nature Communications*, *5*, 3103. https://doi.org/10.1038/ncomms4103.

Robertson, D., Boynton, J. E., & Gillham, N. W. (1990). Cotranscription of the wild-type chloroplast *atpE* gene encoding the CF_1/CF_o ε-subunit with the 3' half of the rps7 gene in *Chlamydomonas reinhardtii* and characterization of frameshift mutations in *atpE*. *Molecular and General Genetics*, *221*(2), 155–163. https://doi.org/10.1007/bf00261715.
Ross, S. A., Zhang, M. X., & Selman, B. R. (1995). Role of the *Chlamydomonas reinhardtii* coupling factor-1 γ-subunit cysteine bridge in the regulation of ATP synthase. *Journal of Biological Chemistry*, *270*(17), 9813–9818.
Rott, M., Martins, N. F., Thiele, W., Lein, W., Bock, R., Kramer, D. M., et al. (2011). ATP synthase repression in tobacco restricts photosynthetic electron transport, CO_2 assimilation, and plant growth by overacidification of the thylakoid lumen. *Plant Cell*, *23*(1), 304–321. https://doi.org/10.1105/tpc.110.079111.
Rott, R., Levy, H., Drager, R. G., Stern, D. B., & Schuster, G. (1998). 3'-Processed mRNA is preferentially translated in Chlamydomonas reinhardtii chloroplasts. *Molecular and Cellular Biology*, *18*(8), 4605–4611. https://doi.org/10.1128/mcb.18.8.4605.
Ryrie, I. J., & Jagendorf, A. T. (1971). An energy-linked conformational change in the coupling factor in chloroplasts. Studies with hydrogen exchange. *Journal of Biological Chemistry*, *246*(11), 3771–3774.
Salewskij, K., Rieger, B., Hager, F., Arroum, T., Duwe, P., Villalta, J., et al. (2020). The spatio-temporal organization of mitochondrial F_1F_o ATP synthase in cristae depends on its activity mode. *Biochimica et Biophysica Acta (BBA): Bioenergetics*, *1861*(1), 148091. https://doi.org/10.1016/j.bbabio.2019.148091.
Saroussi, S., Schushan, M., Ben-Tal, N., Junge, W., & Nelson, N. (2012). Structure and flexibility of the c-ring in the electromotor of rotary F_oF_1-ATPase of pea chloroplasts. *PLoS One*, 7(9). e43045https://doi.org/10.1371/journal.pone.0043045.
Scharff, L. B., Ehrnthaler, M., Janowski, M., Childs, L. H., Hasse, C., Gremmels, J., et al. (2017). Shine-Dalgarno sequences play an essential role in the translation of plastid mRNAs in tobacco. *Plant Cell*, *29*(12), 3085–3101. https://doi.org/10.1105/tpc.17.00524.
Schemidt, R. A., Qu, J., Williams, J. R., & Brusilow, W. S. (1998). Effects of carbon source on expression of F_o genes and on the stoichiometry of the c subunit in the F_1F_o ATPase of *Escherichia coli*. *Journal of Bacteriology*, *180*(12), 3205–3208.
Schumann, J., Richter, M. L., & McCarty, R. E. (1985). Partial proteolysis as a probe of the conformation of the γ subunit in activated soluble and membrane-bound chloroplast coupling factor-1. *Journal of Biological Chemistry*, *260*(21), 11817–11823.
Schwarz, O., Schurmann, P., & Strotmann, H. (1997). Kinetics and thioredoxin specificity of thiol modulation of the chloroplast H^+-ATPase. *Journal of Biological Chemistry*, *272*(27), 16924–16927.
Schwassmann, H. J., Rexroth, S., Seelert, H., & Dencher, N. A. (2007). Metabolism controls dimerization of the chloroplast F_oF_1 ATP synthase in *Chlamydomonas reinhardtii*. *FEBS Letters*, *581*(7), 1391–1396. https://doi.org/10.1016/j.febslet.2007.02.057.
Seelert, H., Poetsch, A., Dencher, N. A., Engel, A., Stahlberg, H., & Muller, D. J. (2000). Structural biology. Proton-powered turbine of a plant motor. *Nature*, *405*(6785), 418–419. https://doi.org/10.1038/35013148.
Selman-Reimer, S., Merchant, S., & Selman, B. R. (1981). Isolation, purification, and characterization of coupling factor 1 from *Chlamydomonas reinhardi*. *Biochemistry*, *20*(19), 5476–5482. https://doi.org/10.1021/bi00522a020.
Sjöholm, J., Bergstrand, J., Nilsson, T., Šachl, R., Ballmoos, C. v., Widengren, J., et al. (2017). The lateral distance between a proton pump and ATP synthase determines the ATP-synthesis rate. *Scientific Reports*, 7(1), 2926. https://doi.org/10.1038/s41598-017-02836-4.
Staehelin, L. A. (2003). Chloroplast structure: From chlorophyll granules to supra-molecular architecture of thylakoid membranes. *Photosynthesis Research*, *76*(1-3), 185–196. https://doi.org/10.1023/A:1024994525586.

Steele, J. A., Durbin, R. D., Uchytil, T. F., & Rich, D. H. (1978). Tentoxin. An uncompetitive inhibitor of lettuce chloroplast coupling factor 1. *Biochimica et Biophysica Acta (BBA): Bioenergetics*, *501*(1), 72–82. https://doi.org/10.1016/0005-2728(78)90096-8.
Steele, J. A., Uchytil, T. F., Durbin, R. D., Bhatnagar, P., & Rich, D. H. (1976). Chloroplast coupling factor 1: A species-specific receptor for tentoxin. *Proceedings of the National Academy of Sciences of the United States of America*, *73*(7), 2245–2248. https://doi.org/10.1073/pnas.73.7.2245.
Stern, D. B., Radwanski, E. R., & Kindle, K. L. (1991). A 3' stem/loop structure of the Chlamydomonas chloroplast *atpB* gene regulates mRNA accumulation *in vivo*. *Plant Cell*, *3*(3), 285–297. https://doi.org/10.1105/tpc.3.3.285.
Sternweis, P. C. (1978). The ε-subunit of *Escherichia coli* coupling factor 1 is required for its binding to the cytoplasmic membrane. *Journal of Biological Chemistry*, *253*(9), 3123–3128.
Strenkert, D., Schmollinger, S., Gallaher, S. D., Salome, P. A., Purvine, S. O., Nicora, C. D., et al. (2019). Multiomics resolution of molecular events during a day in the life of Chlamydomonas. *Proceedings of the National Academy of Sciences of the United States of America*, *116*(6), 2374–2383. https://doi.org/10.1073/pnas.1815238116.
Strotmann, H., & Bickel-Sandkotter, S. (1977). Energy-dependent exchange of adenine nucleotides on chloroplast coupling factor (CF_1). *Biochimica et Biophysica Acta (BBA): Bioenergetics*, *460*(1), 126–135.
Strotmann, H., Hesse, H., & Edelmann, K. (1973). Quantitative determination of coupling factor CF_1 of chloroplasts. *Biochimica et Biophysica Acta (BBA): Bioenergetics*, *314*(2), 202–210.
Strotmann, H., Shavit, N., & Leu, S. (1998). Assembly and function of the chloroplast ATP synthase. In J. D. Rochaix, M. Goldschmidt-Clermont, & S. Merchant (Eds.), *The molecular biology of chloroplasts and mitochondria in Chlamydomonas* (pp. 477–500). Dordrecht: Springer Netherlands.
Takagi, D., Amako, K., Hashiguchi, M., Fukaki, H., Ishizaki, K., Goh, T., et al. (2017). Chloroplastic ATP synthase builds up a proton motive force preventing production of reactive oxygen species in photosystem I. *The Plant Journal*, *91*(2), 306–324. https://doi.org/10.1111/tpj.13566.
Tanigawara, M., Tabata, K. V., Ito, Y., Ito, J., Watanabe, R., Ueno, H., et al. (2012). Role of the DELSEED loop in torque transmission of F_1-ATPase. *Biophysical Journal*, *103*(5), 970–978. https://doi.org/10.1016/j.bpj.2012.06.054.
Thuswaldner, S., Lagerstedt, J. O., Rojas-Stutz, M., Bouhidel, K., Der, C., Leborgne-Castel, N., et al. (2007). Identification, expression, and functional analyses of a thylakoid ATP/ADP carrier from Arabidopsis. *Journal of Biological Chemistry*, *282*(12), 8848–8859. https://doi.org/10.1074/jbc.M609130200.
Viola, S., Cavaiuolo, M., Drapier, D., Eberhard, S., Vallon, O., Wollman, F. A., et al. (2019). MDA1, a nucleus-encoded factor involved in the stabilization and processing of the *atpA* transcript in the chloroplast of Chlamydomonas. *Plant Journal*, *98*(6), 1033–1047. https://doi.org/10.1111/tpj.14300.
Voon, C. P., Guan, X., Sun, Y., Sahu, A., Chan, M. N., Gardestrom, P., et al. (2018). ATP compartmentation in plastids and cytosol of Arabidopsis thaliana revealed by fluorescent protein sensing. *Proceedings of the National Academy of Sciences of the United States of America*, *115*(45), E10778–E10787. https://doi.org/10.1073/pnas.1711497115.
Walker, J. E., Saraste, M., Runswick, M. J., & Gay, N. J. (1982). Distantly related sequences in the α- and β-subunits of ATP synthase, myosin, kinases and other ATP-requiring enzymes and a common nucleotide binding fold. *EMBO Journal*, *1*(8), 945–951.
Wallach, D., Bar-Nun, S., & Ohad, I. (1972). Biogenesis of chloroplast membranes. IX. Development of photophosphorylation and proton pump activities in greening *Chlamydomonas reinhardi* y-I as measured with an open-cell preparation. *Biochimica et Biophysica Acta (BBA): Bioenergetics*, *267*(1), 125–137. https://doi.org/10.1016/0005-2728(72)90144-2.

Weber, A. P., & Linka, N. (2011). Connecting the plastid: Transporters of the plastid envelope and their role in linking plastidial with cytosolic metabolism. *Annual Review of Plant Biology*, *62*, 53–77. https://doi.org/10.1146/annurev-arplant-042110-103903.

Wietrzynski, W., Schaffer, M., Tegunov, D., Albert, S., Kanazawa, A., Plitzko, J. M., et al. (2020). Charting the native architecture of Chlamydomonas thylakoid membranes with single-molecule precision. *eLife9*, e53740. https://doi.org/10.7554/eLife.53740.

Witt, H. T., Schlodder, E., & Graber, P. (1976). Membrane-bound ATP synthesis generated by an external electrical field. *FEBS Letters*, *69*(1), 272–276. https://doi.org/10.1016/0014-5793(76)80702-8.

Woessner, J. P., Masson, A., Harris, E. H., Bennoun, P., Gillham, N. W., & Boynton, J. E. (1984). Molecular and genetic analysis of the chloroplast ATPase of Chlamydomonas. *Plant Molecular Biology*, *3*(3), 177–190. https://doi.org/10.1007/BF00016065.

Wu, G., & Ort, D. R. (2008). Mutation in the cysteine bridge domain of the γ-subunit affects light regulation of the ATP synthase but not photosynthesis or growth in Arabidopsis. *Photosynthesis Research*, *97*(2), 185–193.

Yin, L., Lundin, B., Bertrand, M., Nurmi, M., Solymosi, K., Kangasjarvi, S., et al. (2010). Role of thylakoid ATP/ADP carrier in photoinhibition and photoprotection of photosystem II in Arabidopsis. *Plant Physiology*, *153*(2), 666–677. https://doi.org/10.1104/pp.110.155804.

Yoshida, K., Hara, A., Sugiura, K., Fukaya, Y., & Hisabori, T. (2018). Thioredoxin-like2/2-Cys peroxiredoxin redox cascade supports oxidative thiol modulation in chloroplasts. *Proceedings of the National Academy of Sciences of the United States of America*, *115*(35), E8296–E8304. https://doi.org/10.1073/pnas.1808284115.

Yoshida, K., Matsuoka, Y., Hara, S., Konno, H., & Hisabori, T. (2014). Distinct redox behaviors of chloroplast thiol enzymes and their relationships with photosynthetic electron transport in *Arabidopsis thaliana*. *Plant & Cell Physiology*, *55*(8), 1415–1425. https://doi.org/10.1093/pcp/pcu066.

Younis, H. M., Winget, G. D., & Racker, E. (1977). Requirement of the δ subunit of chloroplast coupling factor-1 for photophosphorylation. *Journal of Biological Chemistry*, *252*(5), 1814–1818.

Zhou, J. M., & Boyer, P. D. (1992). MgADP and free Pi as the substrates and the Mg^{2+} requirement for photophosphorylation. *Biochemistry*, *31*(12), 3166–3171. https://doi.org/10.1021/bi00127a017.

Zhou, J. M., Xue, Z. X., Du, Z. Y., Melese, T., & Boyer, P. D. (1988). Relationship of tightly bound ADP and ATP to control and catalysis by chloroplast ATP synthase. *Biochemistry*, *27*(14), 5129–5135. https://doi.org/10.1021/bi00414a027.

CHAPTER FOUR

Rotation of the γ-subunit in single membrane-bound H^+-ATP synthases from chloroplasts during ATP synthesis

Roland Bienert[a], Paola Turina[b], Michael Börsch[c], and Peter Gräber[a,*]
[a]Department of Physical Chemistry, Albert-Ludwigs-University of Freiburg, Freiburg im Breisgau, Germany
[b]Department of Pharmacy and Biotechnology, University of Bologna, Bologna, Italy
[c]Jena University Hospital, Friedrich Schiller University Jena, Jena, Germany
*Corresponding author: e-mail address: peter.graeber@physchem.uni-freiburg.de

Contents

Abstract

Single-molecule fluorescence resonance energy transfer (smFRET) is used to investigate the movement of the γ-subunit within the H^+-ATPsynthase from chloroplasts during

Advances in Botanical Research, Volume 96
ISSN 0065-2296
https://doi.org/10.1016/bs.abr.2020.07.002
© 2020 Elsevier Ltd
All rights reserved.

proton transport-coupled ATP synthesis. The FRET donor fluorophore is attached to residue γC322 and the FRET acceptor is non-covalently bound to a non-catalytic nucleotide binding site. The labeled enzyme is integrated into liposomes and ATP synthesis is started by an acid-base transition. A custom-built confocal microscope with two spectral detection channels is used to measure smFRET with freely diffusing proteoliposomes. Analysis of the smFRET time traces reveals stepwise distance changes between the two fluorophores. During catalysis the γ-subunit interacts with equal probability with each αβ-pair. In ADP-inhibited enzymes, the γ-subunit interacts with only one specific αβ-pair. Comparison of the smFRET data with cryo-EM data (Hahn, Vonck, Mills, Meier, & Kühlbrandt, 2018, *Science, 360*, pp. 620–628) indicated that the transition from the ADP-inhibited state to the active state is accompanied by a large rotational movement of the γ-subunit.

1. Introduction

F-type H^+-ATPsynthases are membrane-bound enzymes occurring in energy transducing membranes of biological cells and couple a transmembrane proton transport with synthesis and hydrolysis of ATP (Mitchell, 1961). They feature a remarkable rotary mechanism to bring about the coupling between three catalytic nucleotide binding sites located in the hydrophilic F_1-part (subunits $\alpha_3\beta_3\gamma\delta\varepsilon$) and proton translocation through the membrane integrated F_0-part (subunits ab_2c_{10} in *E. coli*) (Abrahams, Leslie, Lutter, & Walker, 1994; Boyer, 1997; Weber & Senior, 2000). Functionally, the subunits $\gamma\varepsilon c_{10}$ form the rotor, and the subunits $\alpha_3\beta_3ab_2$ form the stator. During ATP hydrolysis at high ATP concentrations, the γ-subunit rotates in 120 degree steps, thereby opening and closing sequentially the three catalytic nucleotide binding sites located mainly on the β-subunits. The strong interaction of the γ-subunit with the ε-subunit and the c_{10}-complex leads to a joint rotation of all rotor subunits. This rotational catalysis was proposed on the basis of biochemical experiments (Duncan, Bulygin, Zhou, Hutcheon, & Cross, 1995; Sabbert, Engelbrecht, & Junge, 1996) and has been shown convincingly with a single-molecule fluorescence imaging experiment. The F_1-part from the thermophilic bacterium PS3 was immobilized at a glass slide, a fluorophore-labeled actin filament was attached to the γ-subunit, and an unidirectional stepped rotation was observed after addition of ATP (Noji, Yasuda, Yoshida, & Kinosita Jr., 1997). The same technique was used to investigate the movement of the γ-subunit in the F_1-part from *E. coli* (Noji et al., 1999; Omote et al., 1999). Attachment of the label to the c-subunit allowed the observation of the rotation of the c-subunit ring of F_oF_1 upon addition of ATP (Sambongi et al., 1999).

This method has been further refined by attachment of light-scattering beads to the γ-subunit, so that the improved signal-to-noise ratio allowed to detect rotational substeps in the ms-range (Bilyard et al., 2013; Spetzler et al., 2006; Yasuda, Noji, Yoshida, Kinosita, & Itoh, 2001). Most of the current knowledge of the coupling between rotation of the γ-subunit and ATP hydrolysis has been obtained with that method (Junge & Nelson, 2015).

F-type H^+-ATPsynthases occur in bacteria, chloroplasts and mitochondria. Single-molecule experiments have been carried out mainly with the stable $\alpha_3\beta_3\gamma$-fragments from bacterial enzymes, since they can be easily genetically modified and do not dissociate at very low concentrations, allowing immobilization on glass slides and labeling of the γ-subunit. It is generally assumed that all F-type ATPases use the same mechanism of rotational catalysis in their F_1-parts. On the other hand, the activity of the membrane integrated F_oF_1 holoenzyme is controlled by different subunits to accommodate the distinct functional requirements of bacteria, chloroplasts and mitochondria (Lapashina & Feniouk, 2018; Turina, 1994). Especially, the activity of the chloroplast enzyme is strongly regulated to adapt its activity to the day-night cycle and to avoid the thermodynamically favored ATP hydrolysis.

To accomplish this regulation, the γ-subunit of CF_oF_1 contains a disulfide bridge which is not present in the bacterial and mitochondrial enzymes. The activity of CF_oF_1 is regulated by a ΔpH-dependent activation reaction and a redox reaction. Energization of the membrane transforms the inactive CF_oF_1 into the active state and a tightly bound ADP is released (Gräber, Schlodder, & Witt, 1977). In the absence of membrane energization, the active state is metastable, P_i stabilizes it, and rebinding of ADP leads to inactivation. The redox reaction is the formation of a disulfide bond between γC199 and γC205 under oxidizing conditions and its reduction to two HS-groups by dithiothreitol (in vitro) (Nalin & McCarty, 1984) or by thioredoxin (in vivo) (Mills, Mitchell, & Schürmann, 1980). The inactivation/activation conversion occurs in the oxidized as well as in the reduced enzyme. The activation of the oxidized enzyme requires a ΔpH higher than that necessary for the catalytic reaction, whereas the activation of the reduced enzyme occurs at a ΔpH lower than that required for the catalytic reaction. Therefore under the night-time oxidizing conditions, CF_oF_1 is in the inactive state and does not hydrolyze ATP. Conversely, under the daytime reducing conditions, it can be activated at low membrane energization, and in the active state it is able to catalyze ATP synthesis and hydrolysis according to the energetics of this reaction (Gräber & Schlodder, 1981; Junesch & Gräber, 1987; Strelow & Rumberg, 1993). This regulation has been observed also in plants (Joliot & Joliot, 2008; Kramer et al., 1990; Kramer & Crofts, 1989; Ort & Oxborough, 1992).

The first single-molecule investigation with the chloroplast enzyme was carried out with immobilized CF_1 parts (Häsler, Engelbrecht, & Junge, 1998). A fluorophore-labeled actin filament was attached to the reduced cysteines of the disulfide bridge and, in the presence of ATP a stepped rotary movement of the actin filament was observed, similarly as found with the bacterial enzymes (Hisabori, Kondoh, & Yoshida, 1999). In addition, the redox regulating part of CF_1 was introduced into the bacterial F_1 showing an increased rate upon reduction. In this chimeric F_1-part the attached beads show a modulation of rotation. In the oxidized state more frequent long pauses between the rotational steps were observed (Bald, Noji, Yoshida, Hirono-Hara, & Hisabori, 2001; Hisabori, Konno, Ichimura, Strotmann, & Bald, 2002; Hisabori, Sunamura, Kim, & Konno, 2013).

Our interest in the coupling between proton transport and ATP synthesis led us to develop a different technique for investigating subunit movements in single molecules during catalysis (Börsch, Diez, Zimmermann, Reuter, &Gräber, 2002; Diez et al., 2004; Zimmermann, Diez, Zarrabi, Gräber, & Börsch, 2005). In the following, the general aspects of this technique are described and data obtained with the chloroplast enzyme are presented (Bienert, Rombach-Riegraf, Diez, & Gräber, 2009; Bienert, Zimmermann, Rombach-Riegraf, & Gräber, 2011).

A remark should be added to the nomenclature used in this work. The fluorescence label used in our work was attached to the second to last amino acid of the C-terminal helix of the γ-subunit, residue C322. In the recent structure of CF_OF_1 (Hahn et al., 2018) the numbering of the γ-subunit with its pre-sequence was adopted, so that residue C322 is now numbered C363 and the disulfide bridge is formed now between C240 (C199) and C246 (C205). In the present work we used the amino acid numbering of the γ-subunit as reported in Miki, Maeda, Mukohata, and Futai (1988).

2. Single-molecule FRET measurements in solution with confocal microscopy

2.1 The confocal set-up and detection of single molecules

Intramolecular movements of subunits can be investigated with single-molecule fluorescence energy transfer (smFRET). The efficiency of energy transfer, E_{FRET}, between two fluorescent molecules depends on the distance between them, and conversely, the distance between the two fluorophores can be determined by measuring such efficiency E_{FRET} (Förster, 1948;

Stryer & Haugland, 1967). Recent developments in optical components for microscopy and detection efficiency for recording of single photons allowed the measurement of FRET in single molecules (Deniz et al., 1999; Ha et al., 1996).

Based on these developments we built a confocal microscope setup, shown schematically in Fig. 1A (Börsch et al., 2002). The sample droplet (80 μL) with fluorescent molecules is placed on a cover glass and the laser beam is focused nearly diffraction-limited into the solution. The fluorescence from the sample is collected by the same objective with high numerical aperture, passed through a dichroic mirror and focused on a 100 μm pinhole. The intensity distribution of the laser in the excitation focus is depicted in green, the detection focus defined by the pinhole is shown in yellow. Both foci are adjusted to form the confocal volume, which can be adjusted

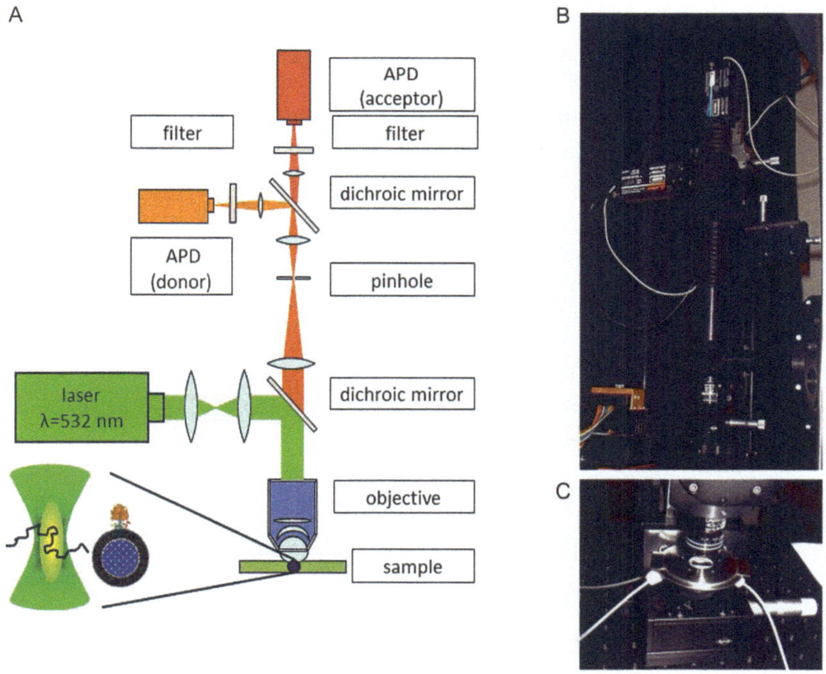

Fig. 1 Confocal set-up for single-molecule FRET measurements in solution. (A) Scheme of the confocal set-up with detectors (APD) and optical filters. Beam path for laser excitation in green, and fluorescence detection from FRET donor in yellow and from FRET acceptor in red. The confocal volume is defined by the superposition of laser focus (green) and pinhole focus (yellow) and is placed within the sample chamber. The schematic shows an arbitrary trajectory of a freely diffusing proteoliposome through the confocal volume. (B) Photo of the set-up. (C) Photo of the covered sample chamber with tubing for buffer mixing.

between 0.1 and 20 fL by using different pinhole sizes in combination with a corresponding laser beam waist underfilling the back aperture of the objective. The second dichroic mirror and appropriate interference filters are used to separate the collected fluorescence of the FRET donor fluorophore from that of the acceptor fluorophore. Fluorescence photons are detected by two single photon counting avalanche photodiodes (APD), counted in parallel with a multichannel scaler (or by TCSPC electronics, respectively) and displayed as fluorescence intensity time traces. In the following the FRET donor intensity traces are displayed in green, the FRET acceptor intensity traces in red.

The detection efficiency of our setup was determined previously (Bienert et al., 2011). The laser beam (wavelength 532 nm, continuous-wave) was adjusted to emit an intensity of $3 \cdot 10^{14}$ photons per second or 100 μW, respectively. The beam was focused by a 40× water immersion objective with a high numerical aperture of 1.15. Approximating the diffusion coefficient of the fluorophore ATTO532 in water at 21 °C, $D_{(21°C)} = 3.5 \cdot 10^{-6}\ \mathrm{cm^2\,s^{-1}}$, and using a measured diffusion time $\tau_D = 0.5$ ms for ATTO532 through the confocal detection volume, we recalculated the radial focus radius ω_0. This radius ω_0 is defined from the center of the focus to a distance where the maximum of the Gaussian intensity distribution has dropped to $1/e^2$. Using the relation for $\omega_0{}^2 = \tau_D 4D$, ω_0 was 0.84 μm. This resulted in a focal area $A = \omega_0{}^2\pi/2$ (Börsch et al., 1998) of approximately $1.1 \cdot 10^{-12}\ \mathrm{m^2}$ and, correspondingly, the mean light intensity in this area was $I = 2.7 \cdot 10^{26}\ \mathrm{m^{-2}\,s^{-1}}$. The fluorophore ATTO532 has an absorption coefficient of $\varepsilon(532) = 115{,}000\ \mathrm{M^{-1}\,cm^{-1}}$, and the maximum number of absorbed photons per second by one ATTO532 molecule is $6 \cdot 10^6\ \mathrm{s^{-1}}$, as estimated from Eq. (1):

$$\frac{dN(abs)}{dt} = \frac{\varepsilon(532)}{N_A} \cdot I = \frac{11{,}500\ \mathrm{m^2\,mol^{-1}} \cdot 2.7 \cdot 10^{26}\ \mathrm{m^{-2}\,s^{-1}}}{6 \cdot 10^{23}\ \mathrm{mol^{-1}}} = 5.2 \cdot 10^6\ \mathrm{s^{-1}} \quad (1)$$

With the quantum yield of ATTO532 ($\Phi = 0.4$) the maximum number of emitted photons per second results in:

$$dN(em)/dt = 2.1 \cdot 10^6\ \mathrm{s^{-1}} \quad (2)$$

Not all photons are collected by the objective since fluorescence is emitted isotropically and the objective collects only those in an angle defined by the numerical aperture. In addition, photons are lost at the diverse optical components (dichroic mirrors, filters and spectral quantum efficiency of the APDs). When a single ATTO532 molecule diffuse through the confocal volume the maximum number of detected photons was

$$dN(detected)/dt = 2 \cdot 10^5 \text{ s}^{-1} \quad (3)$$

Thus the apparent detection efficiency of the setup (detected photons/emitted photons) is $D \approx 0.09$. For visualizing intensity time traces, photons are accumulated with a binning interval of 1 ms. Therefore, if one fluorophore is in the confocal volume, the highest fluorescence count rate is approximately 200 photons per ms, i.e., 200 kHz. In smFRET experiments, the sum of donor and acceptor fluorescence intensities cannot exceed this value (Hellenkamp et al., 2018). Observation of higher fluorescence intensities indicates the presence of more than one FRET donor fluorophore in the confocal volume. Fig. 1B shows a photo of our confocal setup.

2.2 Single-molecule fluorescence, FRET efficiency and subunit movements in EF_oF_1

Investigations of subunit movements during proton transport-coupled ATP synthesis with single enzymes require the use of membrane integrated F_oF_1 where the enzyme has access to two sides with different proton concentrations. In earlier work, liposomes with purified membrane integrated F_oF_1 from chloroplasts (Schmidt & Gräber, 1985), *E. coli* (Fischer et al., 1994) and mitochondria (Förster et al., 2010) had been prepared and energization of the membrane with an acid-base transition resulted in ATP synthesis rates up to 50% of the rates obtained with the different enzymes in their natural membranes.

For measurements of intramolecular smFRET, the enzyme was labeled with one FRET donor and one acceptor fluorophore at appropriate sites and one $F_oF_1$on average was reconstituted into the liposome. These doubly-labeled ("FRET-labeled") F_oF_1-liposomes were diluted to 20 pM, i.e., the mean number of F_oF_1-liposomes in the confocal volume (~10 fL) was 0.12. This implies that most of the time only background count rates (noise) are recorded, since no fluorophores are present in the confocal volume. The appearance of a liposome with a labeled F_oF_1 results in a burst of photons from the FRET donor and acceptor fluorophores, and the duration of the burst indicates the diffusion time of the F_oF_1 liposome through the confocal volume.

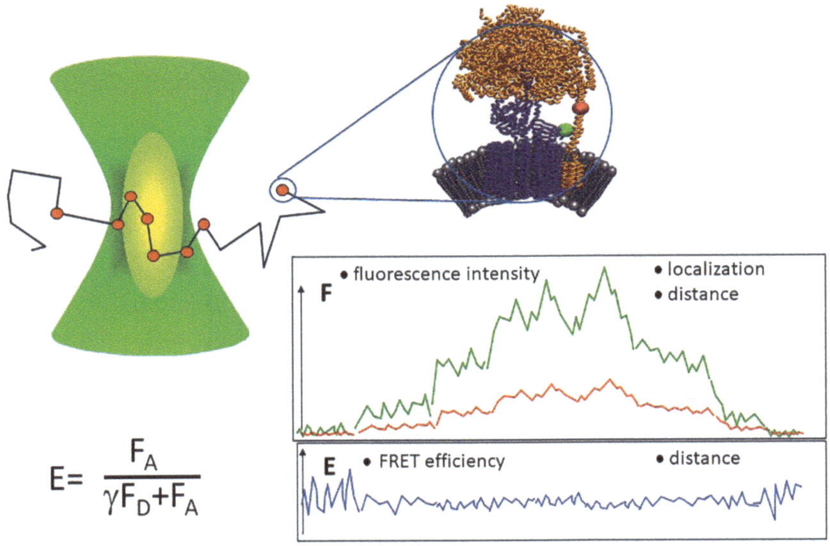

Fig. 2 Interpretation of a photon burst for single-molecule FRET analysis. (Top) Example of a diffusion trajectory of a proteoliposome. (Bottom) Measured fluorescence intensity time traces of FRET donor (F_D, green trace) and FRET acceptor (F_A, red trace). Relative intensities depend on the actual localization within the confocal volume and on the intramolecular distance of the two FRET fluorophores in the enzyme. The FRET efficiency (E, blue trace) is calculated from F_D and F_A per time bin.

Fig. 2 shows schematically the origin and interpretation of a photon burst. Within the confocal volume, the light intensity of the excitation laser depends on the exact position according to Gaussian distributions in x-, y-, z-direction as indicated schematically by the green color intensity in the figure. Therefore, the emitted fluorescence intensity changes, according to the localization of the fluorophore within the confocal volume, from zero outside this volume to its maximal value in the center. This is indicated by the green fluorescence intensity time trace of the donor, F_D, and correspondingly also by the red fluorescence intensity trace of the acceptor F_A. For F_D and F_A, the mean background count rates were subtracted. With these data, the FRET efficiency, E_{FRET} (or E in the following) was calculated from Eq. (4)

$$E = \frac{F_A}{\gamma F_D + F_A} \tag{4}$$

where $\gamma = 0.7$ is a specific correction factor comprising the quantum yields of donor and acceptor fluorophores and the spectral detection efficiencies of our setup. The FRET efficiency trace is depicted in the blue trace. This trace does not depend on the intensity of the exciting light or on the localization of the proteoliposome in the confocal volume. It depends only on the distance and relative orientation of donor- and acceptor fluorophores. The Förster theory allows then the determination of the donor-acceptor distance.

In earlier work, the movements of the rotor subunits (γ- or ε-subunit) in F_oF_1 from *E. coli* were investigated relative to the external stalk (b-subunits) with smFRET. In this case, cysteine residues were introduced at appropriate sites in these subunits. Cysteine residues can be labeled selectively under physiological conditions (pH 7) with maleimide derivatives of the fluorophores. Tetramethylrhodamine maleimide was bound either to the γ- or to the ε-subunit and acted as FRET donor. A bifunctional fluorophore (Cy5 bis-maleimide) was used as FRET acceptor. This fluorophore crosslinked the two b-subunits in F_oF_1, resulting in a unique position relative to the two b-subunits in the enzyme (Diez et al., 2004; Zimmermann et al., 2005; Zimmermann, Diez, Börsch, & Gräber, 2006). These investigations on freely diffusing membrane integrated F_oF_1 molecules led to the following results: First, the γ-subunit and the ε-subunit move stepwise through three positions at saturating ATP concentrations or high energization, and this movement can be described by 120 degree steps around a central axis; second, the direction of the movement during ATP hydrolysis is opposite to that during ATP synthesis; and third, the average rates calculated from the dwell times (dwell time is the time between two subsequent steps) correspond to the biochemically measured bulk rates of ATP hydrolysis and ATP synthesis under the same conditions.

3. Single-molecule FRET measurements with F_oF_1 from chloroplasts

3.1 Catalytic and non-catalytic binding sites of CF_oF_1

CF_oF_1 has three catalytic and three non-catalytic nucleotide binding sites and in earlier work we determined the properties of these binding sites. The non-catalytic binding sites bind ATP, with different binding affinities. The non-catalytic ATP binding site 4 shows the strongest binding of ATP, the non-catalytic site 5 medium binding and the non-catalytic site 6 the weakest. The nucleotides bound to the non-catalytic sites do not exchange with the medium during catalysis and therefore, it is possible to bind labeled nucleotides to these sites, which will remain bound while the enzyme is carrying out catalysis. The fluorescence acceptor AMPPNP-ATTO655 was bound to non-catalytic binding site 4. That site usually contains a tightly bound ATP, which exchanges with much less frequency than the two other sites (Possmayer, Hartog, Berden, & Gräber, 2000a, 2000b, 2001). The ATP on site 4 must be first hydrolyzed to ADP, which can then be exchanged against ATP or AMPPNP-ATTO655. The relation between the structural properties of the three non-catalytic binding sites and their functional properties is not clearly known. It is plausible hypothesis that the properties of one αβ-pair are changed by the binding of the external stalk, and it is that particular αβ-pair which shows the very tight binding of ATP. Therefore, we will assume that the AMPPNP-ATTO655 was bound to the αβ-pair which binds the external stalk, although this assumption is not essential for the interpretation of the smFRET data.

3.2 CF_oF_1-liposomes and ATP synthesis

CF_oF_1 was isolated as described and reconstituted into phosphatidylcholine/phosphatidic acid liposomes (Fromme, Boekema, & Gräber, 1987; Turina, Samoray, & Gräber, 2003). To determine the enzyme activity, an acid-base transition was carried out with the proteoliposomes, generating a transmembrane ΔpH and Δφ. The proteoliposomes were incubated in an acidic buffer (buffer I, 20 mM succinic acid-NaOH pH 4.8, 5 mM NaH_2PO_4, 0.6 mM KOH, 2.5 mM $MgCl_2$ and 10 μM valinomycin) for 3 min and then rapidly mixed with a basic buffer (buffer II, 200 mM tricine-NaOH pH 8.7, 5 mM NaH_2PO_4, 160 mM KOH, 2.5 mM $MgCl_2$ and 0.1 mM ADP). This generated an initial transmembrane energization of ΔpH = 3.9 and Δφ = 120 mV. The concentration of synthesized ATP was determined with luciferin/luciferase as a function of time after mixing and the initial ATP synthesis rate was measured. This initial rate was $(140 \pm 15)\ s^{-1}$ in accordance with earlier results. After labeling the residue γC322 with ATTO532, the rate was $(115 \pm 5)\ s^{-1}$. The occupancy of the non-catalytic site 4 with AMPPNP-ATTO655 decreased the rate to $(65 \pm 10)\ s^{-1}$ and a similar rate was observed when the binding procedure was carried out without AMPPNP-ATTO655. We concluded that the procedure for cysteine labeling and not the binding of the fluorescent nucleotide decreased the enzyme activity. After labeling CF_oF_1 with both FRET fluorophores, an ATP synthesis rate of $(24 \pm 6)\ s^{-1}$ was found.

3.3 Labeling of rotor and stator subunit of CF_oF_1

We investigated subunit movements in oxidized CF_oF_1 during proton transport-coupled ATP synthesis with single enzymes. Such measurements require that the liposomes contain only one doubly-labeled enzyme. We used phosphatidylcholine/phosphatidic acid liposomes with a diameter of 150 nm and such liposomes contain $2.2 \cdot 10^5$ lipid molecules. For reconstitution of CF_oF_1 (15 nM) a lipid concentration of 10.5 mM (corresponding to a final liposome concentration of 48 nM) was used. Therefore, if all enzymes are integrated into the liposome membrane, only approximately 30% of the liposomes will contain one CF_oF_1 on average. Furthermore, liposomes with two or more labeled enzymes can be identified by a significantly increased fluorescence count rate compared to that measured with one fluorophore ($2 \cdot 10^5\ s^{-1}$ at maximum).

CF_oF_1 contains several endogenous cysteine residues. Subunit ε and subunit I (b) contain each 1 cysteine, subunit γ contains 4 cysteines, two of which, γC199 and γC205, form a disulfide bridge under oxidizing conditions. In addition to the modification of these cysteine residues, there are also other possibilities for selective labeling, like the lysine residue K378 of one α-subunit, which can be labeled with fluorophores carrying a vinylsulfon group (Nalin, Snyder, & McCarty, 1985). A completely different approach is the binding of a fluorescent nucleotide to a non-catalytic binding site which remains bound during catalysis. For selective labeling of rotor and stator subunits of CF_oF_1, we investigated the following possibilities: CF_oF_1-γC322-ATTO532 and noncat-AMPPNP-ATTO655; CF_oF_1-γC322-ATTO532 and noncat-ATP-ATTO655; CF_oF_1-γC199/205-ATTO532 and noncat-ATP-ATTO655; CF_oF_1-γC322-Cy5 and αK378-ATTO532; CF_o-IC159-Cy5 and CF1-γC322-ATTO532. Most experiments were carried out with the following system: CF_oF_1-γC322-ATTO532 and noncat-AMPPNP-ATTO655, i.e., the second to last residue of the C-terminal helix of the γ-subunit, γ-C322, was labeled with the fluorophore ATTO532. AMPPNP was labeled at the ribose moiety either at position 3′ or 4′ with ATTO655 and was bound selectively to the non-catalytic ATP binding site 4 (Bienert et al., 2009).

3.4 Fluorescence time traces under different conditions

For the confocal smFRET measurements, proteoliposomes with doubly-labeled CF_oF_1 were reconstituted in liposome membranes. The proteoliposomes were energized as described above and the fluorescence time traces of FRET donor and acceptor were measured as described (Bienert et al., 2009). A microscopic flow chamber was inserted into the confocal setup (see Fig. 1C) and measurements were started after mixing of buffer II with the proteoliposomes which were pre-incubated in buffer I. The proteoliposome concentration was adjusted to 10 pM. Each measurement started with registration of the background count rates and only when a proteoliposome entered the confocal volume, photon bursts with up to 200 photons per ms were observed. The initial energization decreased after mixing, due to proton efflux and, corresponding to the decreasing rate of bulk ATP synthesis, no ATP synthesis activity could be detected after 1 min. Therefore, after each mixing, time traces were registered for 50 s.

Fig. 3 displays typical time traces of photon bursts during ATP synthesis. The FRET donor fluorescence (F_D, green trajectories) and the acceptor fluorescence (F_A, red trajectories) show large intensity fluctuations whereas the calculated FRET efficiency is constant during the photon burst (E, blue trajectories). Approximately 80% of the bursts display such a constant FRET efficiency, i.e., the relative distance and orientation of the FRET donor and acceptor fluorophore does not change during this burst (Fig. 3A–D). Presumably, these traces represent enzymes which are not catalyzing ATP synthesis, i.e., enzymes for which the transmembrane energization has declined so that ATP synthesis is no longer possible. Therefore, we refer to traces obtained under such conditions as "after ATP synthesis," and we call the state of such enzymes "resting state." The high fraction of enzyme displaying a constant FRET efficiency agrees with the fact that during the 50 s time interval in which smFRET traces were registered, the ATP synthesis bulk rate decreased continuously.

About 20% of the bursts show rapid FRET changes indicating a stepwise relative movement of the FRET donor and acceptor. FRET levels changed either with the sequence $L \rightarrow M \rightarrow H$ (see Fig. 3E and F) or with the sequence $L_1 \rightarrow L_2 \rightarrow M \rightarrow H_1 \rightarrow H_2$, i.e., with a splitting of the FRET levels L and H into sublevel (see Fig. 3G and H). Photon bursts with these sequences are observed during ATP synthesis in the presence of 100 μM ADP as well as in the presence of 15 μM ADP. We call the state of such enzymes "active state."

3.5 FRET efficiency histograms and conversion to distance histograms

For further discussion the histograms of the FRET levels have been generated. The data from bursts with the sequence $L \rightarrow M \rightarrow H$ in the presence of 100 μM ADP are collected in Fig. 4A and they show three distinct peaks. These peaks are classified according the maxima of the FRET efficiency states L (Low), M (Medium) and H (High) and they are fitted with

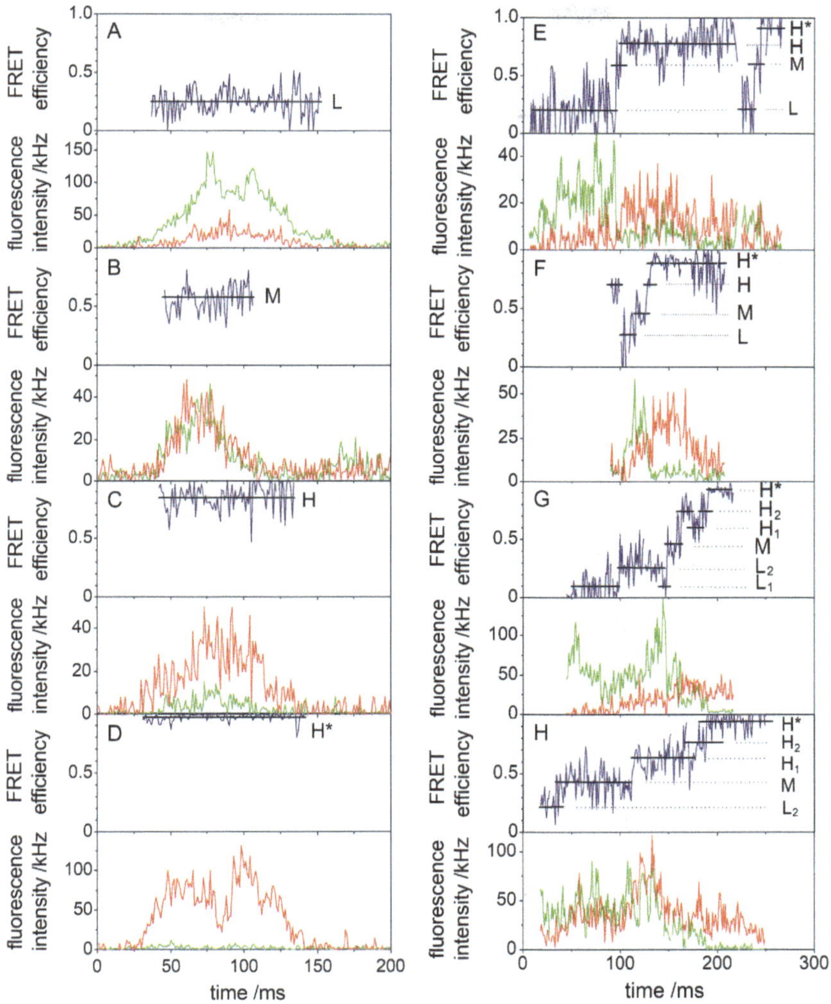

Fig. 3 Photon bursts from liposomes with FRET-labeled CF_oF_1 during ATP synthesis. Corrected fluorescence intensity traces of the donor and acceptor are shown in green and red, respectively. The corresponding FRET efficiencies are shown as blue traces. For each FRET level, the arithmetic mean value is depicted (black line). Time traces measured 2 min after the acid-base transition (after ATP synthesis) show constant FRET efficiencies. The FRET states are labeled L, M, H and H* (left column, panels A–D). Time traces measured within 1 min after the acid-base transition (during ATP synthesis) show stepwise changes of the FRET efficiencies. The FRET levels show the sequence L → M → H → L (right column, panels E and F). Approximately half of the traces indicate a splitting into sublevels showing a FRET sequence $L_1 \rightarrow L_2 \rightarrow M \rightarrow H_1 \rightarrow H_2 \rightarrow H^*$ (right column, panels G and H).

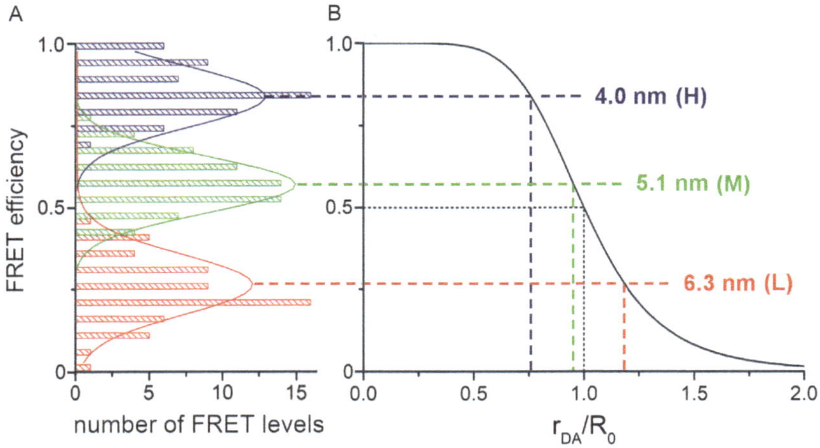

Fig. 4 Histograms of the FRET efficiencies (L, M, H) and determination of the average FRET donor-acceptor distance. The mean FRET efficiency is calculated for each observed FRET level (see black lines in Fig. 3E) and the number of observations is plotted as a function of FRET efficiency. (A) FRET efficiencies during ATP synthesis with 100 μM ADP. The histograms are fitted by Gaussian distributions. (B) For determination of the FRET donor-acceptor distance (r_{DA}) the maxima of the Gaussians on the left are used.

Gaussian functions indicating maxima at FRET efficiencies of 0.26 ± 0.09 (L), 0.57 ± 0.08 (M) and 0.84 ± 0.08 (H). According to the Förster theory, the donor-acceptor distance, r_{DA}, can be calculated from the FRET efficiency, E(FRET), using the following relation:

$$r_{DA} = R_0 \left(\frac{1}{E(FRET)} - 1 \right)^{1/6} \tag{5}$$

In this equation, R_0 is the Förster radius for 50% energy transfer comprising parameters of the FRET donor and acceptor fluorophores. After determining these parameters under our experimental conditions, we calculated $R_0 = 53\,\text{Å}$. In Fig. 4B the relation between FRET efficiency and donor-acceptor distance is depicted. Accordingly, the average distances are calculated from the maxima of the FRET efficiencies resulting in (6.3 ± 0.5) nm for FRET states L, (5.1 ± 0.3) nm for M and (4.0 ± 0.4) nm for H.

3.6 Dwell time histograms and rate of ATP synthesis

Analysis of the time courses of the FRET efficiency indicates that they are kinetically compatible with the rotational movements of the γ-subunit (FRET donor) occurring within the membrane integrated CF_oF_1 during ATP synthesis, relative to the stator (FRET acceptor at non-catalytic site 4). The FRET efficiency time trace in the photon burst is characterized by stepwise changes between the different constant FRET levels, which occur faster than 1 ms (i.e., faster than the 1-ms time resolution of our confocal setup). The time between two FRET levels (dwell time) corresponds to the time for undocking of the γ-subunit from one αβ-pair and docking to the next one. Based on the observation that a full rotation of the γ-subunit corresponds to the hydrolysis of three ATP molecules (Yasuda, Noji, Kinosita, & Yoshida, 1998), a full rotation under ATP synthesis conditions should synthesize three ATP molecules, i.e., one dwell represents the time for the synthesis of one ATP molecule.

The shortest assignable dwell time is 5 ms, since dwells shorter than 5 ms cannot be distinguished from FRET level fluctuations due to photon counting noise. The duration of the longest dwell time depends on the duration of the burst and the additional requirement of three distinguishable FRET level within the photon burst. The first and the last FRET level cannot be used for dwell time histograms, because it remains unknown for how long the enzyme was already in that FRET state before entering the confocal detection volume, or how long it stayed after diffusing out of the detection volume. Histograms of the intermediary dwell times during ATP synthesis with 100 μM ADP and 15 μM ADP are depicted in Fig. 5. The dwell times are fitted with monoexponential decay functions, and average dwell times of (31 ± 6) ms in the presence of 100 μM ADP and (48 ± 9) ms in the presence of 15 μM ADP are obtained.

The reciprocal of the dwell time is the rate of ATP synthesis measured with smFRET. The average dwell time during ATP synthesis in the presence of 100 μM ADP is 31 ms, corresponding to a rate of $34\,s^{-1}$. This is in accordance with the initial rate of $24\,s^{-1}$ measured in bulk. In the presence of 15 μM ADP the average dwell time is 48 ms yielding a rate of $21\,s^{-1}$, again in accordance with the dependence of the initial bulk rate on the ADP concentration.

In addition to bursts with the FRET level sequence L → M → H, some bursts with the opposite sequence H → M → L are observed (17% in presence of 100 μM ADP, 5% in presence of 15 μM ADP). Presumably, the presence of such opposite sequences results from the fact that we consider

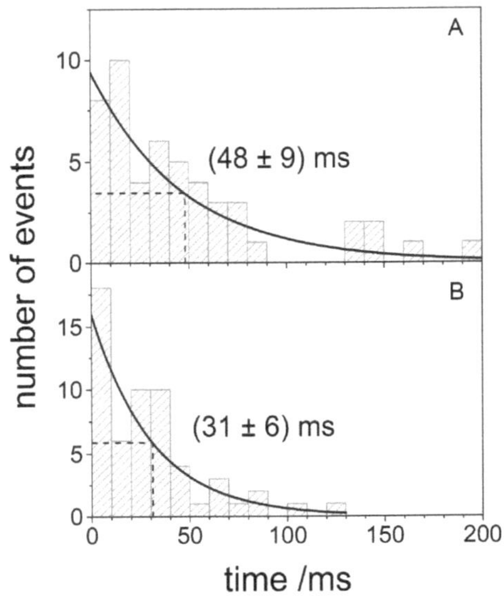

Fig. 5 Histograms of the dwell times from time traces with the FRET level sequence L-M-H during ATP synthesis and monoexponential decay fits. (A) Dwell times in the presence of 15 μM ADP with average dwell time (48 ± 9) ms. (B) Dwell times in the presence of 100 μM ADP with average dwell time (31 ± 6) ms.

a dwell as a separate FRET level only when its duration is longer than 5 ms. A dwell shorter than 5 ms will not be considered as a separate level and missing of one shorter FRET level will lead automatically to the opposite FRET sequence. Assuming an exponential distribution of dwell times:

$$\frac{N}{N_0} = \left(1 - e^{-t/\tau}\right) \tag{6}$$

we can estimate the fraction of dwells shorter than 5 ms from the average dwell times τ. Using $t = 4$ ms and $\tau = 31$ ms (average dwell time in the presence of 100 μM ADP), the fraction of missed dwells results in 12%, and with $\tau = 48$ ms (average dwell time in the presence of 15 μM ADP), that fraction results in 8%. We conclude that the apparently opposite direction of the sequence results mainly from short dwells which are not detected in our smFRET measurements.

3.7 FRET efficiency histograms during ATP synthesis and after catalysis

Histograms of the FRET levels under different conditions are collected in Fig. 6. The results from measurements during ATP synthesis at 15 and

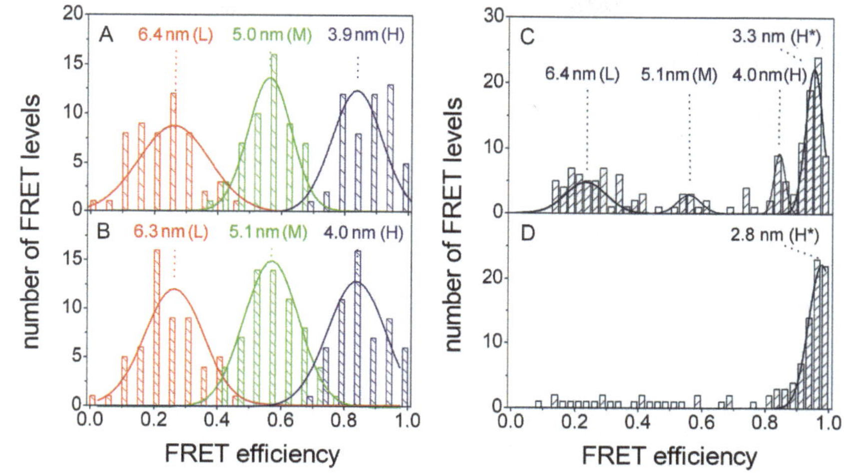

Fig. 6 Histograms of the FRET efficiencies and mean FRET donor-acceptor distances of time traces with FRET levels L, M and H under different conditions. (A) During ATP synthesis with 15 μM ADP. (B) During ATP synthesis with 100 μM ADP. (C) After ATP synthesis in the presence of 100 μM ADP. (D) In the presence of 1 mM AMPPNP. The average FRET donor-acceptor distances are determined as shown in Fig. 4.

100 μM ADP (data from Fig. 4) are shown in Fig. 6A and B. The histograms for both conditions are similar. They both show the three different FRET states L, M and H. All states are equally populated and the maxima of the Gaussian distributions occur at the same FRET efficiencies, indicating the same average donor-acceptor distance.

Fig. 6C shows a histogram of the bursts with overall constant FRET efficiency (see examples in Fig. 3A–D). This histogram displays four different peaks with different populations L (34%), M (7%), H (12%) and H* (42%). The FRET efficiencies and the average distances of L, M and H are the same as observed for the bursts with stepwise changes shown in Fig. 6A and B. It appears that even though these enzymes do not catalyze ATP synthesis, the γ-subunit displays the same three positions as found during catalysis. In addition to these three states a new FRET state H* is observed with the shorter donor-acceptor distance of 3.3 nm. This FRET state is found only after ATP synthesis and its percentage increases with time after the acid-base transition. We interpret H* as an inactive state of the enzyme. It is observed in the presence of ADP after ATP synthesis has declined, and we call it "ADP-inhibited state." The shorter distance of 3.3 nm might be due either to a further rotational movement of the γ-subunit or to a different conformation of the γ-subunit/αβ-pair complex bearing the fluorescent ATP derivative.

In addition to the smFRET measurements during catalysis, photon bursts have been investigated also in the presence of 1 mM AMPPNP. AMPPNP is a not hydrolysable ATP analog which binds to the enzyme and inhibits it competitively (Nalin & Cross, 1982). In agreement with the formation of an inhibited state of the enzyme, all photon bursts show a constant FRET efficiency. The FRET level histogram is depicted in Fig. 6D. Only few bursts with low FRET efficiencies are observed, while the majority of bursts (74%) are found in the highest FRET state H^*_{AMPPNP}, corresponding to a donor-acceptor distance of 2.8 nm. We call this state "AMPPNP-inhibited state." We conclude that the inhibited state induced by AMPPNP either features a rotary position of the γ-subunit relative to the bound fluorophore-labeled ATP, which differs from each of the three orientations during catalysis, or such a shorter FRET distance is due to a γ/α/β conformation similar to that of the H* state observed in the presence of ADP.

Photon bursts with FRET changes can be subdivided into bursts showing the simple sequence L → M → H (32% in the presence of 15 μM ADP and 46% in the presence of 100 μM ADP) and bursts showing an additional splitting of the low and high FRET levels L and H resulting in the sequence $L_1 \rightarrow L_2 \rightarrow M \rightarrow H_1 \rightarrow H_2$. In addition forward and backward steps between the sublevels are also observed (see Fig. 3G and H). Fig. 7 shows FRET level histograms in the presence of 100 μM ADP comprising FRET levels L_1 (26%), L_2 (27%), M (25%), H_1 (19%) and H_2 (16%), i.e., the five FRET levels are populated similarly, with sublevels L_1 and L_2 slightly more than the others. The average donor-acceptor distances were 7.7 nm for L_1, 6.4 nm for L_2, 5.4 nm for M, 4.4 nm for H_1 and 3.6 nm for H_2. By comparing Figs. 5 and 7, it is shown that level L_2 and H_2 (Fig. 7) are similar to L and H,

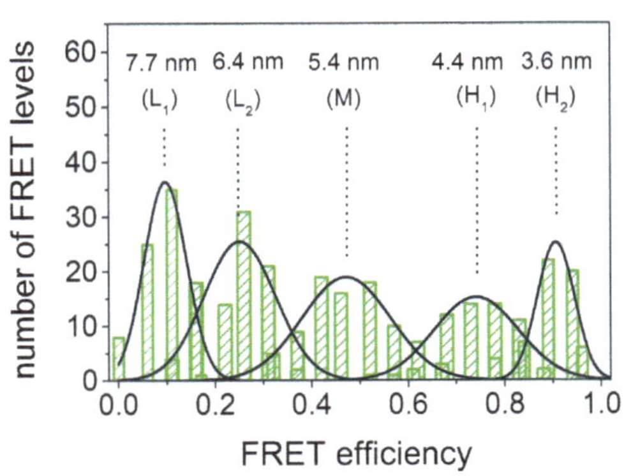

Fig. 7 Histograms of the FRET efficiencies and mean FRET donor-acceptor distances of time traces with five FRET levels during ATP synthesis with 100 μM ADP. The average FRET donor-acceptor distances are determined as described in Fig. 4.

respectively (Fig. 5). This splitting leads to two new conformational states of the enzyme: L_1, which has a longer donor-acceptor distance than L, and H_1, which has a longer donor-acceptor distance than H. Level M does not show a splitting. All these data are collected in Table 1.

Table 1 Statistics and donor-acceptor distances from FRET levels with stepwise transitions, constant FRET levels and structural data.

ATP synthesis 100 μM, (5 states) active enzymes	L_1	L_2	M	H_1	H_2	
Fraction of levels (%)	20	25	23	17	15	
E (FRET)	0.11	0.27	0.47	0.74	0.88	
Distance (nm)	7.7	6.4	5.4	4.4	3.6	
ATP synthesis 15 μM, (3 states) active enzymes		L	M		H	
Fraction of levels (%)		34	34		32	
E (FRET)		0.24	0.57		0.86	
Distance (nm)		6.4	5.0		3.9	
ATP synthesis 100 μM, (3 states) active enzymes		L	M		H	
Fraction of levels (%)		32	36		32	
E (FRET)		0.26	0.57		0.84	
Distance (nm)		6.3	5.1		4.0	
After ATP synthesis, (4 states) resting enzymes, inactive enzyme		L	M		H	H*
Fraction of levels (%)		32	8		16	44
E (FRET)		0.24	0.56		0.84	0.95
Distance (nm)		6.4	5.1		4.0	3.3
Structure		6FKI	6FKH			6FKF
Fraction of levels (%)		7.6	8			83
Distances of labeled positions (nm) 363 g@ca:ATP@C3′		5.2	4.6			4.3
Fluorophore distances (nm); data from (Fig. 9)		6.7	5.7			4.6

Compared to the three-stepped rotation sequence of the γ-subunit found in membrane integrated F_oF_1 from *E. coli*, two additional features are observed in CF_oF_1 (Bienert et al., 2009; Diez et al., 2004). First, in some traces the FRET levels L and H show a splitting into substeps, and second, forward and backward substeps are observed before transition to the next FRET level. At low ATP concentrations, substeps have been observed in bacterial F_1-parts and identified as ATP waiting steps and catalytic steps (Yasuda et al., 2001). Also during ATP synthesis in CF_oF_1, waiting steps for ADP, P_i and energizing protons from the internal phase of the liposomes might in principle be expected. It is also possible that the splitting is due to two different stable positions of the fluorophore without any mechanistic effect. According to our interpretation, the substeps seen here are due to regulatory movements (see Section 5.3).

4. Functional and conformational states of CF_oF_1 observed with smFRET

4.1 The different enzyme states

The H^+-ATPsynthase from chloroplast is a strongly regulated enzyme and the results obtained with smFRET leads to the detection of different enzyme states and this requires some discussion.

Enzymes in the "active state": When the rate of ATP synthesis is measured in bulk with CF_oF_1 in liposomes, the initial rate after an acid-base transition decreases exponentially and no activity is observed after 1 min. In the smFRET measurements, we registered fluorescence time traces for up to 50 s after the acid-base transition at regular time intervals. Therefore, during that time, it was expected that the enzyme molecules floating one after another into the confocal regions would be progressively more and more silent in catalysis, due to the decreasing energization. Indeed, during the measuring time, almost 80% of the photon bursts showed a constant FRET efficiency, and only 20% showed stepwise FRET efficiency change. We conclude that only these 20% of bursts are related to enzymes which are active in ATP synthesis (Fig. 6A). Three FRET levels are observed, indicating that the labeled γ-subunit carries out a three-stepped movement relative to the labeled non-catalytic nucleotide binding site. The transition between these states are faster than 1 ms and the three states are equally populated as expected from the binding change mechanism. The rate of ATP synthesis and its dependence on ADP concentration are in accordance with the rates calculated from the average dwell times of all three states.

Enzymes in the "resting state": Almost 80% of the photon bursts measured in the 50 s time span after the acid-base transition show a constant FRET efficiency (Fig. 6C). Even though they are not equally populated, the three peaks, labeled H, M and L, indicate the same average distances as observed for the photon bursts with stepwise changes, shown in Fig. 6A and B. Therefore, it appears that, under conditions of bulk ATP synthesis, a large fraction of enzyme molecules spend long times with the same γ-subunit positions as the active molecules, but without rotational movements. As mentioned above, it is plausible that such molecules are waiting for H^+ to bind from the inner side or to be released to the outer side, and, as a consequence, also for ADP and P_i to bind to their catalytic sites.

Enzymes in the "AMPPNP-inhibited state": In the presence of the inhibitor AMPPNP, the enzyme molecules are mostly inactive and, correspondingly, mainly bursts with constant FRET efficiencies (donor-acceptor distance 2.8 nm) are observed (Fig. 6D). As mentioned, the shorter donor-acceptor distance may be due to a different rotary position of the γ-subunit relative to the bound fluorophore-labeled ATP, or to a different γ/α/β conformation, which would change the distance between the γ-C-terminus and the bound fluorophore-labeled ATP.

Enzymes in the "ADP-inhibited state": In addition to the resting enzymes with the same FRET donor-acceptor distances as the active enzymes (L, M, H), a fourth FRET level (H*) is observed with a shorter donor-acceptor distance (3.3 nm). This level represents the largest fraction of enzymes (42%) without FRET level changes. We assume that this conformation H* represents the stable inactive state of the enzyme in the presence of ADP. Differently from the case of the enzyme inhibited by AMPPNP, where only one conformation is observed (H^*_{AMPPNP}), the histogram in Fig. 6C indicates that in this case a mixture of enzymes is present, which are either in the resting states with FRET donor-acceptor distances L, M and H, or in the inactive state with donor-acceptor distance H*. It is interesting to note that the difference in distance between H* and H^*_{AMPPNP} is consistent with the fact that occupation of the nucleotide binding sites by ATP or AMPPNP leads to a decrease of the volume of the enzyme (Börsch et al., 1998; Böttcher, Bertsche, Reuter, & Gräber, 2000).

4.2 Relation between the enzyme states

During catalysis three equally populated FRET states with different donor-acceptor distances are observed L (32%), M (36%), H (32%), i.e., active enzymes occur in three different conformations and the γ-subunit interacts on average equally with each αβ-pair. This is in accordance with the rotational binding change mechanism. When catalysis stops, e.g., when ΔpH decreases below the thermodynamic threshold, the enzyme switches to the resting state. The three resting states have the same FRET donor-acceptor distances as the active states but they are not equally populated, i.e., L (34%), M (7%), H (12%). This indicates that the transition from the active state into the resting state (or vice versa) does not occur with the same probability at each αβ-pair.

The processes of activation and inactivation can be observed directly with smFRET. The FRET transition H → H* indicates the conformational change from an active into the inactive state (see Fig. 3E and F). When the state H*, which corresponds to the ADP-inhibited state, is reached, in all examined traces the enzyme remains in this state until the end of the photon burst. Therefore, it appears that inactivation of the enzyme occurs only when the γ-subunit C-terminus is rotated closest to the tightly bound nucleotide at site 4, and appears to bring these two positions even closer. Direct steps from the M-state to H* and from the L-state to H* have not been observed. According to these data, the activation of the enzyme is not just the direct back reaction of inactivation, rather it involves a different conformational pathway, i.e., the activation step is accompanied by a large rotational movement of the γ-subunit, i.e., H* → L. Accordingly, most photon bursts displaying catalytic FRET changes start in state L. The relations between catalysis, activation and inactivation discussed here are summarized schematically in Fig. 8.

If the active-to-inactive transition is connected with deenergization of the enzyme, it should be possible that it occurs from all FRET states L, M and H. However, inactivation (to the ADP-inhibited state H*) has been observed only from state H. Such preference shows that in the resting state the three γ-positions are not energetically equivalent, and, in addition, that, at low energization and in the presence of bound ADP, there is only one αβ-pair to which the γ-subunit will preferentially bind.

The observation that in the inactive state one of the three γ-positions has a higher interaction energy with a given αβ-pair implies a breaking of the threefold symmetry which seems to characterize the γ-αβ interaction during catalysis (Fig. 5A and B). An obvious element of asymmetry in the αβ barrel is the external stalk δbb′. Therefore, it is possible that, in the inactive state, the γ-subunit is attached to the αβ-pair which interacts strongly with the external stalk.

5. CF_oF_1 structure and relation to the smFRET data

5.1 The localization of the fluorophores in the structure of CF_oF_1

The structure of the CF_1-part with a resolution of 3.2 nm has been reported (Groth & Pohl, 2001). The first cryo-electron microscopic investigation of the CF_oF_1 was carried out in the presence of AMPPNP and resulted in a map with 20 Å resolution showing the internal and external stalk. Fitting of the map with the structure of mitochondrial F_1 (Abrahams et al., 1994) indicated that the oxidized, inactive enzyme in the presence of AMPPNP adopts a unique resting position (Mellwig & Böttcher, 2003). This structural observation is in accordance with the data shown in Fig. 6D, where, in the presence of AMPPNP, only one γ-position is found relative to the fluorophore-labeled nucleotide in the non-catalytic site 4. Recently, the structure of the oxidized CF_oF_1 in lipid nanodiscs has been solved in the absence of added nucleotides using cryo-electron microscopy, with a resolution of up to 2.9 Å for the CF_1-part. The dataset revealed three distinct conformations of the enzyme, with PDB numbers 6FKF, 6FKH and 6FKI (Hahn et al., 2018). These three conformations are similar,

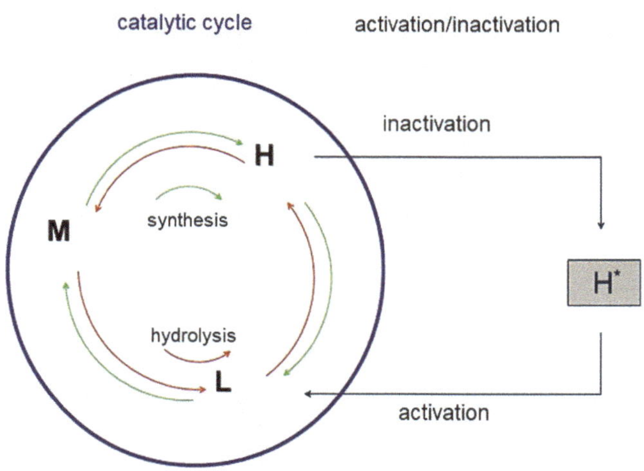

Fig. 8 Scheme of relation between catalysis, activation and inactivation based on smFRET data. Comparison of smFRET data with the structural data yields the following assignments: 6FKF = H*, 6FKI = L, 6FKH = M.

and rotary positions of the central stalk $\gamma\varepsilon c_{14}$ are separated by 103, 112 and 145 degrees. The three different central stalk positions are interpreted as the three conformational states during rotational catalysis.

The structures of the different conformations of CF_oF_1 have been used to compare the smFRET donor-acceptor distances. The FRET acceptor AMPPNP-ATTO663 was bound to the non-catalytic binding site 4 and we have assumed that the $\alpha\beta$-pair bearing this site is the one which interacts with the external stalk. It should be noted, however, that, due to the three-fold pseudo-symmetry of the $(\alpha\beta)_3$ barrel, such distances would not change by positioning the fluorescent ATP analog in one of the other three $\alpha\beta$-pairs. As mentioned, the ATP bound to site 4 is not readily exchanged during catalysis. In the cryo-EM structures 6FKF, 6FKH and 6FKI, this $\alpha\beta$-pair is constituted by the α-chain "A" and the β-chain "F," and an ATP molecule (called "ATP A") is bound at the A/F interface. In AMPPNP-ATTO655, the ATTO655 fluorophore is covalently bound either to the O3′ or O4′ atoms of the ribose moiety of the nucleotide. The position of residue γC322, to which the FRET donor ATTO532 was covalently bound, is well resolved in the structure. However, the exact position and orientation of the fluorophores in the protein structure is not known. In general, fluorophores have bulky aromatic systems and are attached via a linker to that residue. They must fit into the structure avoiding clashes between fluorophore and protein. We tried to map the CF_oF_1 structure with smFRET data using two approaches. First, to avoid the problem of positioning of the fluorophores we just determined the distance between the residues selected for labeling. The distances between the C_α of γ-C322 and the C3′ at the ribose moiety of ATP bound at the non-catalytic binding site 4 (ATP A) are 4.3 nm (6FKF), 4.6 nm (6FKH) and 5.2 nm (6FKI). Second, to approximate the position of the fluorophore moiety in the cryo-EM structures, we have manually placed the ATTO655 structure in the crevice at the A/F interface, close to the O3′ of the ATP A molecule. Accordingly, the distance between the center of the ATTO655 and the O3′ of ATP A resulted in 0.7 nm. To guide the manual positioning of the ATTO532 moiety, first of all an approximate γ-subunit rotation axis was drawn through the center of, and perpendicular to, the plane defined by the three C_α's of residues γ146 in 6FKF, 6FKH and 6FKI. The ATTO532 moiety was then placed in each of the structures 6FKF/H/I on a plane passing through the C_α's of residues γ363 in the 6FKF/H/I structures and perpendicular to that axis. Then, the three ATTO532 moieties were placed in the dimple formed by the six β-barrel C-terminal domains of the three $\alpha\beta$-pairs, in such a way as to avoid clashes with the protein backbone, and so that the distances and orientations, relative to the γ363-Cα's, were similar in the three structures 6FKF/H/I. The distances from the center of ATTO532 to C_α's of γ363 resulted 1.2 nm. Finally, the positions of the three ATTO532 moieties were adjusted so that the three distances to the ATTO655 fluorophores at the A/F interface would satisfy the rotary FRET sequence H $\rightarrow$ L $\rightarrow$ M in the clockwise direction (i.e., ATP synthesis direction, seen from the CF_o-part). As a result, the expected FRET donor-acceptor distances were 4.7 nm (6FKF), 6.3 nm (6FKH), and 5.7 nm (6FKI), in fair agreement with the H, L, M distances calculated from the FRET efficiencies. The structural analysis was carried out with UCSF chimera (Pettersen et al., 2004), the data are listed in Table 1.

5.2 Donor-acceptor distances from structure and from smFRET

The CF_oF_1 structures with these modifications are shown in Fig. 9 in side and top views. The AMPPNP part of the FRET acceptor is located at the same position as the ATP in the structure and is shown as ball and sticks. The FRET acceptor ATTO655 label and the FRET donor ATTO532 label are shown with their van der Waals volume and enclosed in an ellipsoid (red for the acceptor, green for the donor). The mass centers of the ATTO655 and ATTO 532 moieties were used to determine the distances between the FRET donor and acceptor.

The donor-acceptor distances obtained with smFRET and those estimated from the structure are collected in Table 1. The fair agreement between smFRET data and those from the structures is the result of combined structural restraints (explained above) and arbitrary degrees of freedom (due to unknowns, as the exact conformations of the linkers, and the exact position of the transition dipole moments of the FRET fluorophores). However, the overall fitting of the fluorophores into the structure shows that the three FRET distances measured in the active enzyme are consistent with the newest high resolution cryo-EM structures and with their three different γ-subunit positions.

5.3 Relation between structural conformations and enzyme states from smFRET

The cryo-EM measurements were carried out in the absence of added nucleotides and of membrane energization, therefore their states can be compared to the ADP-inactivated state and/or to the resting states, as

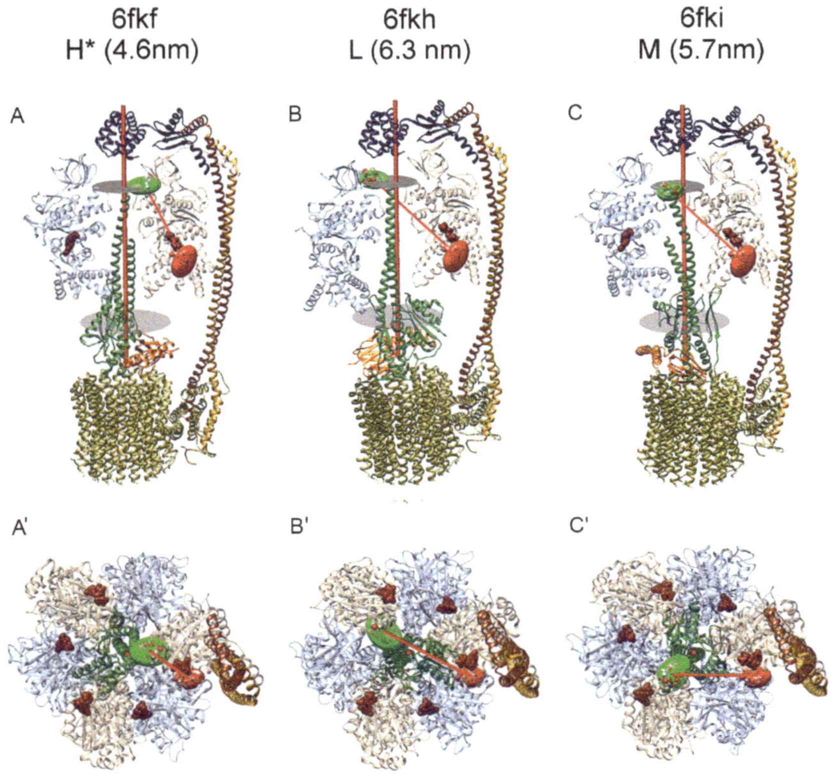

Fig. 9 Visualization of the fluorophore positions for FRET donor-acceptor distances in the CF_oF_1 cryo-EM structures (Hahn et al., 2018) with PDB entries 6FKF, 6FKH, 6FKI. (A–C) side views, (A′–C′) corresponding top views (seen from the CF_1 side of the membrane). Details see text.

defined in the present work (Fig. 6C). As discussed above, the enzyme state under the conditions of Fig. 6C (no energization) is a mixture of one inactive state (H*, 42%), and of the three resting states L (34%), M (7%), H (12%), where the conformations in the resting states have the same FRET donor-acceptor distances as the active enzymes. By comparison, the cryo-EM data reveal three CF_oF_1 conformations with the following population, 6FKF (83%), 6FKH (8%) and 6FKI (7.6%). Given the common condition of absence of energization in both cases, it is plausible that the highest populated state in the cryo-EM experiment (6FKF, 83%), and the highest populated state in the smFRET experiment (H*, 42%) coincide, and therefore, that the 6FKF structure represents the structure of the ADP-inhibited state of the oxidized CF_oF_1, as identified by smFRET.

During ATP synthesis the smFRET level sequence is L → M → H and the rotary sequence proposed from the structural data is 6FKH → 6FKI → 6FKF. The state L is observed frequently since during catalysis most bursts start in this state and consequently, L is also frequently observed in the resting state. Therefore, we conclude that 6FKH is the structure of state L and 6FKI that of M. As mentioned, the L- and H-states show a splitting of the FRET levels into sublevels, whereas in the M-state such splitting is not observed. According to our model (Fig. 8) the M-state is involved only in the catalytic cycle whereas the L-state and H-state are involved in two reactions, in the catalytic cycle and in the activation/inactivation reaction. Possibly, different rotational positions or different orientations of γC322-ATTO532 are favored in the two reaction pathways, even though it cannot be excluded that two different stable positions of the fluorophore exist, leading to a level splitting without any mechanistic effect.

Our hypothesis on the relation between the functionally characterized FRET states is shown in Fig. 8. The catalytic cycle in the ATP synthesis direction L → M → H is depicted by green arrows, the backward reaction ATP hydrolysis with red arrows. The inactivation process H → H* and the activation step plus the rotational step H* → L (or L_1/L_2) have been observed only in one direction (black arrows). According to the assignment of the smFRET states to the structural conformations, the transition 6FKH → 6FKI corresponds to the movement of the γ-subunit in the catalytic cycle, the transition 6FKI → 6FKF corresponds to catalysis and inactivation, and the transition 6FKF → 6FKH corresponds to activation and catalysis.

The rotary angles of the γ-subunit for the transitions change between the different states: for 6FKH → 6FKI by 103°, for 6FKI → 6FKF by 145° and for 6FKF → 6FKH by 112° (Hahn et al., 2018). Based on our assignments, only the first transition is a purely catalytic transition, in which the 103° movement of the γ-subunit would correspond to the movement of $4H^+$ per ATP, in accordance with measurements of the H^+/ATP ratio observed with CF_oF_1 (Turina et al., 2003). The transitions to and from 6FKF include both a step in the catalytic cycle and an activation/inactivation step. While, to our knowledge, it has never been investigated whether the inactivation step involves any proton translocation (e.g., in direction of ATP hydrolysis), the activation step has been shown to require additional protons (Junesch & Gräber, 1987). This might be the reason for the different rotary step sizes of the γ-subunit as observed in the CF_oF_1 structural analysis.

6. Conclusion and future perspective

Proton transport-coupled ATP synthesis and the activation-inactivation of oxidized CF_0F_1 were investigated with smFRET. Data analysis resulted in a model describing the relation between the three active enzyme conformations, H, M and L during catalysis, which is in accordance with the rotating binding change mechanism. Only one inactive conformation H* was observed after catalysis. H* differs from the three catalytic conformations and presumably represents the ADP-inhibited state of the deenergized enzyme. Comparison of these data with the structural conformations of the oxidized CF_0F_1 (Hahn et al., 2018) indicated that one conformation (6FKF) corresponds to the inactive state and two conformations (6FKH and 6FKI) correspond to active (resting) states. It is known that the activation reaction is connected with the release of tightly bound ADP and requires energization of the enzyme. Since the data of the present work indicate that activation and inactivation are coupled with rotation of the c-ring, we propose that additional rotational movements of rotor subunits take place, possibly coupled with proton transport, which differ from the rotation due to proton transport-coupled ATP synthesis.

Mitochondrial and bacterial enzymes are also regulated by different processes, including ADP binding, however, investigations of such mechanisms have been carried out mainly with F_1-parts. The interaction between regulation of the enzyme activity and proton translocation is an interesting aspect of the membrane integrated enzymes and their function in vivo.

References

Abrahams, J. P., Leslie, A. G., Lutter, R., & Walker, J. E. (1994). Structure at 2.8 Å resolution of F1-ATPase from bovine heart mitochondria. *Nature*, *370*, 621–628.

Bald, D., Noji, H., Yoshida, M., Hirono-Hara, Y., & Hisabori, T. (2001). Redox regulation of the rotation of F1-ATP synthase. *The Journal of Biological Chemistry*, *276*, 39505–39507.

Bienert, R., Rombach-Riegraf, V., Diez, M., & Gräber, P. (2009). Subunit movements in single membrane-bound H+-ATPsynthases from chloroplasts during ATP synthesis. *The Journal of Biological Chemistry*, *284*, 36240–36247.

Bienert, R., Zimmermann, B., Rombach-Riegraf, V., & Gräber, P. (2011). Time-dependent FRET with single enzymes: Domain motions and catalysis in H^+-ATPsynthases. *ChemPhysChem*, *12*, 510–517.

Bilyard, T., Nakanishi-Matsui, M., Steel, B. C., Pilizota, T., Nord, A. L., Hosokawa, H., et al. (2013). High-resolution single-molecule characterization of the enzymatic states in Escherichia coli F1-ATPase. *Philosophical Transactions of the Royal Society B: Biological Sciences*, *368*, 20120023.

Börsch, M., Diez, M., Zimmermann, B., Reuter, R., & Gräber, P. (2002). Stepwise rotation of the γ-subunit of EF0F1-ATP synthase observed by intramolecular single-molecule fluorescence resonance energy transfer. *FEBS Letters*, *527*, 147–152.

Börsch, M., Turina, P., Eggeling, C., Fries, J. R., Seidel, C. A. M., Labahn, A., et al. (1998). Conformational changes of the H+-ATPase from Escherichia coli upon nucleotide binding detected by single molecule fluorescence. *FEBS Letters*, *437*, 251–254.

Böttcher, B., Bertsche, I., Reuter, R., & Gräber, P. (2000). Direct visualisation of conformational changes in EF0F1 by electron microscopy. *Journal of Molecular Biology*, *296*, 449–457.

Boyer, P. D. (1997). The ATP synthase—A splendid molecular machine. *Annual Review of Biochemistry*, *66*, 717–749.

Deniz, A. A., Dahan, M., Grunwell, J. R., Ha, T. J., Faulhaber, A. E., Chemla, D. S., et al. (1999). Single-pair fluorescence resonance energy transfer on freely diffusing molecules: Observation of Forster distance dependence and subpopulations. *Proceedings of the National Academy of Sciences of the United States of America*, *96*, 3670–3675.

Diez, M., Zimmermann, B., Börsch, M., König, M., Schweinberger, E., Steigmiller, S., et al. (2004). Proton-powered subunit rotation in single membrane-bound F0F1-ATP synthase. *Nature Structural & Molecular Biology*, *11*, 135–141.

Duncan, T. M., Bulygin, V. V., Zhou, Y., Hutcheon, M. L., & Cross, R. L. (1995). Rotation of subunits during catalysis by Escherichia coli F1-ATPase. *Proceedings of the National Academy of Sciences of the United States of America*, *92*, 10964–10968.

Fischer, S., Etzold, E., Turina, P., Deckers-Hebestreit, D., Altendorf, K., & Gräber, P. (1994). ATP synthesis catalyzed by the ATP synthase of Escherichia coli reconstituted into liposomes. *European Journal of Biochemistry*, *225*, 167–172.

Förster, T. (1948). Zwischenmolekulare Energiewanderung und Fluoreszenz. *Ann Phys*, *2*, 55–70.

Förster, K., Turina, P., Drepper, F., Haehnel, W., Fischer, S., Gräber, P., et al. (2010). Proton transport coupled ATP synthesis by the purified yeast H+-ATPsynthase in proteoliposomes. *Biochimica et Biophysica Acta*, *1797*, 1828–1837.

Fromme, P., Boekema, E. J., & Gräber, P. (1987). Isolation and characterisation of a supramolecular complex of subunit III of the ATP-synthase from chloroplasts. *Zeitschrift für Naturforschung*, *42c*, 1239–1245.

Gräber, P., & Schlodder, E. (1981). Regulation of the rate of ATP synthesis/hydrolysis by ΔpH and Δψ. In G. Akoyunoglou (Ed.), *Vol. II. Proceedings of the 5th International Congress of Photosynthesis, Kallithea, Greece, 1980* (pp. 867–880). Philadelphia: Balaban International Science Service.

Gräber, P., Schlodder, E., & Witt, H. T. (1977). Conformational change of chloroplast Atpase induced by a transmembrane electric-field and its correlation to phosphorylation. *Biochimica et Biophysica Acta*, *461*, 426–440.

Groth, G., & Pohl, E. (2001). The structure of the chloroplast F1-ATPase at 3.2 resolution. *The Journal of Biological Chemistry*, *276*, 1345–1352.

Ha, T., Enderle, T. H., Ogletree, D. F., Chemla, D. S., Selvin, P. R., & Weiss, S. (1996). Probing the interaction between two single molecules: Fluorescence resonance energy transfer between a single donor and a single acceptor. *Proceedings of the National Academy of Sciences of the United States of America*, *93*, 6264–6268.

Hahn, A., Vonck, J., Mills, D. J., Meier, T., & Kühlbrandt, W. (2018). Structure, mechanism, and regulation of the chloroplast ATP synthase. *Science*, *360*, 620–628.

Häsler, K., Engelbrecht, S., & Junge, W. (1998). Three-stepped rotation of subunits gamma and epsilon in single molecules of F-ATPase as revealed by polarized, confocal fluorometry. *FEBS Letters*, *426*, 301–304.

Hellenkamp, B., Schmid, S., Doroshenko, O., Kühnemuth, R., Rezaei, A. S., Ambrose, B., et al. (2018). Precision and accuracy of single-molecule FRET measurements—A multi-laboratory benchmark study. *Nature Methods*, *15*, 669–676.

Hisabori, T., Kondoh, A., & Yoshida, M. (1999). The γ subunit in chloroplast F1-ATPase can rotate in a unidirectional and counter-clockwise manner. *FEBS Letters*, *463*, 35–38.

Hisabori, T., Konno, H., Ichimura, H., Strotmann, H., & Bald, H. (2002). Molecular devices of chloroplast F1-ATP synthase for the regulation. *Biochim Biophys Acta*, *1555*, 140–146.

Hisabori, T., Sunamura, E., Kim, Y., & Konno, H. (2013). The chloroplast ATP synthase features the characteristic redox regulation machinery. *Antioxidants & Redox Signaling*, *19*, 1846–1854.

Joliot, P., & Joliot, A. (2008). Quantification of the electrochemical proton gradient and activation of ATP synthase in leaves. *Biochimica et Biophysica Acta*, *1777*, 676–683.

Junesch, U., & Gräber, P. (1987). Influence of the redox state and the activation of the chloroplast Atp synthase on proton-transport-coupled Atp synthesis hydrolysis. *Biochimica et Biophysica Acta*, *893*, 275–288.

Junge, W., & Nelson, N. (2015). ATP synthase. *Annual Review of Biochemistry*, *84*, 631–657.

Kramer, D. M., & Crofts, A. R. (1989). Activation of the chloroplast ATPase measured by the electrochromic change in leaves of intact plants. *Biochimica et Biophysica Acta*, *976*, 28–41.

Kramer, D. M., Wise, R. R., Frederick, J. R., Alm, D. M., Hesketh, J. D., Ort, D. R., et al. (1990). Regulation of coupling factor in field-grown sunflower: A redox model relating coupling factor activity to the activities of other thioredoxin-dependent chloroplast enzymes. *Photosynthesis Research*, *26*, 213–222.

Lapashina, A. S., & Feniouk, B. A. (2018). ADP-inhibition of H^+-F_oF_1-ATP synthase. *Biochemistry (Moscow)*, *83*, 1141–1160.

Mellwig, C., & Böttcher, B. (2003). A unique resting position of the ATP-synthase from chloroplasts. *The Journal of Biological Chemistry*, *278*, 18544–18549.

Miki, J., Maeda, M., Mukohata, Y., & Futai, M. (1988). The y-subunit of ATP synthase from spinach chloroplasts. Primary structure deduced from the cloned cDNA sequence. *FEBS Letters*, *232*, 221–226.

Mills, J. D., Mitchell, P., & Schürmann, P. (1980). Modulation of coupling factor ATPase activity in intact chloroplasts: The role of the thioredoxin system. *FEBS Letters*, *112*, 173–177.

Mitchell, P. (1961). Coupling of phosphorylation to electron and hydrogen transfer by a chemiosmotic type of mechanism. *Nature*, *191*, 144–148.

Nalin, C. M., & Cross, R. L. (1982). Adenine nucleotide binding sites on beef heart F1-ATPase. *The Journal of Biological Chemistry*, *257*, 2874–2881.

Nalin, C. M., & McCarty, R. E. (1984). Role of a disulfide bond in the gamma-subunit in activation of the Atpase of chloroplast coupling factor-I. *The Journal of Biological Chemistry*, *259*, 7275–7280.

Nalin, C. M., Snyder, B., & McCarty, R. E. (1985). Selective modification of an α-subunit of chloroplast coupling factor 1. *Biochemistry*, *24*, 2318–2324.

Noji, H., Häsler, K., Junge, W., Kinosita, K., Yoshida, M., & Engelbrecht, S. (1999). Rotation of Escherichia coli F_1-ATPase. *Biochemical and Biophysical Research Communications*, *260*, 597–599.

Noji, H., Yasuda, R., Yoshida, M., & Kinosita, K., Jr. (1997). Direct observation of the rotation of F1-ATPase. *Nature*, *386*, 299–302.

Omote, H., Sambonmatsu, N., Saito, K., Sambongi, K., Iwamoto-Kihara, A., Yanagida, T., et al. (1999). The γ-subunit rotation and torque generation in F_1-ATPase from wild-type or uncoupled mutant Escherichia coli. *Proceedings of the National Academy of Sciences of the United States of America*, *96*, 7780–7784.

Ort, D. R., & Oxborough, K. (1992). In situ regulation of chloroplast coupling factor activity. *Annual Review of Plant Physiology and Plant Molecular Biology*, *43*, 269–291.

Pettersen, E. F., Goddard, T. D., Huang, C. C., Couch, G. S., Greenblatt, D. M., Meng, E. C., et al. (2004). UCSF chimera-A visualizationsystem for exploratory research and analysis. *Journal of Computational Chemistry, 25*, 1605–1612.
Possmayer, F. E., Hartog, A. F., Berden, J. A., & Gräber, P. (2000a). Covalent modification of the catalytic sites of the H+-ATPase from chloroplasts with 2-nitreno-ADP. Modification of the catalytic site 1 (tight) and catalytic sites 1 and 2 together impairs both uni-site and multi-site catalysis of ATP synthesis and ATP hydrolysis. *Biochimica et Biophysica Acta, 1459*, 202–217.
Possmayer, F. E., Hartog, A. F., Berden, J. A., & Gräber, P. (2000b). Covalent modification of the catalytic sites of the H+-ATPase from chloroplasts, CF_oF_1, with 2-azido-[alpha-P-32]ADP: Modification of the catalytic site 2 (loose) and the catalytic site 3 (open) impairs multi-site, but not uni-site catalysis of both ATP synthesis and ATP hydrolysis. *Biochimica et Biophysica Acta, 1456*, 77–98.
Possmayer, F. E., Hartog, A. F., Berden, J. A., & Gräber, P. (2001). Covalent modification of the non-catalytic sites of the H+-ATPase from chloroplasts with 2-azido-[alpha-P-32] ATP and its effect on ATP synthesis and ATP hydrolysis. *Biochimica et Biophysica Acta, 1510*, 378–400.
Sabbert, D., Engelbrecht, S., & Junge, W. (1996). Intersubunit rotation in active F-ATPase. *Nature, 381*, 623–625.
Sambongi, Y., Iko, Y., Tanabe, M., Omote, H., Iwamoto-Kihara, A., Ueda, I., et al. (1999). Mechanical rotation of the c subunit oligomer in ATP synthase (F0F1): Direct observation. *Science, 286*, 1722–1724.
Schmidt, G., & Gräber, P. (1985). The rate of ATP-synthesis by reconstituted CF_oF_1 liposomes. *Biochimica et Biophysica Acta, 808*, 46–51.
Spetzler, D., York, J., Daniel, D., Fromme, R., Lowry, D., & Frasch, W. (2006). Microsecond time scale rotation measurements of single F1-ATPase molecules. *Biochemistry, 45*, 3117–3124.
Strelow, F., & Rumberg, B. (1993). Kinetics and energetics of redox regulation of ATP synthase from chloroplasts. *FEBS Letters, 323*, 19–22.
Stryer, L., & Haugland, R. P. (1967). Energy transfer: A spectroscopic ruler. *Proceedings of the National Academy of Sciences of the United States of America, 58*, 719–726.
Turina, P. (1994). Influence of the transmembrane electrochemical proton gradient on catalysis and regulation of the H+-ATP synthase from Rhodobacter capsulatus. *Bioelectrochemistry and Bioenergetics, 33*, 31–43.
Turina, P., Samoray, D., & Gräber, P. (2003). H^+/ATP ratio of proton transport-coupled ATP synthesis and hydrolysis catalysed by CF_oF_1-liposomes. *The EMBO Journal, 22*, 418–426.
Weber, J., & Senior, A. E. (2000). ATP synthase: What we know about ATP hydrolysis and what we do not know about ATP synthesis. *Biochimica et Biophysica Acta, 1458*, 300–309.
Yasuda, R., Noji, H., Kinosita, K., & Yoshida, M. (1998). F1-ATPase is a highly efficient molecular motor that rotates with discrete 120 degree steps. *Cell, 93*, 1117–1124.
Yasuda, R., Noji, H., Yoshida, M., Kinosita, K., & Itoh, H. (2001). Resolution of distinct rotational substeps by submillisecond kinetic analysis of F_1-ATPase. *Nature, 410*, 898–904.
Zimmermann, B., Diez, M., Börsch, M., & Gräber, P. (2006). Subunit movement in membrane-integrated F_0F_1 during ATP synthesis detected by single-molecule spectroscopy. *Biochimica et Biophysica Acta, 1757*, 311–319.
Zimmermann, B., Diez, M., Zarrabi, N., Gräber, P., & Börsch, M. (2005). Movements of the ε-subunit during catalysis and activation in single membrane-bound H^+-ATP synthase. *The EMBO Journal, 24*, 2053–2063.

CHAPTER FIVE

P700 oxidation suppresses the production of reactive oxygen species in photosystem I

Riu Furutani[a], Kentaro Ifuku[b], Yuji Suzuki[c], Ko Noguchi[d], Ginga Shimakawa[e], Shinya Wada[a], Amane Makino[f], Takayuki Sohtome[g], and Chikahiro Miyake[a,h,]*

[a]Department of Applied Biological Science, Faculty of Agriculture, Graduate School for Agricultural Science, Kobe University, Kobe, Japan
[b]Graduate School of Biostudies, Kyoto University, Kyoto, Japan
[c]Faculty of Agriculture, Iwate University, Morioka, Japan
[d]School of Life Sciences, Tokyo University of Pharmacy and Life Sciences, Tokyo, Japan
[e]Research Center for Solar Energy Chemistry, Osaka University, Osaka, Japan
[f]Graduate School of Agricultural Science, Tohoku University, Sendai, Japan
[g]Department of System Development, Bunkoukeiki Co., Ltd, Tokyo, Japan
[h]Core Research for Environmental Science and Technology (CREST), Japan Science and Technology Agency, Tokyo, Japan
*Corresponding author: e-mail address: cmiyake@hawk.kobe-u.ac.jp

Contents

Advances in Botanical Research, Volume 96
ISSN 0065-2296
https://doi.org/10.1016/bs.abr.2020.08.001
© 2020 Elsevier Ltd
All rights reserved.

Abstract

The main production site of reactive oxygen species (ROS) in photosynthetic organisms is photosystem (PS) I of thylakoid membranes. Unless the suppression mechanism of ROS production functions, PSI easily suffers from oxidative damages by ROS attack. Miyake group has elucidated the production and suppression mechanisms of ROS in PSI. The reaction center chlorophyll, P700, in PSI functions in P700 photo-oxidation reduction cycle. The photoexcited P700, P700*, can donate electron to O_2 producing superoxide radical, ROS, with oxidized to $P700^+$. The accumulation of the $P700^+$ decreases the probability of the presence of P700* not to produce ROS. The present review describes the molecular mechanism to oxidize P700 and to accumulate $P700^+$ in PSI. Tight coupling between the light and the dark reactions in photosynthesis accumulates H^+ in the luminal side and electron in plastoquinone pool of thylakoid membranes on exposure to the environmental stress, which lowers the electron transport activity of Cyt b_6/f-complex and suppresses the electron flux to PSI with $P700^+$ accumulated. We discuss the molecular mechanisms to accumulate e^- and H^+ and its relationship with ATP synthase activity from the aspect of P700 oxidation in PSI.

1. Overview

P700 oxidation in photosystem I (PSI) is a robust mechanism that suppresses the production of reactive oxygen species (ROS) and is indispensable for the growth of photosynthetic organisms. In this review, we aimed to clarify the molecular mechanism underlying P700 oxidation based on the robust relationship between the light and dark reactions of photosynthesis. The light reaction is driven by the dark reaction in a tightly coupled manner. The dark and light reactions are suppressed simultaneously under abiotic stresses, increasing the probability of electron accumulation at the acceptor side of PSI to reduce O_2 level and produce ROS level. In response to the suppression of the dark reaction, P700 is oxidized to $P700^+$. This P700 oxidation is accompanied by both proton (H^+) accumulation in the luminal space and electron (e^-) accumulation in the plastoquinone (PQ) pool of thylakoid membranes, thus suppressing the activity of Cyt b_6/f PQ oxidation. This leads to the reduction of $P700^+$, which is therefore the rate-determining step (RdS) of the P700 oxidation/reduction cycle. Electron accumulation at the acceptor side of PSI is then greatly alleviated to suppress ROS production.

2. How photosynthetic organisms survive in the field under exposure to severe sunlight intensity

Generally, the rate of photosynthetic CO_2 fixation in photosynthetic organisms saturates at approximately 1/4 to 1/2 of sunlight intensity (Asada,

1999, 2000, 2006), indicating that photosynthesis occurs under excess light energy. In other words, photosynthetic CO_2 fixation functions even at low photosynthetic efficiency. In this situation, it is easy for e^- to accumulate in the photosynthetic electron transport system. Fig. 1 shows the general mechanism underlying photosynthesis, where light energy is absorbed by photosystem II (PSII) and PSI and then stored in the chemical compounds NADPH and ATP. Sejima, Takagi, Fukayama, Makino, and Miyake (2014) described the method by which e^- accumulate at the acceptor side of PSI by applying a light pulse (light intensity, 20,000 μmol photons m^{-2} s^{-1}; pulse duration, 400 ms; pulse frequency, 0.1 Hz) to the intact leaves of C3 plants. This repetitive short-pulse (rSP) illumination treatment induced ROS production at the acceptor side of PSI; the ROS attacked and inactivated PSI (Miyake, 2020; Sejima et al., 2014; Shimakawa & Miyake, 2018b). The electron carriers A_0, A_1, F_X, and F_A/F_B at the acceptor side of PSI are all molecular candidates for O_2 reduction to superoxide (O_2^-) (Charepanov, Milanovsky, Petrova, Tikhonov, & Semenov, 2017; Khorobrykh, Havurinne, Mattila, & Tyystjarvi, 2020; Takahashi & Asada, 1982, 1983). Compared with the lifespan of the reduced forms of A_0 and A_1, that of F_X and F_A/F_B are longer, and thus, the chances of F_X and F_A/F_B reacting with O_2 to produce O_2^- are higher.

As described above, photosynthetic organisms are exposed to ROS attacks. Miyake's group has elucidated the molecular mechanism underlying the suppression in ROS production in PSI of the P700 oxidation system (Miyake, 2020; Sejima et al., 2014; Shimakawa & Miyake, 2018b; Shimakawa, Shaku, & Miyake, 2016). PSI inactivation by rSP illumination treatment was suppressed under actinic light (AL) illumination (Sejima et al., 2014). The extent of this suppression increased with increasing AL intensity, indicating that the light-induced inactivation of PSI was prevented by light itself (Miyake, 2020). AL illumination suppressed PSI inactivation by the induction and accumulation of oxidized P700. Increasing the AL intensity resulted in $P700^+$ accumulation in the P700 photo-oxidation reduction cycle. That is, the prevalence of P700* in the P700 photo-oxidation reduction cycle decreased, thereby decreasing the probabilities of the electron carriers A_0, A_1, F_X, and F_A/F_B and Fd to reduce O_2 to O_2^- at the acceptor side of PSI (Miyake, 2020; Shimakawa & Miyake, 2018b). This is the molecular mechanism underlying the suppression of ROS production in PSI occurring in the field.

The oxidation of P700 has been evidenced as essential for the growth of photosynthetic organisms (Miyake, 2020; Shimakawa & Miyake, 2018b; Shimakawa et al., 2016, 2017). Numerous researchers have reported P700 oxidation in the PSI of photosynthetic organisms exposed to

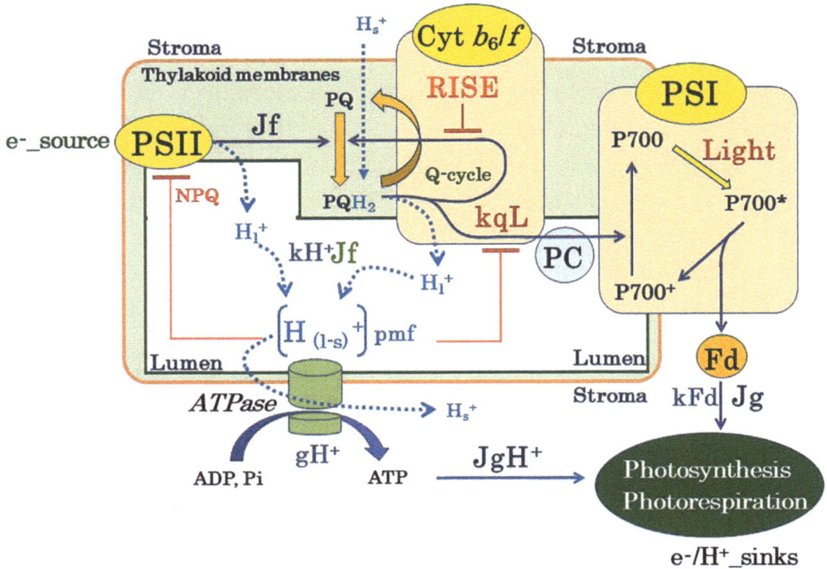

Fig. 1 Tight coupling of the light and dark reactions of photosynthesis in C3 angiosperms. Blue arrows: electron (e^-) flow; blue dotted arrows: proton (H^+) flow; Hs^+: stroma H^+; Hl^+: lumen H^+; $[H(l-s)^+]$: difference in H^+ concentration (ΔpH) between the stroma and the lumen, i.e., the proton motive force (pmf). Photosystem II (PSII): electron source. In the photo-oxidation reduction cycle, P700 catalyzes the electron flow from plastocyanin (PC) to ferredoxin (Fd) in photosystem I (PSI). P700 is photoexcited to P700*, which donates e^- to Fd through the electron carriers A_0, A_1, F_A/F_B, and F_X to produce $P700^+$. In turn, $P700^+$ is reduced by e^- in PSII through plastoquinone (PQ), cytochrome (Cyt) b_6/f complex, and PC. The Q cycle in the Cyt b_6/f complex transfers e^- from reduced PQ along the PC and oxidized PQ routes. Enhanced reduction of PQ suppresses the reduction of oxidized PQ in the Cyt b_6/f complex, which slows down the activity of the Q cycle and the electron flux from the Cyt b_6/f complex to $P700^+$ in PSI. H^+ conductance (gH^+) is an apparent rate constant that depends on the concentrations of ADP, Pi, ATP synthase, and the catalytic constant ($k_{(ATPase)}$). The non-photochemical quenching (NPQ) of chlorophyll fluorescence is activated by the acidification of the lumenal space of the thylakoid membrane, as reflected by the increase of pmf. NPQ suppresses the quantum yield of PSII, as indicated by the electron flux in the photosynthetic linear electron flow (Jf); pmf also suppresses the oxidation activity of reduced PQ by the Cyt b_6/f complex. Tight coupling of the light reaction (that involves generation of e^- and H^+) to the dark reaction (that involves the consumption of e^- and H^+ in both photosynthesis and photorespiration) yields a robust relationship, in which Jf is equal to the electron flux in electron sinks (photosynthesis and photorespiration) expressed as Jg. The pmf is the driving force to synthesize ATP by ATP synthase using ADP and Pi, and the pmf-dissipation rate (vH^+) is expressed as $gH^+ \times pmf$. Thus, $Jf = vFd = Jg$ (Kadota et al., 2019), where vFd is the oxidation rate of reduced Fd, $kJf \times Jf = gH^+ \times pmf$ (Avenson, Cruz, Kanazawa, & Kramer, 2005; Avenson, Cruz, & Kramer, 2004; Wada, Suzuki, & Miyake, 2020), and $vH^+ = gH^+ \times pmf = JgH^+$ (Wada et al., 2020). Jg and JgH^+ mutually determine each other. The pH of the lumenal space of the thylakoid membrane decreases during the photo-oxidation of water in PSII and the Q cycle in the Cyt b_6/f complex. This decrease in pH (i.e., increased pmf) acts as the driving force for ATP production via ATP synthase. *Reproduced from Miyake, C. (2020). Molecular mechanism of oxidation of P700 and suppression of ROS production in photosystem I in response to electron-sink limitations in C3 plants.* Antioxidants, *9(3).*

conditions that reduce photosynthesis efficiency, such as high AL intensity, low temperature, and drought (Golding & Johnson, 2003; Golding, Joliot, & Johnson, 2005; Harbinson & Foyer, 1991; Harbinson, Genty, & Baker, 1989, 1990; Harbinson, Genty, & Foyer, 1990; Harbinson & Hedley, 1993; Miyake, Horiguchi, et al., 2005; Miyake, Miyata, Shinzaki, & Tomizawa, 2005). Our research clarified the physiological importance of P700 oxidation in the PSI of photosynthetic organisms.

Plants, including those in *Arabidopsis*, barley, cucumber, and potato, may display PSI photoinhibition upon exposure to low temperature (Havaux & Davaud, 1994; Ivanov, Morgan, Gray, Velitchkova, & Huner, 1998; Teicher, Möller, & Scheller, 2000; Terashima, Funayama, & Sonoike, 1994; Tjus, Möller, & Scheller, 1998; Tjus, Möller, & Scheller, 1999; Tjus, Scheller, Andersson, & Möller, 2001). Interestingly, the photoinhibition of PSI in *Arabidopsis* spp. and barley occurs with the photoinhibition of PSII (Tjus et al., 1998; Zhang & Scheller, 2004). Furthermore, before the above facts were reported, the photoinactivation of PSI had already been identified by Japanese researchers (Inoue et al., 1989; Inoue, Sakurai, & Hiyama, 1986; Satoh, 1970a, 1970b, 1970c, 1982; Satoh & Fork, 1982), who suggested that PSII e^- are required for the photoinhibition of PSI. They observed PSI photoinhibition under severe reduced conditions using intact chloroplasts, revealing that although photosynthetic organisms can oxidize P700 to $P700^+$; thus, PSI may be photoinactivated in field conditions.

The photoinhibition of PSI lowers the photosynthetic activity of intact plant leaves (Sejima et al., 2014; Zivcak, Brestic, Kunderlikova, Sytar, & Allakhverdiev, 2015; Zivcak, Kalaji, Shao, Olsovska, & Brestic, 2014). Moreover, the recovery of PSI from photoinhibition requires several weeks (Sonoike, 2011; Zivcak et al., 2015). Thus, PSI photoinhibition in the field may lead to suppressed crop growth and reduced grain yield. To understand the molecular mechanism underlying ROS production and oxidative damages in PSI under field conditions, it is indispensable to clarify the oxidation mechanism of P700 in PSI. As described, the reaction-center chlorophyll P700 of PSI drives the photo-oxidation/reduction cycle (Fig. 1). That is, the production and suppression of ROS in PSI depend on the RdS of the cycle. If the RdS was the oxidation of $P700^*$, e^- would accumulate at the acceptor side of PSI to enhance ROS production. In contrast, if the RdS was the reduction of $P700^+$, the production of ROS would be suppressed. In the present review, the molecular mechanism underlying the selection of the RdS of the photo-oxidation/reduction cycle and that underlying the oxidation of P700 in response to the fluctuation in the photosynthesis activity of C3 plants (Miyake, 2020) have been described.

3. Molecular mechanism underlying P700 oxidation in PSI

In the P700 photo-oxidation/reduction cycle, both P700* oxidation and $P700^+$ reduction reactions limit the turnover rate of the cycle. For example, in the induction phase of photosynthesis, P700* oxidation limits the turnover rate of the cycle immediately after AL illumination initiation; then, with the initiation of photosynthesis, $P700^+$ reduction starts to limit the turnover rate of the cycle (Miyake, Suzuki, Yamamoto, Amako, & Makino, 2012). Furthermore, under high AL intensity or low CO_2 partial pressure, $P700^+$ reduction limits the P700 photo-oxidation/reduction cycle (Wada et al., 2020). That is, the transition of the RdS of the cycle depends on the ability of photosynthesis. Therefore, it crucial to understand the interaction between the light and dark reactions of photosynthesis (Fig. 1) for clarifying the regulatory mechanism of P700 oxidation in the P700 photo-oxidation/reduction cycle. Notably, the suppression of the $P700^+$ reduction reaction of the cycle would greatly contribute to the oxidation of P700. Thus, understanding the trigger to change the RdS of the cycle from P700* oxidation to $P700^+$ reduction and the magnitude of the limitation of $P700^+$ reduction are indispensable to clarify the physiological function of P700 oxidation in PSI. The following description is as presented by Miyake (2020).

3.1 Tight coupling between the light and dark reactions of photosynthesis

Much evidence has accumulated for a robust tight coupling between the light and dark reactions of photosynthesis in C3 plants (Fig. 1) (Miyake, 2020). For instance, Genty, Harbinson, Briantais, and Baker (1990) reported a positive linear relationship between the photosynthetic linear electron flow rate (Jf) in the light reaction and the photosynthetic electron consumption rate (Jg) based on NADPH consumption in the dark reaction. Consistently, a positive linear relationship was also found between Jf in the light reaction and Jg-based on both NADPH and reduced Fd consumptions—in the dark reaction, where photosynthesis and photorespiration occur (Eq. 1) (Driever & Baker, 2011; Genty et al., 1990; Ghashghaie & Cornic, 1994; Ruuska, Badger, Andrews, & von Caemmerer, 2000). These results indicate that Jf is the sole driver of both photosynthesis and photorespiration. In other words,

$$Jf = Jg \tag{1}$$

Recently, a positive linear relationship was noted between the reduced Fd oxidation rate (vFd) and Jf (Eq. 2) (Kadota et al., 2019). These results indicate that electron flow via Fd is driven by both photosynthesis and photorespiration. In other words,

$$vFd = kFd \times Fd^{-} \tag{2}$$

where kFd is the apparent constant rate of reduced Fd oxidation and Fd^{-} is the amount of reduced Fd.

A positive linear relationship was reported between Jf and the proton motive force (pmf) consumption rate (Eq. 3).

$$vH^{+} = gH^{+} \times pmf \tag{3}$$

where gH^{+} is proton conductance (Avenson et al., 2004, 2005; Kanazawa & Kramer, 2002; Kanazawa et al., 2017). These results indicate that Jf supports the consumption of pmf (Eq. 4), which can be indicated using the following formula:

$$kH^{+} \times Jf = vH^{+} \tag{4}$$

where the rate constant kH^{+} reflects H^{+} accumulation efficiency in the lumenal space of thylakoid membranes, which depends on H_2O oxidation in PSII and on reduced PQ oxidation in the Q cycle of the Cyt b_6/f complex.

Furthermore, a positive linear relationship was noted between vH^{+} and the H^{+} usage rate (JgH^{+}) owing to the regeneration of ATP required for both photosynthesis and photorespiration (Eq. 5) (Sejima et al., 2016). These results indicate that vH^{+} also reflects the pmf consumption rate during photosynthesis and photorespiration:

$$vH^{+} = JgH^{+} \tag{5}$$

A positive linear relationship was also found between JgH^{+} and Jg (=Jf) (Eqs. 6 and 7) (Sejima et al., 2016; Tikhonov, 2014), indicating that JgH^{+} is determined by Jg based on the consumption of both reduced Fd and NADPH:

$$JgH^{+} = kJg \times Jg \tag{6}$$

$$= kH^{+} \times Jf \tag{7}$$

Eqs. (6) and (7) indicate the tight coupling between the light and dark reactions of photosynthesis and photorespiration. Constants kjg and kH^{+}

are the coefficients that link electron flow with H^+ flow during photosynthesis and photorespiration. These findings indicate that the light reaction is driven by the dark reaction and vice versa and provide a sound explanation for the observed decrease in Jf under environmental stresses, when the photosynthetic rate decreases because of dark reaction rate limitation (i.e., when JgH^+ and Jg decrease).

Tight coupling between the light and dark reactions, as shown in Eqs. (6) and (7), may result in P700* oxidation being the RdS of the P700 photo-oxidation/reduction cycle. This implies that the production of ROS in PSI would be enhanced because the photo-excitation efficiency of converting ground state P700 to excited P700 (i.e., P700*) is constant under constant light intensity. For example, Jg and Jf, in which electron flux are similar, both decrease when photosynthetic activity decreases under abiotic stress. However, the photo-excitation efficiency of P700 to P700* does not decrease under constant light intensity. Then, P700* and e^- accumulate, increasing the probability of ROS production in PSI. Counteracting this mechanism should increase the safe use of light energy to fix CO_2 in photosynthetic organisms (Shimakawa & Miyake, 2018a, 2018b).

However, it is well known that e^- do not accumulate at the acceptor side of PSI and that P700 is oxidized to a greater extent to $P700^+$ under stress conditions than that under no stress conditions (Harbinson & Foyer, 1991; Harbinson et al., 1989; Miyake, Miyata, et al., 2005). Based on the model for the photo-oxidation reduction cycle of P700, the reduction of $P700^+$ must be the RdS to accumulate oxidized P700 in PSI under stress conditions. Thus, the electron flux to oxidized P700 should be suppressed such that P700 is oxidized. This indicates that photosynthetic organisms have a regulatory system to oxidize P700 in PSI.

3.2 Protons and electrons accumulate in both the lumenal space of thylakoid membranes and PQ pool in response to dark reaction suppression

Generally, it is proposed that the electron flux from PQ to P700 in PSI is regulated by the oxidation activity of the reduced PQ (PQH_2) of the Cyt b_6/f complex (Anderson, 1992; Baker, Harbinson, & Kramer, 2007; Golding et al., 2005; Hope, 2000; Hope, Matthews, & Valente, 1994; Hope, Valente, & Matthews, 1994; Rantala et al., 2020; Rott et al., 2011; Shaku, Shimakawa, Hashiguchi, & Miyake, 2016; Shimakawa, Shaku, & Miyake, 2018; Tikhonov, 2014, 2018; Zhang et al., 2019). The PQH_2 oxidation activity of the Cyt b_6/f complex decreases as pH decreases, with a H^+ gradient (ΔpH) formation across thylakoid membranes. Therefore, the accumulation of H^+ in the lumenal space of thylakoid membranes suppresses the electron flow from PQH_2 to the oxidized P700 in PSI, which is a photosynthesis control mechanism (Baker et al., 2007; Rantala et al., 2020; Tikhonov, 2018). Furthermore, the over-reduction of PQ inhibits the Q cycle in the Cyt b_6/f complex and contributes to the oxidation of P700 in PSI, which is known as the reduction-induced suppression of electron flow (RISE) (Shaku et al., 2016; Shimakawa et al., 2018). We propose that RISE is also involved in photosynthesis control (Miyake, 2020; Shaku et al., 2016; Shimakawa et al., 2018; Wada et al., 2020).

To clarify the molecular mechanism underlying P700 oxidation in PSI upon exposure to the limitation of photosynthesis, we must understand how H^+ and e^- accumulate in both the lumenal space and PQ pool of the photosynthetic electron transport system in response to both photosynthesis and photorespiration.

The mechanism underlying the induction of ΔpH formation across the thylakoid membranes by the accumulation of H^+ in the lumenal space of thylakoid membranes is described below. ΔpH formation is observed with an increase in pmf (Avenson et al., 2005, 2004; Johnson & Ruban, 2014; Kramer, Avenson, & Edwards, 2004). In general, ΔpH is the main factor influencing pmf at low photosynthetic efficiency, e.g., high light and low CO_2 conditions (Wada et al., 2020). The rate of change in pmf [d(pmf)/dt] is determined by the difference between the pmf production rate and the pmf consumption rate. The pmf production rate depends on the Jf (Eq. 4), and the pmf consumption rate depends on the ATP consumption rate, which is driven by both photosynthesis and photorespiration (Eqs. 5–7). Thus, the velocity of change in pmf is defined as follows (Eqs. 8 and 9):

$$\mathrm{d(pmf)/dt} = \mathrm{kH^+} \times \mathrm{Jf} - \mathrm{vH^+} \tag{8}$$

$$= \mathrm{kH^+} \times \mathrm{Jf} - \mathrm{gH^+} \times \mathrm{pmf} \tag{9}$$

At the steady state, $\mathrm{d(pmf)/dt} = 0$, and the pmf generation rate, $\mathrm{kH^+} \times \mathrm{Jf}$, is equal to the pmf consumption rate, $\mathrm{vH^+}$. The pmf generation rate depends on H_2O oxidation in PSII and on the Q cycle rotation of the Cyt b_6/f complex. The $\mathrm{vH^+}$ can be replaced with $\mathrm{JgH^+}$ as follows (Eq. 10):

$$\mathrm{d(pmf)/dt} = \mathrm{kH^+} \times \mathrm{Jf} - \mathrm{JgH^+} \tag{10}$$

In a steady state, where d(pmf)/dt = 0, and based on Eqs. (4) and (5), Eq. (11) can be devised:

$$kH^+ \times Jf = JgH^+ \quad (11)$$

Eqs. (9) and (10) show that Jf in the light reaction is associated with photosynthetic and photorespiration activities in the dark reaction through pmf generation and consumption. The dependence of pmf on Jf, gH+, and JgH is as follows (Eqs. 12 and 13):

$$pmf = (kH^+ \times Jf)/gH^+ \quad (12)$$
$$= JgH^+/gH^+ \quad (13)$$

Based on these model equations, the molecular mechanism underlying H^+ accumulation in the lumenal space of thylakoid membranes in response to photosynthesis and photorespiration can be further elucidated. For example, if the extent of the decrease in the dark reaction rate, shown as decreases in both Jg and JgH^+, is smaller than that in gH^+, pmf would increase (Wada et al., 2020). This is how H^+ accumulates in the lumenal space of thylakoid membranes at a lower efficiency of photosynthesis.

The accumulation of e^- in the photosynthetic electron transport system can be reflected by the redox states of both PQ and Fd. The reduced state of PQ is evaluated as 1 − qL (Kramer, Johnson, Kiirats, & Edwards, 2004; Miyake, Amako, Shiraishi, & Sugimoto, 2009), in which qL is the oxidation state of PQ in thylakoid membranes. A higher value of 1 − qL indicates higher accumulation of e^- in the PQ pool. The oxidation rate of reduced PQ can be expressed as kqL × (1 − qL), and the oxidation rate of Fd^- is expressed as kFd × Fd^-. Both kqL and kFd are apparent rate constants. In the light reaction, the electron fluxes of the oxidations of both reduced PQ and reduced Fd are equal to Jf and Jg, as shown in Eqs. (14) and (15) (Kadota et al., 2019; Shimakawa & Miyake, 2018a, 2018b).

$$Jf = kqL \times (1 - qL) \quad (14)$$
$$= kFd \times Fd^- \quad (15)$$
$$= Jg \quad (1)$$

The redox states of both PQ and Fd can then be determined by Jg, kqL, and kFd, as shown in Eqs. (16) and (17), respectively:

$$1 - qL = Jg/kqL \quad (16)$$
$$Fd^- = Jg/kFd \quad (17)$$

kqL has a positive value that is expected to decrease with H^+ accumulation in the lumenal space of thylakoid membranes, i.e., the photosynthetic control mechanism (Foyer, Furbank, Harbinson, & Horton, 1990; Schöttler & Toth, 2014; Tikhonov, 2014), and with RISE (Shaku et al., 2016; Shimakawa et al., 2018). Generally, a decrease in Jg increases the value of $1-qL$ (Miyake et al., 2009; Miyake, Miyata, et al., 2005). Thus, in response to the decrease in electron sink activities (photosynthesis and photorespiration), the extent of the decrease in kqL is larger than that in Jg. From Eq. (16), we can clearly understand why $1-qL$ increases, which explains how e^- accumulate in the PQ pool of thylakoid membranes.

However, a decrease in Jg does not induce a reduction of Fd (Kadota et al., 2019). For example, lowering the intercellular partial pressure of CO_2 decreases photosynthesis rate and Jg, but the reduced level of Fd may not change or may decrease, and the extent of the decrease in kFd is almost the same as that in Jg. kFd depends on both the activity of $Fd-NADP^+$ oxidoreductase (FNR) and the efficiency of $NADP^+$ regeneration. The decrease in the photosynthesis rate lowers the $NADP^+$ regeneration rate, which decreases kFd to an extent that is similar to that of Jg.

3.3 ATP synthase activity is associated with pmf in thylakoid membranes

The robust relationship between the light and dark reactions allows determining pmf as the ratio of $(kH^+ \times Jf)$ to gH^+ (Eq. 12). A miss-sense mutation on the g-subunit of ATP synthase in *Arabidopsis* spp. (*hope*2 mutant) induced a decrease in the quantity of ATP synthase and increased H^+ conductivity (gH^+) for the synthesis of ATP (Takagi et al., 2017). Interestingly, this mutation affected the phenotype both qualitatively and quantitatively. Generally, a decrease in ATP synthase leads to the accumulation of H^+ in the lumenal space of thylakoid membranes, that is, increases pmf (Kanazawa et al., 2017; Kramer, Cruz, & Kanazawa, 2003; Takizawa, Cruz, Kanazawa, & Kramer, 2007; Takizawa, Kanazawa, & Kramer, 2008). Enhancing gH^+ decreased pmf, as revealed by ΔpH formation across thylakoid membranes under high light intensity (Takagi et al., 2017), as well as suppressed P700 oxidation in PSI. This response of P700 in PSI to pmf is illustrated by the P700 oxidation model shown in Fig. 2. Compared with that of wild type individual, the PSI of *hope2* mutant was photoinhibited by the rSP illumination treatment under the steady-state AL condition. Moreover, *hope2* mutant could not maintain the P700 oxidized state, and

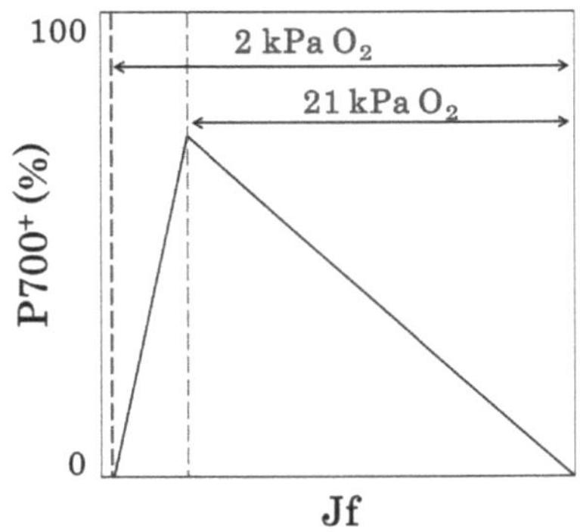

Fig. 2 Oxidized P700 ($P700^+$) induction depends on the electron flux in the photosynthetic linear electron flow in intact leaves. Generally, P700 is oxidized with decreasing electron flux in the photosynthetic linear electron flow (Jf) by lowering CO_2 partial pressure. At 21 kPa O_2, when both photorespiration and photosynthetic CO_2 fixation are occurring, $P700^+$ level is at the maximum value at the CO_2 compensation point, and Jf shows the minimum value (blue dotted line). The horizontal arrow at 21 kPa O_2 shows the range over which Jf is driven by both photosynthetic CO_2 fixation and photorespiration. At 2 kPa O_2, when no detectable photorespiration occurs and photosynthetic CO_2 fixation functions normally, $P700^+$ starts to decrease at the CO_2 compensation point, after reaching its maximum value, and Jf reaches its minimum value (red dotted line). The horizontal arrow at 2 kPa O_2 shows the range over which Jf is driven by photosynthetic CO_2 fixation. *Reproduced from Miyake, C. (2020). Molecular mechanism of oxidation of P700 and suppression of ROS production in photosystem I in response to electron-sink limitations in C3 plants.* Antioxidants, *9(3).*

ROS production was enhanced. These findings indicate that the quantity of ATP synthase that is required for best photosynthesis performance and for ROS production suppression in the PSI of thylakoid membranes can be determined.

3.4 How accumulated H^+ in the lumenal space and accumulated e^- in the PQ pool contribute to P700 oxidation in PSI

The decrease in Jf (=Jg) has a negative linear relationship with $P700^+$ increase (Fig. 2) (Shimakawa & Miyake, 2018b; Shimakawa et al., 2019). That is, the decrease in photosynthesis efficiency induces the oxidation of P700 in PSI. As described above, the suppression of photosynthesis enhances both pmf formation and PQ reduction by increasing $1 - qL$, which would lead $P700^+$ reduction to be the RdS in the photo-oxidation reduction cycle of P700 in PSI. The decrease in Jg reduces Jf in the tight coupling of the light reaction with the dark reaction. However, the decreases in both gH^+ and kqL are larger than that in Jg; thus, both pmf and 1 − qL increase.

In addition, the magnitude of Jf can be decreased by both the formation of pmf and the reduction of PQ, such that the $P700^+$ reduction reaction becomes the RdS in the photo-oxidation reduction cycle of P700 in PSI. Thus, Jf must be lower than the theoretical value of Jg obtained by the suppression of photosynthesis. For example, if Jf decreases to the same extent as the decrease in Jg, in response to the decrease in the intercellular partial pressure of CO_2, the photoexcited P700 (P700*) would accumulate because the efficiency of the photo-excitation of P700 to P700* does not change under constant AL intensity. However, this is contradicted by the experimental results as suppressing photosynthesis induces the oxidation of P700.

RISE induced by the increased reduction of PQ may suppress the electron flow in the light reaction and oxidize P700 in the PSI of cyanobacteria (Shaku et al., 2016; Shimakawa et al., 2018). In these organisms, flavodiiron protein (FLV) catalyzes the reduction of O_2 to H_2O using NADPH/Fd (Helman et al., 2003; Shimakawa et al., 2015). The electron flux in the FLV reaction is so high that it alternates the electron flux in photosynthesis under CO_2-deficient conditions (Hayashi et al., 2014). FLV-deficient cyanobacteria show suppressed photosynthesis, with the accumulation of e^- in the PQ pool (Shaku et al., 2016). Furthermore, saturated-pulse (SP) illumination reduces PQ almost transiently with the suppression of O_2 reactions under steady-state conditions (Shaku et al., 2016). Thus, FLV can oxidize PQ, even at high photosynthetic activity, and it keeps PQ oxidized to prevent RISE at maximum photosynthetic activity if enough CO_2 is supplied to cyanobacteria.

We also found that the induction of both pmf and RISE oxidizes P700 in the PSI of the leaves of wheat, a C3 plant (Shimakawa & Miyake, 2018b; Shimakawa et al., 2019). A decrease in the intercellular partial pressure of CO_2 reduces the photosynthetic rate and increases the photorespiration rate. However, the photorespiration rate cannot reach the value theoretically derived using Rubisco kinetics at low CO_2 pressure (Miyake & Yokota, 2000). The induction of both pmf and RISE reduces the activity of the light reaction, which leads to the suppression of the dark reaction (i.e., of photorespiration rate), even though the potential activity of the dark reaction exceeds the suppressed activity of the light reaction. Thus, the oxidation of P700 requires the suppression of the electron flux in the light reaction more than it requires the same in the dark reaction; moreover, the suppressed activity in the dark reaction is lower than the theoretical potential activity. As a result, the reduction of $P700^+$ becomes the RdS in the P700 photo-oxidation reduction cycle. This is the essence of the molecular mechanism underlying P700 oxidation.

However, the induction of both pmf and RISE does not always oxidize P700 in response to the limitation of photosynthesis. Unless the electron flux of the light reaction reaches a value below the potential electron flux of the dark reaction, as theoretically expected from the limited activity of photosynthesis, the PSI acceptor-side limitation is so strong that P700 is not oxidized (Fig. 2). When photosynthesis is severely limited, with the suppression of both photosynthesis and photorespiration, P700 cannot be oxidized because the oxidation of P700* becomes the RdS in the P700 photo-oxidation reduction cycle.

The molecular mechanism underlying P700 oxidation in the PSI of chloroplasts is summarized in Fig. 3. Both PSI and PSII absorb photon energy to excite the reaction centers P680 (PSII) and P700 (PSI). Similar to P700 in PSI, P680 undergoes a photo-oxidation reduction cycle. The ground state of P680 is photoexcited to P680*, and this donates e^- to PQ through the electron carriers in PSII. The e^- then start to flow from the reduced PQ to PSI through the Cyt b_6/f complex and plastocyanin (PC). During the electron flow, H^+ from both H_2O oxidation in PSII and PQH_2 (plastoquinol) oxidation of the Cyt b_6/f complex accumulates in the lumenal space of thylakoid membranes. This accumulated H^+ is the pmf that drives ATP synthase to produce ATP in the stroma. Simultaneously, NADPH is produced by FNR at the acceptor side of PSI, and ATP, NADPH, and Fd^- are used for the regeneration of ribulose 1,5-bisphosphate (RuBP) in both photosynthesis and photorespiration. The regeneration rates of $NADP^+$, oxidized Fd, and ADP are determined by the activities of both photosynthesis and photorespiration.

The suppression of photosynthesis from (I) large electron/H^+ sinks to (II) small electron/H^+ sinks decreases both Jg and JgH^+. Simultaneously, the regeneration rates of ADP, Pi, and $NADP^+$ decrease. That is, gH^+ decreases to increase the pmf, and kqL decreases to increase $1-qL$, which induces RISE, as shown in Eqs. (12), (13), and (16). Both pmf and RISE lower kqL, as reflected by the suppressed activity of the reduced PQ oxidation by the Cyt b_6/f complex. The NPQ induced by the pmf also decreases the quantum yield of PSII. In situation (II), the suppressed electron flow from PSII to the oxidized P700 enhances the reduction of $P700^+$ in the photo-oxidation reduction cycle, which becomes the RdS, leading to the oxidation of P700.

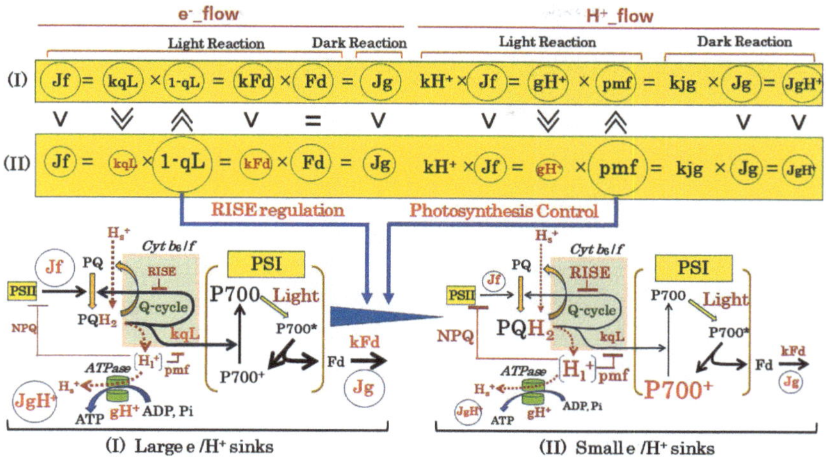

Fig. 3 P700 oxidation in PSI by sensing both electron accumulation in the PQ pool and H^+ accumulation in the luminal space of thylakoid membranes. Electron accumulation in the PQ pool induces RISE regulation, which suppresses the Q cycle in the Cyt b_6/f complex. H^+ accumulation in the luminal space of thylakoid membranes induces photosynthesis control, which suppresses reduced PQ (PQH_2) oxidation by the Cyt b_6/f complex. Both RISE regulation and photosynthesis control result in $P700^+$ reduction being the RdS of the P700 photo-oxidation reduction cycle, leading to the accumulation of $P700^+$. Black arrows indicate electron flow, and red dotted arrows indicate H^+ flow. The width of both the arrows reflects the magnitude of the apparent rate constant in each elementary reaction. In situation (I), both electron and H^+ flows are driven by both photosynthesis and photorespiration, which have large electron and H^+ sinks. That is, photosynthetic linear electron flow activity is high. In situation (II), the photosynthetic linear electron flow activity is lower, with lower electron and H^+ sinks than those in situation (I). Both electron and H^+ fluxes (Jg and JgH^+); the rate constants kqL, kFd, and gH^+; and the reduced state of PQ, $1-qL$, were compared between situations (I) and (II). The magnitude of the circles reflects the magnitude of values. The magnitude of circles (Jf, kqL, kFd, and Jg in electron flow and gH^+ in H^+ flow) is lower in situation (II) than that in situation (I), as indicated by the ∨ symbol. The decreases in both kqL and gH^+ are larger than the decrease in Jf (=Jg), as indicated by the double ∨ symbol, and the magnitudes of $1-qL$ and pmf in situation (II) are larger than those in situation (I), as indicated by the double ∧ symbol. The enhanced $1-qL$ and pmf trigger RISE regulation and photosynthesis control, which enhances the RdS of the $P700^+$ reduction reaction to further oxidize P700 in response to small electron/H^+ sinks. The redox level of Fd does not change in response to smaller electron/H^+ sinks, as indicated by the =symbol. This is because the extent of kFd decrease is close to that of Jf (=Jg) decrease in the transition of situation (I) to situation (II).

4. P700 oxidation contributes to the maintenance of Fd redox state

Generally, the reduced state of the electron acceptor side of PSI and of both Fd and NADPH has been considered to increase when photosynthesis efficiency decreases under stress conditions (Asada, 1999, 2000, 2006). However, Kadota et al. (2019) found that reducing photosynthetic activity by decreasing the partial pressure of CO_2 did not increase the reduced level of Fd in wheat leaves. Simultaneously, the oxidation level of P700 increased (Kadota et al., 2019). Furthermore, during the induction of photosynthesis, approximately 90% of Fd was reduced just after AL treatment was initiated; in addition, the reduced level of Fd decreased with the activation of photosynthesis, in which P700 oxidation occurred (Kadota et al., 2019). These facts show that the oxidation of P700 is negatively related to the reduction of Fd. As described in the model of P700 oxidation (Fig. 3), both pmf and RISE suppress the $P700^+$ reduction reaction and maintain the reduced level of Fd at a constant value, rather than the redox level of Fd. The suppression of Fd reduction contributes to lowering the chance of reduced Fd to donate an electron to O_2, thereby producing ROS around the PSI of thylakoid membranes (Asada, 2000; Furbank, Badger, & Osmond, 1982, 1983). The homeostasis of Fd redox level also shows that the ratio of NADPH to $NADP^+$ ($NADPH/NADP^+$) is kept constant because the reduction and oxidation reactions are determined by the dark reaction in photosynthesis (Eq. 3, Fig. 1) (Kadota et al., 2019). In addition, Fd^- is an electron donor for the reduction reaction of $NADP^+$ catalyzed by FNR. If NADPH is accumulated, almost all Fd would be reduced. Thus, in response to the photosynthesis efficiency, the oxidation of P700 would suppress the over-reduction of both Fd and $NADP^+$ and would maintain $NADPH/NADP^+$ at the constant value.

5. P700 oxidation ensures ATP-supply limiting photosynthesis

Those interested in the stoichiometry of the usage of both ATP and NADPH, that is, H^+ and e^-, in both photosynthesis and photorespiration have stated that alternative electron flow, especially Fd-dependent cyclic electron flow, supplies ATP because photosynthetic linear electron flow alone cannot fulfill the amount of ATP necessary to drive both reactions

(Heber, 2002; Heber & Walker, 1992; Munekage et al., 2002; Shikanai, 2016; Wada et al., 2018; Yamamoto & Shikanai, 2019; Yamamoto, Takahashi, Badger, & Shikanai, 2016; Yamori, Makino, & Shikanai, 2016; Yamori, Sakata, Suzuki, Shikanai, & Makino, 2011; Yamori & Shikanai, 2016; Yamori, Shikanai, & Makino, 2015). ATP is synthesized by ATP synthase in thylakoid membranes, which uses pmf. Photosynthesis linear electron flow accumulates H^+ in the lumenal space of thylakoid membranes at the H^+ flow/electron flow ratio of 3. The ratio of H^+ usage to electron usage in both photosynthesis and respiration is $(4.67/2)*(3+3.5\ \phi)/(2+2\ \phi)$, where ϕ is the ratio of RuBP oxygenation rate to RuBP carboxylation rate in the Rubisco reaction (Miyake, Miyata, et al., 2005; Seelert et al., 2000; Wada et al., 2020). In C3 photosynthesis, the ratio of JgH^+ to ATP consumption rate (vATP) is 4.67 because ATP synthase uses 4.67 H^+ ions for the synthesis of one molecule of ATP (Seelert et al., 2000). Thus, photosynthetic linear electron flow alone cannot provide the necessary H^+ for the synthesis of ATP to drive both photosynthesis and photorespiration at any value of ϕ. Therefore, Heber and Walker (1992) argued that extra ATP is required to match the ratio of H^+ and electron supplies in the light reaction to H^+ and electron consumptions in the dark reaction, and proposed that Fd-cyclic electron flow (CEF) would provide the extra ATP for driving both photosynthesis and photorespiration. However, the activity of Fd-CEF is negligible, compared to that of electron flux in the photosynthetic linear electron flow in vivo (Kadota et al., 2019). That is, only the photosynthetic linear electron flow can drive both photosynthesis and photorespiration (Eqs. 6 and 7).

To understand how both photosynthesis and photorespiration proceed in photosynthetic organisms it should be discussed why the production ratio of H^+ to e^- in the light reaction should correspond to the consumption ratio of H^+ to e^- in the dark reaction. The dark reaction can indeed proceed at ATP-supply limiting photosynthesis. If both photosynthesis and photorespiration function at ATP-supply limiting photosynthesis, both NADPH and Fd^- are produced in excess for the dark reaction, which could cause a dangerous situation, as discussed in the previous sections. However, photosynthetic organisms have a P700 oxidation system to suppress the surplus accumulation of e^- at the acceptor side of PSI. This is also a physiological function of P700 oxidation in PSI. Furthermore, even though P700 was not oxidized at sub-saturated photosynthesis against light intensity, pmf accumulated (Kanazawa et al., 2017; Kanazawa, Blanchard, Szabo, Ralph, & Kramer, 2014; Kanazawa & Kramer, 2002). This is because the potential rate formation of pmf in the light reaction exceeds the potential consumption rate in the dark reaction, that is, the dark reaction limits the usage of ATP under lower light intensity. Furthermore, as described above, at high light intensity and low CO_2, pmf increases by H^+ accumulation in the lumenal space of thylakoid membranes. We should recognize how pmf is formed in photosynthesis. If the supply of ATP to the dark reaction is limited, the formation of pmf is not observed under low and saturated light intensity.

6. CEF around PSI: An artifact

Conventionally, the electron flux in PSI has been evaluated by the method of SP illumination-dependent apparent quantum yield of PSI (Baker et al., 2007) (Fig. 4). In principle, P700 is excited by SP-illumination (LED pulse, 200–800 ms, 10,000–20,000 μmol photons $m^{-2} s^{-1}$) to the excited state (P700*), and then P700* is oxidized to $P700^+$ by the primary electron acceptor (A0) localized at the acceptor side of PSI. The time required for the transition from P700 to $P700^+$ ranges from ns to μs. The increased LED pulse illuminated the leaves within μs, which is shorter than the reduction rate of $P700^+$ by the reduced PC. Under AL illumination, where some P700 is oxidized to $P700^+$, P700 is excited by SP-illumination without the reduction of the already existing $P700^+$. Then, the apparent quantum yield of PSI is expected to be correctly determined, as described in Fig. 4A. However, the production of $P700^+$ by SP-illumination requires about 3–10 ms (Fig. 4B). Furthermore, the magnitude of SP-illumination-dependent oxidation of P700 [$Y(I) = (Pm' - P700^+)/Pm'$] depends on the redox state of PC before SP-illumination (Fig. 4B). In the dark, where both PC and P700 are reduced, SP-illumination produces Pm′(dark1). Far-Red (FR) light illumination oxidizes both PC and P700 and then SP-illumination produces Pm′(FR). Thus, Pm′(FR) is larger than Pm′(dark1) because the turnover rate of P700 in the photo-oxidation reduction cycle of P700 is limited by the reduction rate of $P700^+$ by the reduced PC. Then, Pm′ under FR-illumination is enhanced, compared to Pm′ under dark condition. After FR-illumination is turned off, $P700^+$ is rapidly reduced, compared to the reduction of the oxidized PC. SP-illumination produces Pm′(PCox) in mush higher levels than Pm′(dark1) but the same amount of Pm′(dark2) and Pm′(dark1). These facts clearly show that SP-illumination produces Pm′ on a magnitude that depends on the redox state of PC. If PC is in the oxidized state, the turnover rate of P700 in the photo-oxidation reduction cycle of P700 is limited by the reduction rate of $P700^+$ by the reduced PC. Then, during SP-illumination, the extra P700 is photo-oxidized

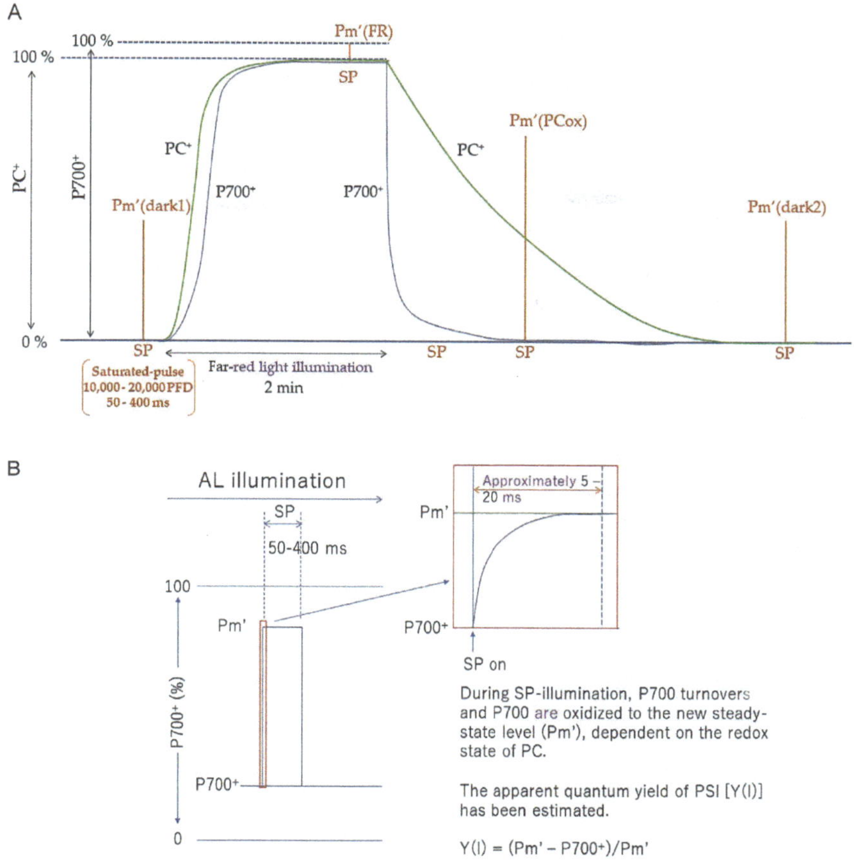

Fig. 4 Overestimation of the apparent quantum yield of PSI [Y(I)] under plastocyanin (PC) oxidation. (A) Saturated-pulse (SP) light (10,000–20,000 μmol photons $m^{-2}\,s^{-1}$, 50–400 ms) is illuminated to the leaves of plants in the dark. Then, P700 is oxidized to the level of Pm′(dark1). Far-red (FR) light is illuminated to plant leaves for 2 min, and both P700 and PC are oxidized. Then, SP-light is illuminated to obtain Pm′(FR). Pm′(FR) reflects the total amount of P700 in PSI. After FR light is turned off and P700 is fully reduced, and while PC is still oxidized, SP-light is illuminated to obtain Pm′ (PCox). After PC is fully reduced in the dark, SP-light is illuminated to get Pm′(dark2). The magnitudes of Pm′ are ordered as follows: Pm′(FR) > Pm′(PCox) > Pm′(dark1) = Pm′ (dark2). (B) Under actinic light (AL) illumination, P700 is oxidized to $P700^+$. Then, SP-light is illuminated to obtain Pm′. P700 is excited to $P700^+$ (Pm′) during SP-illumination, and its level depends on the redox state of PC under AL illumination as shown in (A). The turnover rate of P700 in the P700 photo-oxidation reduction cycle is limited by the reduction rate of $P700^+$, as shown in the accumulation of $P700^+$ under AL illumination. Under such conditions, $P700^+$ gradually increases to the new steady-state level during SP-illumination leading to excess oxidation of P700. Then, Y(I) becomes larger than the apparent quantum yield of PSII [Y(II)]. This is the cause for the overestimation of P700, which would be the ground state of P700 in PSI.

producing excess Pm′, as shown in Fig. 4B. This issue points out that extra attention is required in the evaluation of Y(I). Under the steady state of AL illumination, if P700 is oxidized to $P700^+$, Pm′ is over-estimated; excess Y(I) is then obtained and Y(I) becomes larger than the apparent quantum yield of PSII, Y(II). These cases have been previously reported (Miyake, Shinzaki, Miyata, & Tomizawa, 2004; Miyake, Horiguchi, et al., 2005; Miyake, Miyata, et al., 2005; Yamamoto et al., 2006) and used as evidence to support the functioning of CEF around PSI under low photosynthetic efficiency (Shikanai, 2016; Yamori & Shikanai, 2016). In fact, conflicting data and contradicting arguments were reported by Shikanai's group. PGR5-deficient mutants of Arabidopsis and rice showed the suppressed oxidation of P700 and the lower Y(I) value than the Y(II) value (Yamamoto et al., 2016; Wada et al., 2018). However, FLV-overexpressed PGR5-deficient mutants showed the oxidation of P700 again and Y(I) values were recovered and larger than Y(II) values. That is, if CEF existed, its activity CEF would be recovered without PGR5. Thus, CEF is an artifact in the estimation of Y(I). At present, we do not have precise methods to evaluate PSI activity.

7. Conclusions and perspectives

Photosynthetic organisms convert photon energy to both electron and proton energy flows for the reduction of CO_2 to sugar in photosynthesis. The energy of both electrons and protons is stored as NADPH and ATP in the light reaction, and these compounds drive the Calvin-Benson cycle to fix CO_2 in the dark reaction. The light and dark reactions function in a tightly coupled manner. Therefore, the limitation of the dark reaction under environmental stress could easily suppress the light reaction thereby increasing the reduction of O_2 to O_2^-, a ROS, in PSI. However, photosynthetic organisms have an excellent mechanism to suppress ROS production. The stagnation of energy flows is correlated with the accumulation of both electrons and protons in the Cyt b_6/f complex in thylakoid membranes: electrons, RISE; protons, pmf. Both RISE and pmf suppress the activity of plastoquinol oxidation by the Cyt b_6/f-complex and cause P700 oxidation in PSI by limiting $P700^+$ reduction. The oxidation of P700 inhibits the accumulation of electrons at the acceptor side of PSI, which suppresses ROS production. The activity of ATP synthase is linked to the plastoquinol oxidation activity of the Cyt b_6/f complex through both RISE and pmf; this central management system and the P700 oxidation system control the tight coupling of the light and dark reactions of photosynthesis and ROS production in PSI. The amount of PSI is larger than that required for driving photosynthesis (Shimakawa & Miyake, 2019). Probably, the excess amount of PSI is used to account for PSI damaged by ROS attack. The mechanism underlying the regulation of the amount of ATP synthase remains to be clarified, although it might be related to the amount of PSI for the suppression of ROS as ATP synthase can regulate pmf.

Acknowledgment

The authors are thankful to Editage for English language editing.

Funding

This work was supported by JST CREST (grant number JPMJCR15O3), Japan.

Conflicts of interest

The funders had no role in the design of the study; in the collection, analyses, or interpretation of data; in the writing of the manuscript; and in the decision to publish the results.

References

Anderson, J. M. (1992). Cytochrome b_6/f complex: Dynamic molecular organization, function and acclimation. *Photosynthesis Research, 34*(3), 341–357.

Asada, K. (1999). The water-water cycle in chloroplasts: Scavenging of active oxygens and dissipation of excess photons. *Annual Review of Plant Physiology and Plant Molecular Biology, 50*, 601–639.

Asada, K. (2000). The water-water cycle as alternative photon and electron sinks. *Philosophical Transactions of the Royal Society of London. Series B, Biological Sciences, 355*(1402), 1419–1431.

Asada, K. (2006). Production and scavenging of reactive oxygen species in chloroplasts and their functions. *Plant Physiology, 141*(2), 391–396.

Avenson, T. J., Cruz, J. A., Kanazawa, A., & Kramer, D. M. (2005). Regulating the proton budget of higher plant photosynthesis. *Proceedings of the National Academy of Sciences of the United States of America, 102*(27), 9709–9713.

Avenson, T. J., Cruz, J. A., & Kramer, D. M. (2004). Modulation of energy-dependent quenching of excitons in antennae of higher plants. *Proceedings of the National Academy of Sciences of the United States of America, 101*(15), 5530–5535.

Baker, N. R., Harbinson, J., & Kramer, D. M. (2007). Determining the limitations and regulation of photosynthetic energy transduction in leaves. *Plant, Cell & Environment, 30*(9), 1107–1125.

Charepanov, D. A., Milanovsky, G. E., Petrova, A. A., Tikhonov, A. N., & Semenov, A. Y. (2017). Electron transfer through the acceptor side of photosystem I: Interaction with exogenous acceptors and molecular oxygen. *Biochemistry (Moscow), 82*(11), 1249–1268.

Driever, S. M., & Baker, N. R. (2011). The water-water cycle in leaves is not a major alternative electron sink for dissipation of excess excitation energy when CO_2 assimilation is restricted. *Plant, Cell & Environment, 34*(5), 837–846.

Foyer, C., Furbank, R., Harbinson, J., & Horton, P. (1990). The mechanisms contributing to photosynthetic control of electron transport by carbon assimilation in leaves. *Photosynthesis Research, 25*(2), 83–100.
Furbank, R. T., Badger, M. R., & Osmond, C. B. (1982). Photosynthetic oxygen exchange in isolated cells and chloroplasts of C_3 plants. *Plant Physiology, 70*(4), 927–931.
Furbank, R. T., Badger, M. R., & Osmond, C. B. (1983). Photoreduction of oxygen in mesophyll chloroplasts of C_4 plants: A model system for studying an in vivo Mehler reaction. *Plant Physiology, 73*(4), 1038–1041.
Genty, B., Harbinson, J., Briantais, J. M., & Baker, N. R. (1990). The relationship between non-photochemical quenching of chlorophyll fluorescence and the rate of photosystem II photochemistry in leaves. *Photosynthesis Research, 25*(3), 249–257.
Ghashghaie, J., & Cornic, G. (1994). Effect of temperature on partitioning of photosynthetic electron flow between CO_2 assimilation and O_2 reduction and on the CO_2/O_2 specificity of Rubisco. *Journal of Plant Physiology, 143*, 642–650.
Golding, A. J., & Johnson, G. N. (2003). Down-regulation of linear and activation of cyclic electron transport during drought. *Planta, 218*(1), 107–114.
Golding, A. J., Joliot, P., & Johnson, G. N. (2005). Equilibration between cytochrome *f* and P700 in intact leaves. *Biochimica et Biophysica Acta, 1706*(1–2), 105–109.
Harbinson, J., & Foyer, C. H. (1991). Relationships between the efficiencies of photosystems I and II and stromal redox state in CO_2-free air: Evidence for cyclic electron flow in vivo. *Plant Physiology, 97*(1), 41–49.
Harbinson, J., Genty, B., & Baker, N. R. (1989). Relationship between the quantum efficiencies of photosystems I and II in pea leaves. *Plant Physiology, 90*(3), 1029–1034.
Harbinson, J., Genty, B., & Baker, N. R. (1990). The relationship between CO_2 assimilation and electron transport in leaves. *Photosynthesis Research, 25*(3), 213–224.
Harbinson, J., Genty, B., & Foyer, C. H. (1990). Relationship between photosynthetic electron transport and stromal enzyme activity in pea leaves: Toward an understanding of the nature of photosynthetic control. *Plant Physiology, 94*(2), 545–553.
Harbinson, J., & Hedley, C. L. (1993). Changes in P700 oxidation during the early stages of the induction of photosynthesis. *Plant Physiology, 103*(2), 649–660.
Havaux, M., & Davaud, A. (1994). Photoinhibition of photosynthesis in chilled potato leaves is not correlated with a loss of photosystem-II activity: Preferential inactivation of photosystem I. *Photosynthesis Research, 40*(1), 75–92.
Hayashi, R., Shimakawa, G., Shaku, K., Shimizu, S., Akimoto, S., Yamamoto, H., et al. (2014). O_2-dependent large electron flow functioned as an electron sink, replacing the steady-state electron flux in photosynthesis in the cyanobacterium *Synechocystis* sp. PCC 6803, but not in the cyanobacterium *Synechococcus* sp. PCC 7942. *Bioscience, Biotechnology, and Biochemistry, 78*(3), 384–393.
Heber, U. (2002). Irrungen, Wirrungen? The Mehler reaction in relation to cyclic electron transport in C_3 plants. *Photosynthesis Research, 73*(1–3), 223–231.
Heber, U., & Walker, D. (1992). Concerning a dual function of coupled cyclic electron transport in leaves. *Plant Physiology, 100*(4), 1621–1626.
Helman, Y., Tchernov, D., Reinhold, L., Shibata, M., Ogawa, T., Schwarz, R., et al. (2003). Genes encoding A-type flavoproteins are essential for photoreduction of O_2 in cyanobacteria. *Current Biology, 13*(3), 230–235.
Hope, A. B. (2000). Electron transfers amongst cytochrome *f*, plastocyanin and photosystem I: Kinetics and mechanisms. *Biochimica et Biophysica Acta, 1456*(1), 5–26.
Hope, A. B., Matthews, D. B., & Valente, P. (1994). The kinetics of reactions around the cytochrome b_6/f complex studied in an isolated system. *Photosynthesis Research, 40*(2), 199–206.
Hope, A. B., Valente, P., & Matthews, D. B. (1994). Effects of pH on the kinetics of redox reactions in and around the cytochrome b_6/f complex in an isolated system. *Photosynthesis Research, 42*(2), 111–120.

Inoue, K., Fujii, T., Yokoyama, E., Matsuura, K., Hiyama, T., & Sakurai, H. (1989). The photoinhibition site of photosystem I in isolated chloroplasts under extremely reducing conditions. *Plant & Cell Physiology, 30*, 65–71.
Inoue, K., Sakurai, H., & Hiyama, T. (1986). Photoinactivation sites of photosystem I in isolated chloroplasts. *Plant & Cell Physiology, 27*, 961–968.
Ivanov, A. G., Morgan, R. M., Gray, G. R., Velitchkova, M. Y., & Huner, N. P. (1998). Temperature/light dependent development of selective resistance to photoinhibition of photosystem I. *FEBS Letters, 430*(3), 288–292.
Johnson, M. P., & Ruban, A. V. (2014). Rethinking the existence of a steady-state DY component of the proton motive force across plant thylakoid membranes. *Photosynthesis Research, 119*(1–2), 233–242.
Kadota, K., Furutani, R., Makino, A., Suzuki, Y., Wada, S., & Miyake, C. (2019). Oxidation of P700 induces alternative electron flow in photosystem I in wheat leaves. *Plants, 8*(6).
Kanazawa, A., Blanchard, G. J., Szabo, M., Ralph, P. J., & Kramer, D. M. (2014). The site of regulation of light capture in *Symbiodinium*: Does the peridinin-chlorophyll *a*-protein detach to regulate light capture? *Biochimica et Biophysica Acta, 1837*(8), 1227–1234.
Kanazawa, A., & Kramer, D. M. (2002). In vivo modulation of nonphotochemical exciton quenching (NPQ) by regulation of the chloroplast ATP synthase. *Proceedings of the National Academy of Sciences of the United States of America, 99*(20), 12789–12794.
Kanazawa, A., Ostendorf, E., Kohzuma, K., Hoh, D., Strand, D. D., Sato-Cruz, M., et al. (2017). Chloroplast ATP synthase modulation of the thylakoid proton motive force: Implications for photosystem I and photosystem II photoprotection. *Frontiers in Plant Science, 8*, 719.
Khorobrykh, S., Havurinne, V., Mattila, H., & Tyystjarvi, E. (2020). Oxygen and ROS in photosynthesis. *Plants (Basel), 9*(1).
Kramer, D. M., Avenson, T. J., & Edwards, G. E. (2004). Dynamic flexibility in the light reactions of photosynthesis governed by both electron and proton transfer reactions. *Trends in Plant Science, 9*(7), 349–357.
Kramer, D. M., Cruz, J. A., & Kanazawa, A. (2003). Balancing the central roles of the thylakoid proton gradient. *Trends in Plant Science, 8*, 27–32.
Kramer, D. M., Johnson, G., Kiirats, O., & Edwards, G. E. (2004). New fluorescence parameters for the determination of Q_A redox state and excitation energy fluxes. *Photosynthesis Research, 79*(2), 209.
Miyake, C. (2020). Molecular mechanism of oxidation of P700 and suppression of ROS production in photosystem I in response to electron-sink limitations in C3 plants. *Antioxidants, 9*(3).
Miyake, C., Amako, K., Shiraishi, N., & Sugimoto, T. (2009). Acclimation of tobacco leaves to high light intensity drives the plastoquinone oxidation system—Relationship among the fraction of open PSII centers, non-photochemical quenching of Chl fluorescence and the maximum quantum yield of PSII in the dark. *Plant & Cell Physiology, 50*(4), 730–743.
Miyake, C., Horiguchi, S., Makino, A., Shinzaki, Y., Yamamoto, H., & Tomizawa, K. (2005). Effects of light intensity on cyclic electron flow around PSI and its relationship to non-photochemical quenching of Chl fluorescence in tobacco leaves. *Plant & Cell Physiology, 46*(11), 1819–1830.
Miyake, C., Miyata, M., Shinzaki, Y., & Tomizawa, K. (2005). CO_2 response of cyclic electron flow around PSI (CEF-PSI) in tobacco leaves—relative electron fluxes through PSI and PSII determine the magnitude of non-photochemical quenching (NPQ) of Chl fluorescence. *Plant & Cell Physiology, 46*(4), 629–637.
Miyake, C., Shinzaki, Y., Miyata, M., & Tomizawa, K. (2004). Enhancement of cyclic electron flow around PSI at high light and its contribution to the induction of non-photochemical

quenching of Chl fluorescence in intact leaves of tobacco plants. *Plant and Cell Physiology*, *45*(10), 1426–1433.

Miyake, C., Suzuki, Y., Yamamoto, H., Amako, K., & Makino, A. (2012). O_2-enhanced induction of photosynthesis in rice leaves: The Mehler-ascorbate peroxidase (MAP) pathway drives cyclic electron flow within PSII and cyclic electron flow around PSI. *Soil Science & Plant Nutrition*, *58*(6), 718–727.

Miyake, C., & Yokota, A. (2000). Determination of the rate of photoreduction of O_2 in the water-water cycle in watermelon leaves and enhancement of the rate by limitation of photosynthesis. *Plant & Cell Physiology*, *41*(3), 335–343.

Munekage, Y., Hojo, M., Meurer, J., Endo, T., Tasaka, M., & Shikanai, T. (2002). PGR5 is involved in cyclic electron flow around photosystem I and is essential for photoprotection in Arabidopsis. *Cell*, *110*(3), 361–371.

Rantala, S., Lempiainen, T., Gerotto, C., Tiwari, A., Aro, E. M., & Tikkanen, M. (2020). PGR5 and NDH-1 systems do not function as protective electron acceptors but mitigate the consequences of PSI inhibition. *Biochimica et Biophysica Acta—Bioenergetics*, *1861*(3), 148154.

Rott, M., Martins, N. F., Thiele, W., Lein, W., Bock, R., Kramer, D. M., et al. (2011). ATP synthase repression in tobacco restricts photosynthetic electron transport, CO_2 assimilation, and plant growth by overacidification of the thylakoid lumen. *Plant Cell*, *23*(1), 304–321.

Ruuska, S. A., Badger, M. R., Andrews, T. J., & von Caemmerer, S. (2000). Photosynthetic electron sinks in transgenic tobacco with reduced amounts of Rubisco: Little evidence for significant Mehler reaction. *The Journal of Experimental Botany*, *51*(Suppl. 1), 357–368.

Satoh, K. (1970a). Mechanism of photoinactivation in photosynthetic systems. I. The dark reaction in photoinactivation. *Plant & Cell Physiology*, *11*, 15–27.

Satoh, K. (1970b). Mechanism of photoinactivation in photosynthetic systems. II. The occurrence and properties of two different types of photoinactivation. *Plant & Cell Physiology*, *11*, 29–38.

Satoh, K. (1970c). Mechanism of photoinactivation in photosynthetic systems. III. Site and mode of photoinactivation in photosystem I. *Plant & Cell Physiology*, *11*, 187–197.

Satoh, K. (1982). Mechanism of photoactivation of electron transport in intact *Bryopsis* chloroplasts. *Plant Physiology*, *70*(5), 1413–1416.

Satoh, K., & Fork, D. C. (1982). Photoinhibition of reaction centers of photosystems I and II in intact *Bryopsis* chloroplasts under anaerobic conditions. *Plant Physiology*, *70*(4), 1004–1008.

Schöttler, M. A., & Toth, S. Z. (2014). Photosynthetic complex stoichiometry dynamics in higher plants: Environmental acclimation and photosynthetic flux control. *Frontiers in Plant Science*, *5*, 188.

Seelert, H., Poetsch, A., Dencher, N. A., Engel, A., Stahlberg, H., & Muller, D. J. (2000). Structural biology. Proton-powered turbine of a plant motor. *Nature*, *405*(6785), 418–419.

Sejima, T., Hanawa, H., Shimakawa, G., Takagi, D., Suzuki, Y., Fukayama, H., et al. (2016). Post-illumination transient O_2 uptake is driven by photorespiration in tobacco leaves. *Physiologia Plantarum*, *156*(2), 227–238.

Sejima, T., Takagi, D., Fukayama, H., Makino, A., & Miyake, C. (2014). Repetitive short-pulse light mainly inactivates photosystem I in sunflower leaves. *Plant & Cell Physiology*, *55*(6), 1184–1193.

Shaku, K., Shimakawa, G., Hashiguchi, M., & Miyake, C. (2016). Reduction-induced suppression of electron flow (RISE) in the photosynthetic electron transport system of *Synechococcus elongatus* PCC 7942. *Plant & Cell Physiology*, *57*(7), 1443–1453.

Shikanai, T. (2016). Regulatory network of proton motive force: Contribution of cyclic electron transport around photosystem I. *Photosynthesis Research*, *129*(3), 253–260.

Shimakawa, G., Matsuda, Y., Nakajima, K., Tamoi, M., Shigeoka, S., & Miyake, C. (2017). Diverse strategies of O_2 usage for preventing photo-oxidative damage under CO_2 limitation during algal photosynthesis. *Scientific Reports*, 7, 41022.
Shimakawa, G., & Miyake, C. (2018a). Changing frequency of fluctuating light reveals the molecular mechanism for P700 oxidation in plant leaves. *Plant Direct*, *2*(7), e00073.
Shimakawa, G., & Miyake, C. (2018b). Oxidation of P700 ensures robust photosynthesis. *Frontiers in Plant Science*, *9*, 1617.
Shimakawa, G., & Miyake, C. (2019). What quantity of photosystem I is optimum for safe photosynthesis? *Plant Physiology*, *179*(4), 1479–1485.
Shimakawa, G., Murakami, A., Niwa, K., Matsuda, Y., Wada, A., & Miyake, C. (2019). Comparative analysis of strategies to prepare electron sinks in aquatic photoautotrophs. *Photosynthesis Research*, *139*(1–3), 401–411.
Shimakawa, G., Shaku, K., & Miyake, C. (2016). Oxidation of P700 in photosystem I is essential for the growth of cyanobacteria. *Plant Physiology*, *172*(3), 1443–1450.
Shimakawa, G., Shaku, K., & Miyake, C. (2018). Reduction-induced suppression of electron flow (RISE) is relieved by non-ATP-consuming electron flow in *Synechococcus elongatus* PCC 7942. *Frontiers in Microbiology*, *9*, 886.
Shimakawa, G., Shaku, K., Nishi, A., Hayashi, R., Yamamoto, H., Sakamoto, K., et al. (2015). FLAVODIIRON2 and FLAVODIIRON4 proteins mediate an oxygen-dependent alternative electron flow in *Synechocystis* sp. PCC 6803 under CO_2-limited conditions. *Plant Physiology*, *167*(2), 472–480.
Sonoike, K. (2011). Photoinhibition of photosystem I. *Physiologia Plantarum*, *142*(1), 56–64.
Takagi, D., Amako, K., Hashiguchi, M., Fukaki, H., Ishizaki, K., Goh, T., et al. (2017). Chloroplastic ATP synthase builds up a proton motive force preventing production of reactive oxygen species in photosystem I. *The Plant Journal*, *91*(2), 306–324.
Takahashi, M., & Asada, K. (1982). Dependence of oxygen affinity for Mehler reaction on photochemical activity of chloroplast thylakoids. *Plant & Cell Physiology*, *23*(8).
Takahashi, M. A., & Asada, K. (1983). Superoxide anion permeability of phospholipid membranes and chloroplast thylakoids. *Archives of Biochemistry and Biophysics*, *226*(2), 558–566.
Takizawa, K., Cruz, J. A., Kanazawa, A., & Kramer, D. M. (2007). The thylakoid proton motive force in vivo. Quantitative, non-invasive probes, energetics, and regulatory consequences of light-induced pmf. *Biochimica et Biophysica Acta—Bioenergetics* *1767*, 1233–1244.
Takizawa, K., Kanazawa, A., & Kramer, D. M. (2008). Depletion of stromal P_i induces high 'energy-dependent' antenna exciton quenching (q_E) by decreasing proton conductivity at CF_O-CF_1 ATP synthase. *Plant, Cell & Environment*, *31*, 235–243.
Teicher, H. B., Möller, B. L., & Scheller, H. V. (2000). Photoinhibition of photosystem I in field-grown barley (*Hordeum vulgare* L.): Induction, recovery and acclimation. *Photosynthesis Research*, *64*, 53–61.
Terashima, I., Funayama, S., & Sonoike, K. (1994). The site of photoinhibition in leaves of *Cucumis* L. at low temperature is photosystem I, not photosystem II. *Planta*, *193*, 300–306.
Tikhonov, A. N. (2014). The cytochrome b_6/f complex at the crossroad of photosynthetic electron transport pathways. *Plant Physiology and Biochemistry*, *81*, 163–183.
Tikhonov, A. N. (2018). The cytochrome b_6/f complex: Biophysical aspects of its functioning in chloroplasts. *Sub-Cellular Biochemistry*, *87*, 287–328.
Tjus, S. E., Möller, B. L., & Scheller, H. V. (1998). Photosystem I is an early target of photoinhibition in barley illuminated at chilling temperatures. *Plant Physiology*, *116*(2), 755–764.
Tjus, S. E., Möller, B. L., & Scheller, H. V. (1999). Photoinhibition of photosystem I damages both reaction centre proteins PSI-A and PSI-B and acceptor-side located small photosystem I polypeptides. *Photosynthesis Research*, *60*, 75–86.

Tjus, S. E., Scheller, H. V., Andersson, B., & Möller, B. L. (2001). Active oxygen produced during selective excitation of photosystem I is damaging not only to photosystem I, but also to photosystem II. *Plant Physiology*, *125*(4), 2007–2015.

Wada, S., Suzuki, Y., & Miyake, C. (2020). Photorespiration enhances acidification of the thylakoid lumen, reduces the plastoquinone pool, and contributes to the oxidation of P700 at a lower partial pressure of CO_2 in wheat leaves. *Plants (Basel)*, *9*(319).

Wada, S., Yamamoto, H., Suzuki, Y., Yamori, W., Shikanai, T., & Makino, A. (2018). Flavodiiron protein substitutes for cyclic electron flow without competing CO_2 assimilation in rice. *Plant Physiology*, *176*(2), 1509–1518.

Yamamoto, H., Kato, H., Shinzaki, Y., Horiguchi, S., Shikanai, T., Hase, T., et al. (2006). Ferredoxin limits cyclic electron flow around PSI (CEF-PSI) in higher plants—Stimulation of CEF-PSI enhances non-photochemical quenching of Chl fluorescence in transplastomic tobacco. *Plant and Cell Physiology*, *47*(10), 1355–1371.

Yamamoto, H., & Shikanai, T. (2019). PGR5-dependent cyclic electron flow protects photosystem I under fluctuating light at donor and acceptor sides. *Plant Physiology*, *179*(2), 588–600.

Yamamoto, H., Takahashi, S., Badger, M. R., & Shikanai, T. (2016). Artificial remodelling of alternative electron flow by flavodiiron proteins in Arabidopsis. *Nature Plants*, *2*, 16012.

Yamori, W., Makino, A., & Shikanai, T. (2016). A physiological role of cyclic electron transport around photosystem I in sustaining photosynthesis under fluctuating light in rice. *Scientific Reports*, *6*, 20147.

Yamori, W., Sakata, N., Suzuki, Y., Shikanai, T., & Makino, A. (2011). Cyclic electron flow around photosystem I via chloroplast NAD(P)H dehydrogenase (NDH) complex performs a significant physiological role during photosynthesis and plant growth at low temperature in rice. *The Plant Journal*, *68*(6), 966–976.

Yamori, W., & Shikanai, T. (2016). Physiological functions of cyclic electron transport around photosystem I in sustaining photosynthesis and plant growth. *Annual Review of Plant Biology*, *67*, 81–106.

Yamori, W., Shikanai, T., & Makino, A. (2015). Photosystem I cyclic electron flow via chloroplast NADH dehydrogenase-like complex performs a physiological role for photosynthesis at low light. *Scientific Reports*, *5*, 13908.

Zhang, M. M., Fan, D. Y., Murakami, K., Badger, M. R., Sun, G. Y., & Chow, W. S. (2019). Partially dissecting electron fluxes in both photosystems in spinach leaf discs during photosynthetic induction. *Plant & Cell Physiology*, *60*(10), 2206–2219.

Zhang, S., & Scheller, H. V. (2004). Photoinhibition of photosystem I at chilling temperature and subsequent recovery in *Arabidopsis thaliana*. *Plant & Cell Physiology*, *45*(11), 1595–1602.

Zivcak, M., Brestic, M., Kunderlikova, K., Sytar, O., & Allakhverdiev, S. I. (2015). Repetitive light pulse-induced photoinhibition of photosystem I severely affects CO_2 assimilation and photoprotection in wheat leaves. *Photosynthesis Research*, *126*(2–3), 449–463.

Zivcak, M., Kalaji, H. M., Shao, H. B., Olsovska, K., & Brestic, M. (2014). Photosynthetic proton and electron transport in wheat leaves under prolonged moderate drought stress. *Journal of Photochemistry and Photobiology B*, *137*, 107–115.

CHAPTER SIX

Regulation of photosynthesis by cyclic electron transport around photosystem I

Toshiharu Shikanai*
Department of Botany, Graduate School of Science, Kyoto University, Kyoto, Japan
*Corresponding author: e-mail address: shikanai@pmg.bot.kyoto-u.ac.jp

Contents

Abstract

Cyclic electron transport (CET) around photosystem I (PSI) was discovered more than 60 years ago but it was 2004 when two pathways of CET were proposed. The main pathway depends on the PROTON GRADIENT REGULATION 5 (PGR5) protein, whereas the minor pathway is mediated by the chloroplast NADH dehydrogenase-like (NDH) complex. Although it is still a topic of debate, as to whether PGR5 is really involved in CET, the Arabidopsis *pgr5* mutant is severely defective in the ΔpH-dependent regulation of photosynthetic electron transport. PGR5 is necessary to induce a ΔpH-dependent component of nonphotochemical quenching (NPQ) of chlorophyll fluorescence at a high light intensity and this qE mechanism dissipates excessively absorbed light energy safely as heat from photosystem II antennae. The lumenal acidification also downregulates the electron transport at the cytochrome b_6f complex. The latter is essential for protecting PSI and the *pgr5* mutant cannot survive the fluctuating light intensity. Although the contribution of the NDH complex to the *pmf* formation is much smaller than that of the PGR5-dependent CET during steady-state photosynthesis, the NDH complex efficiently pumps protons probably in the absence of large *pmf*. In addition to the size of *pmf*, it is also necessary to regulate the partitioning *pmf* components, ΔpH, and membrane

ISSN 0065-2296
https://doi.org/10.1016/bs.abr.2020.07.005
© 2020 Elsevier Ltd
All rights reserved.

potential ($\Delta\Psi$) across the thylakoid membrane. Ion channels and transporters localized to the thylakoid membrane involve the regulation and form a regulatory network of photosynthesis with the CET machinery.

1. Introduction

The light reactions of photosynthesis mediate linear electron transport (LET) from water to $NADP^+$ (Fig. 1). Water splitting in photosystem II (PSII) and the quinone (Q) cycle in the cytochrome (Cyt) b_6f complex contribute to the translocation of protons (H^+) across the thylakoid membrane. Resulting ΔpH forms proton motive force (*pmf*), as well as membrane potential ($\Delta\Psi$) across the thylakoid membrane (Fig. 2). The *pmf* is utilized in ATP synthesis by F_1F_o ATP synthase. ATP and NADPH are utilized in the Calvin-Benson cycle for CO_2 fixation. However, this LET does not satisfy the ATP/NADPH production ratio of >1.67 required by CO_2 fixation and photorespiration (Allen, 2002). Additional ATP is considered to be provided by cyclic electron transport (CET) around photosystem I (PSI).

In CET around PSI, electrons are recycled from ferredoxin (Fd) to the plastoquinone (PQ) pool, generating *pmf* without the net accumulation of

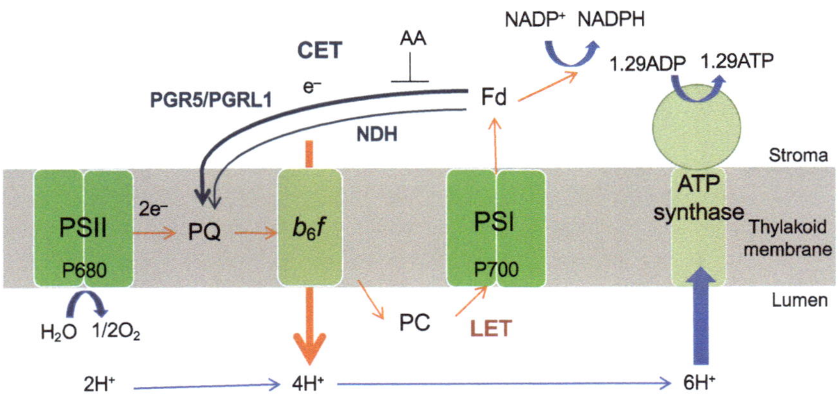

Fig. 1 Schematic representation of photosynthetic electron transport. In LET, the movement of two electrons (e^-) from water to $NADP^+$ is coupled with the translocation of six H^+ across the thylakoid membrane, generating 1.29 molecules of ATP. This ATP/NADPH production ratio does not satisfy the requirement by the Calvin-Benson cycle and photorespiration. Additional ATP has been considered to be supplemented by CET. In angiosperms, CET consists of two pathways. The main pathway depends on PGR5/PGRL1 proteins and is sensitive to antimycin A (AA). The minor pathway is mediated by the chloroplast NDH complex. *b_6f*, Cyt b_6f complex. *PC*, plastocyanin.

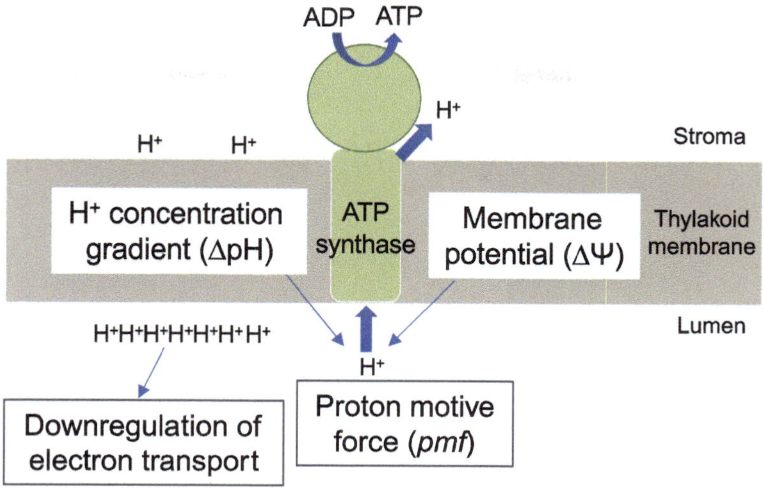

Fig. 2 Two components of *pmf*. Because H^+ has a positive charge, both the H^+ concentration gradient (ΔpH) and membrane potential (ΔΨ) across the thylakoid membrane contribute to ATP synthesis as *pmf*. However, the only ΔpH downregulates electron transport via inducing qE and the donor-side regulation at the Cyt b_6f complex.

NADPH (Yamori & Shikanai, 2016). In angiosperms, CET around PSI consists of two pathways (Fig. 1). The main pathway depends on PROTON GRADIENT REGULATION 5 (PGR5)/PGR5-like Photosynthetic Phenotype 1 (PGRL1) proteins. Since some of the *pgr5* mutant phenotypes can be explained by other ideas, it is still a topic of debate as to whether the PGR5/PGRL1 proteins directly involve CET around PSI (Avenson, Cruz, Kanazawa, & Kramer, 2005; Nandha, Finazzi, Joliot, Hald, & Johnson, 2007; Rantala et al., 2020; Takagi & Miyake, 2018). The chloroplast NADH dehydrogenase-like (NDH) complex mediates the minor route of electron transport. The physiological significance of CET around PSI has long been underestimated mainly because of the lack of methods monitoring the rate of operation in vivo (Bendall & Manasse, 1995). On the basis of the mutant phenotypes, however, CET around PSI has been considered to be the main mechanisms for regulating photosynthetic electron transport, especially under stressful light environments.

In addition to the contribution of *pmf* to ATP synthesis, a ΔpH component of *pmf* downregulates electron transport by acidifying the thylakoid lumen (Shikanai & Yamamoto, 2017) (Fig. 2). The lumenal acidification induces a qE (energization dependent) component of nonphotochemical quenching (NPQ) of chlorophyll fluorescence (Ruban, 2018) (Fig. 3).

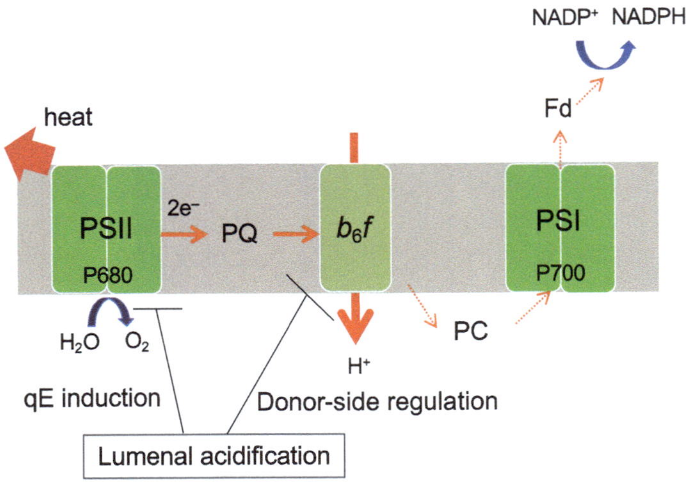

Fig. 3 ΔpH-dependent downregulation of electron transport. Lumenal acidification triggers the thermal dissipation of excessively absorbed light energy from the PSII antennae. This qE mechanism is monitored as the induction of NPQ. Low lumenal pH also downregulates the rate of electron transport through the Cyt b_6f complex (b_6f). This regulation is essential to protect PSI from photodamage by controlling the electron transport toward PSI from the donor side (donor-side regulation).

Under the conditions where light energy limits photosynthesis, absorbed light energy by PSII antennae is efficiently transferred to the PSII reaction center chlorophylls (P680) and used for the charge separation. However, excessively absorbed light energy that cannot be used for the charge separation is safely dissipated as heat from the PSII antennae. This qE mechanism is not essential for the survival of angiosperms even under the constant high light intensity (Li et al., 2000) or the fluctuating light intensity (Tikkanen, Grieco, Kangasjärvi, & Aro, 2010) but is required for the fitness for natural fluctuating light environments (Külheim, Agren, & Jansson, 2002).

The lumenal acidification also downregulates the rate of electron transport through the Cyt b_6f complex (Nishio & Whitmarsh, 1993; Rumberg & Siggel, 1969; Tikhonov, Khomutov, Ruuge, & Blumenfeld, 1981) (Fig. 3). This regulation is important to protect PSI from photodamage, especially under the fluctuating light intensity (Allahverdiyeva, Suorsa, Tikkanen, & Aro, 2015). Despite its physiological significance, it has been still unclear how the lumenal acidification or *pmf* downregulates the activity of the Cyt b_6f complex. One of the most important physiological functions of CET around PSI might be the induction of the ΔpH-dependent downregulation of the Cyt b_6f activity.

In this chapter, I summarize the current progress to understand the contribution of CET around PSI to the regulation of photosynthetic electron transport and also on its collaboration with the machinery for regulating the partitioning of two *pmf* components, ΔpH and ΔΨ.

2. What is CET? The link between the pioneering work and the current genetics

CET around PSI was discovered by a group of Arnon in the 1950s (Arnon, Allen, & Whatley, 1954). CET activity was monitored as ATP production in isolated thylakoid membranes (cyclic phosphorylation) and depended on Fd. Since the discovery, the molecular identity of this electron transport has long been unclear. The Arnon's cyclic phosphorylation was inhibited by antimycin A (Tagawa, Tsujimoto, & Arnon, 1963) and it has been a key to clarify the molecules involving this enigmatic electron transport. Once the idea of LET producing both ATP and NADPH was established, the physiological function of CET around PSI has been underestimated and was often ignored (Bendall & Manasse, 1995).

In 2002, we reported the Arabidopsis (*Arabidopsis thaliana*) *pgr5* mutant which was defective in the qE induction under high light intensity (Munekage et al., 2002). Mainly on the basis of reduced activity of Fd-dependent PQ reduction in ruptured chloroplasts, we concluded that the *pgr5* mutant was defective in CET around PSI. Antimycin A phenocopied the *pgr5* mutant in the Fd-dependent PQ reduction assay and did not further affect the electron flow in the *pgr5* mutant (Munekage et al., 2002). Infiltration of antimycin A into intact leaves of wild-type (WT) plants also mimicked the *pgr5* mutant phenotypes in photosynthesis (Sugimoto et al., 2013). How does antimycin A inhibits CET around PSI? Although antimycin A binds the Qi site of the Cyt bc_1 complex and inhibits the Q cycle, this is not the case for the Cyt b_6f complex (de Wolf, Krab, Visschers, de Waard, & Kraayenhof, 1998). Unexpectedly, the Arabidopsis plants transformed by the *PGR5* gene isolated from loblolly pine (*Pinus taeda*) were resistant to antimycin A in CET around PSI (Sugimoto et al., 2013). We finally discovered that a single amino acid alteration in PGR5 (V3K) conferred the resistance. Simultaneously, a group of Dario Leister showed that PGRL1-dependent reduction of the PQ analog, 2,6-dimethyl-p-benzoquinone in vitro was inhibited by antimycin A (Hertle et al., 2013). Most likely, antimycin A inhibits the function of the PGR5/PGRL1 protein complex in CET around PSI.

To confirm the link between the Arnon's CET and genetically identified PGR5, we reconstructed the classical experiment to monitor cyclic phosphorylation (Wang, Takahashi, & Shikanai, 2018). Rate of ATP synthesis and the *pmf* formation was reduced in the *pgr5* mutant compared to those in the WT plants. Antimycin A phenocopied the *pgr5* mutation although it further affected the *pgr5* mutant unexpectedly. Antimycin A may partially inhibit ATP synthase directly, as well as affecting CET around PSI. The historical Arnon's experiment and our molecular genetics likely observed the same physiological event.

Despite the significant contribution of PGR5 protein in the regulation of photosynthesis (Munekage et al., 2002, 2004), the exact molecular function of PGR5 and PGRL1 proteins has not been clarified yet. For the conclusion that PGR5 involves CET, it would be necessary to directly show what PGR5 and PGRL1 do in CET around PSI.

3. Physiological function of PGR5-dependent CET around PSI

PGR5-dependent CET around PSI contributes to the formation of *pmf* without the net production of NADPH (Fig. 1), facilitating the balancing of the ATP/NADPH production ratio. In addition to ATP synthesis, a ΔpH component of *pmf* also contributes to the downregulation of electron transport via the lumenal acidification (Figs. 2 and 3). In fact, the *pgr5* mutant was isolated on its reduced induction of NPQ at high light intensity (Munekage et al., 2002). The Arabidopsis *nonphotochemical quenching 4* (*npq4*) mutant defective in PsbS, a pH sensor for qE (a ΔpH-dependent component of NPQ) induction, is much more resistant to high light stress than the *pgr5* mutant (Li et al., 2000), especially against fluctuating light intensity (Tikkanen et al., 2010). Apparently, the defect in the qE induction does not explain many of the *pgr5* mutant phenotypes and we have to consider the contribution of CET to other ΔpH-dependent regulation.

In addition to the qE induction, ΔpH also downregulates electron transport through the Cyt b_6f complex (Fig. 3). The process is often called "photosynthetic control" but I prefer to use the term "donor-side regulation" (at the donor-side of PSI) in this chapter for the contrast with the regulation at the acceptor-side of PSI (see below). This donor-side regulation is essential for protecting PSI from photodamage by restricting electron transport toward PSI at the Cyt b_6f complex. In the *pgr5* mutant, PSI is highly reduced by electrons at a high light intensity, leading to severe photodamage (Munekage et al., 2002; Tiwari et al., 2016).

Why are electrons trapped in PSI at the high light intensity in the *pgr5* mutant? Because PGR5-dependent CET functions in ATP synthesis, it may be necessary to supply the sufficient electron acceptors from PSI (oxidized Fd or $NADP^+$) by balancing the ATP/NADPH production ratio (acceptor-side regulation). Originally, we explained the sensitivity of PSI to high light in the *pgr5* mutant by this acceptor limitation (Munekage et al., 2002; Shikanai, 2007). Alternatively, PGR5-dependent CET may control the rate of electron transport toward PSI by the acidification of the thylakoid lumen for downregulating the Cyt b_6f complex. The defect in the induction of this donor-side regulation has been proposed to be the main reason for the *pgr5* phenotype in PSI (Suorsa et al., 2012).

To evaluate the contribution of the donor-side and acceptor-side regulation to the photoprotection of PSI quantitatively, we took an advantage of the Arabidopsis *pgr1* mutant, in which the strength of the donor-side regulation was enhanced (Yamamoto & Shikanai, 2019). The *pgr1* mutant has an amino acid alteration in the Rieske subunit of the Cyt b_6f complex (Munekage et al., 2001). Although the *pgr1* mutation does not affect the stability of the Rieske subunits, it makes the Cyt b_6f complex hypersensitive to the lumenal acidification. In the *pgr1* mutant, the Cyt b_6f complex is downregulated even when the thylakoid lumen is weakly acidified (Jahns, Graf, Munekage, & Shikanai, 2002). In the *pgr1 pgr5* double mutant, the strong donor-side regulation was induced even at relatively high lumenal pH in the *pgr5* mutant background and consequently, PSI was resistant to the fluctuating light intensity (Yamamoto & Shikanai, 2019). This result suggests the preferential contribution of the PGR5-dependent CET to the donor-side regulation. However, too strong donor-side regulation in the *pgr1* mutant makes PSII sensitive to constant high light (Munekage et al., 2001). The sensitivity of the b_6f complex to lumenal acidification (donor-side regulation) is optimized to the natural fluctuating light environments and the lumenal pH is adjusted to the appropriate range by the operation of PGR5-dependent CET.

How does the acceptor-side regulation (balancing of the ATP/NADPH production ratio) induced by the PGR5-dependent CET contribute to the photoprotection of PSI under fluctuating light intensity? Flavodiiron (Flv) proteins accept electrons from PSI probably via Fd or NADPH and reduce molecular oxygen to H_2O (Vicente, Gomes, Wasserfallen, & Teixeira, 2002). It forms a large electron sink from PSI, efficiently protecting PSI from photodamage under fluctuating light intensity in cyanobacteria (Allahverdiyeva et al., 2013) and *Chlamydomonas reinhardtii* (Chaux et al., 2017). Mysteriously, angiosperms lost Flv during their evolution (Ilik et al., 2017). To test the impact of the enhanced acceptor-side regulation, we introduced *FlvA* and *FlvB* genes isolated from a moss (*Physcomitrella patens*) into Arabidopsis WT and *pgr5* mutant plants (Yamamoto, Takahashi, Badger, & Shikanai, 2016). The exogenous Flv does not function during steady-state photosynthesis in the WT background. This is important in the genetic engineering for enhancing the resistance of PSI to the fluctuating light intensity because Flv may compete with the CO_2 fixation for oxidizing NADPH. In the *pgr5* mutant, however, up to 25% of electrons originated by water splitting in PSII are used for Flv-dependent O_2 reduction (Yamamoto, Takahashi, et al., 2016). Because the electron flow starts by the water splitting in PSII and ends by the Flv-dependent O_2 reduction to water, Flv mediates the pseudo-CET (water-water cycle). This pseudo-CET contributes to the *pmf* formation via the Q cycle at the Cyt b_6f complex and complements the function of the PGR5-dependent CET, especially in the oxidation of P700 at high light intensity (Yamamoto, Takahashi, et al., 2016). Unlikely in the water-water cycle depending on the Mehler reaction (Asada, 1999), Flv-dependent electron transport does not produce any reactive oxygen species. Because the contribution of $\Delta\Psi$ to *pmf* was larger in the *pgr5* mutant accumulating Flv than in the WT plants, NPQ induction was mildly affected at the moderate light intensity in the lines. The plants were more resistant to the fluctuating light intensity than the *pgr5* mutant. Notably, the similar effect was also observed in the WT background (Yamamoto, Takahashi, et al., 2016), suggesting that the exogenous Flv system can be used to artificially improve the resistance of PSI to fluctuating light in crops, and this was proved in rice (Wada et al., 2018). Our approach suggests the significant contribution of a large electron sink (acceptor-side regulation) to the resistance of PSI against the fluctuating light intensity. But the Flv-dependent pseudo-CET also induces the donor-side regulation via the lumenal acidification. It is not simple to separately evaluate the donor-side and acceptor-side regulation.

By carefully inspecting the induction kinetics of the donor-side regulation by monitoring the redox state of PSI reaction center chlorophylls (P700), we could discuss the physiological function of the acceptor-side regulation (Fig. 4). Immediately after the shift from low light to high light (<20 s) in the fluctuating light cycle, the acceptors from PSI are transiently limited even in the WT plants (Yamamoto & Shikanai, 2019). This limitation is rapidly

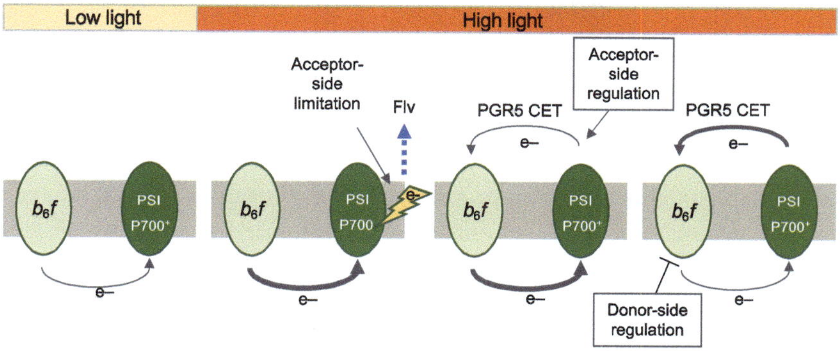

Fig. 4 PGR5-dependent CET protects PSI from photodamage after the shift from low light to high light. Immediately after the shift to high light (<20 s), P700 is transiently reduced by electrons (e^-) from the Cyt b_6f complex (b_6f) (acceptor limitation). Flv efficiently sucks these electrons from P700 but angiosperms do not have Flv. In the presence of PGR5, plants move electrons from P700 to the PQ pool via CET. The operation of the PGR5-dependent CET induces the donor-side regulation via lumenal acidification and consequently, P700 is oxidized.

relaxed by the induction of the donor-side regulation in the WT but this was not the case in the *pgr5* mutant. PGR5-dependent CET is likely important to rapidly relax the acceptor-side limitation by moving electrons through the CET pathway. Notably, some population of LHCII is associated with PSI (state 2 of the state transitions) in low light in Arabidopsis, and this is important to protect PSI immediately after the shift to high light (Grieco, Tikkanen, Paakkarinen, Kangasjärvi, & Aro, 2012). Collaboration between the PGR5-dependent CET and state transitions plays a key role in protecting PSI immediately after the shift from low light to high light. In this acceptor-side regulation, Flv more efficiently sucks up electrons from P700 than the PGR5-dependent CET. The donor-side regulation at the Cyt b_6f complex plays a key role to control the rate of electron transport toward PSI. The acceptor-side regulation is also important to initially move electrons to induce the donor-side regulation by acidifying the thylakoid lumen (Fig. 4). Either the PGR5-dependent CET or the Flv-dependent pseudo-CET function in both processes, whereas Flv is better machinery for the accepter-side regulation. It would be reasonable to conserve both regulatory systems, the Flv system for a safety valve, and the PGR5 system for the steady-state *pmf* formation and this strategy is actually taken by non-angiosperms (Ilik et al., 2017). It has been still a mystery why angiosperms lost Flv.

On the basis of the careful reevaluation of mutant phenotypes, the most important physiological function of PGR5-dependent CET may be the induction of the donor-side regulation via acidification of the thylakoid lumen. Because angiosperms do not have Flv, PGR5-dependent CET is extremely important to protect PSI under the fluctuating light intensity (Suorsa et al., 2012). The sensitivity of PSI to high light is mainly explained by the reduced induction of the donor-side regulation. In the *pgr5* mutant, the restriction of electron transport at the Cyt b_6f complex (donor-side regulation) is not fully induced but LET is saturated at the lower level (Munekage et al., 2002). In the WT plants, the Cyt b_6f complex likely restricts electron transport at high light intensity via the donor-side regulation (Tikhonov, 2014). How is the reduced rate of LET explained in the *pgr5* mutant, in which the Cyt b_6f complex unlikely restricts electron transport? It is still necessary to rely on the original idea of the reduced ATP production ratio to explain all the *pgr5* mutant phenotypes. In the mutant, the ATP level likely limits the rate of LET at the acceptor side from PSI.

4. NDH-dependent CET around PSI

4.1 Structure and activity of the chloroplast NDH complex

The chloroplast NDH complex was first recognized by the determination of two plastid genomes in *Marchantia polymorpha* and *Nicotiana tabacum* (Matsubayashi et al., 1987). Eleven *ndh* genes (*ndhA–ndhK*) encode proteins similar to subunits of respiratory NADH dehydrogenase (complex I) (Table 1). It was a mystery why the chloroplast genome encodes the proteins related to respiratory electron transport. But the cyanobacterial NDH-1 complex was discovered to mediate CET in the analysis of the *M55* mutant (Δ*ndhB*) defective in the growth in the low CO_2 conditions (Ogawa, 1991). The chloroplast NDH may also be a machinery for CET around PSI in chloroplasts. This idea was proved by using a knockout technique of the chloroplast gene (Burrows, Sazanov, Svab, Maliga, & Nixon, 1998; Shikanai et al., 1998).

In the chloroplast genome, four genes (*ndhH–ndhK*) encode proteins forming a core of subcomplex A (SubA = Q module) involving quinone reduction. The remaining seven chloroplast genes (*ndhA–ndhG*) encode subunits forming the Membrane subcomplex (SubM = P module) involving H^+ pumping across the membrane coupled with electron transport in the complex (Table 1). In the respiratory complex I, N module involving NADH oxidation is on the top of the Q module (Parey, Wirth, Vonck, & Zickermann, 2020). However, three genes encoding the N module subunits are missing in the chloroplast genome and the research has been focused on

Table 1 List of NDH subunits

Subcomplex		Subunit	TM	Motifs, etc.
SubA		NdhH		Core of Q module
		NdhI		
		NdhJ		
		NdhK		
		NdhM		
		NdhN		
		NdhO		
	Connecting	NdhL	✓☐	
SubM		NdhA	✓☐	
		NdhB	✓☐	P module
		NdhC	✓☐	
		NdhD	✓☐	
		NdhE	✓☐	
		NdhF	✓☐	
		NdhG	✓☐	
SubB		PnsB1		
		PnsB2		
		PndB3		FeS cluster
		PndB4	✓☐	
		PndB5	✓☐	
	Lumenal side	PnsL3		PsbQ family
SubL		PnsL1		PsbP family
		PnsL2		PsbQ family
		PnsL4		FKBP16-2
		PnsL5		AtCYP20-2
SubED		NdhS		Src homology 3 domain-like fold
		NdhT	✓☐	J domain
		NdhU	✓☐	J-like domain
		NdhV		
Linker		Lhca5	✓☐	LHCI family
		Lhca6	✓☐	LHCI family

Shadowed boxes indicate chloroplast-encoded subunits. *TM*; transmembrane domain the identification of nuclear-encoded subunits involving NADH or NADPH oxidation. Because of the lack of biochemical evidence for the electron donor to the complex, the chloroplast NDH complex has been often referred to as NAD(*P*)H (NADH or NADPH) dehydrogenase. But it has been clear now that the chloroplast NDH accepts electrons from Fd rather than NAD(*P*)H (Schuller et al., 2019). To avoid any confusion, we have proposed to keep using NDH for the complex but the NDH complex stands for the NADH dehydrogenase-like complex (Yamamoto, Peng, Fukao, & Shikanai, 2011). Because this enzyme activity and the structure are also conserved in the cyanobacterial NDH-1 complex, we call this group of enzymes "photosynthetic NDH." Despite the structural similarity of the core (SubA/Q module and SubM/P module), respiratory complex I and photosynthetic NDH have different enzyme activities.

Except for core eleven subunits, the nuclear genome of angiosperms encodes some subunits specific to photosynthetic NDH. Notably, core eleven subunits conserved between respiratory complex I and photosynthetic NDH have not moved from the chloroplast genome to the nuclear genome after the endosymbiosis. In addition to four core subunits, SubA includes three additional nuclear-encoded subunits, NdhM, NdhN, and NdhO (Rumeau et al., 2005). SubA is almost fully assembled in the stroma, the process that is assisted by a series of specific assembly factors in Arabidopsis (Peng, Cai, & Shikanai, 2010; Peng et al., 2011; Peng, Fukao, Fujiwara, & Shikanai, 2012; Yamamoto et al., 2016). After incorporating NdhN, the assembly intermediate is inserted into the membrane part consisting of other subcomplexes (Peng et al., 2012). NdhL and NdhA are transmembrane proteins and are necessary to connect SubA with SubM. Although we previously grouped NdhL and NdhA into SubA and SubM respectively (Ifuku, Endo, Shikanai, & Aro, 2011), both subunits are more appropriate to be categorized into the connecting part between SubA and SubM from a functional point of view (Table 1).

Instead of the N module which oxidizes NADH and transfer two electrons to a FeS cluster present in the Q module in respiratory complex I, photosynthetic NDH possess the electron donor subcomplex (SubED) which supports Fd binding directly to SubA. NdhS was discovered in the proteome analysis of the NDH-PSI supercomplex in Arabidopsis and the knockout of the *ndhS* gene resulted in impaired NDH activity in the *crr31/ndhs* mutant (Yamamoto et al., 2011). In the 3D structure, NdhS is similar to PsaE forming the Fd-binding site of PSI, suggesting that NdhS forms an Fd-binding site of the chloroplast NDH complex. In fact, NdhS was necessary for the efficient binding of Fd to the NDH complex (Yamamoto et al., 2011). We hypothesized the positive surface charge of NdhS is important for interaction with negatively charged Fd based on the impact of the site-directed mutations that changes the surface charge of NdhS (Yamamoto & Shikanai, 2013). However, the surface of NdhS is embedded in SubA in the cryo-electron microscopy structure of cyanobacterial NDH-1 (Laughlin, Bayne, Trempe, Savage, & Davies, 2019; Schuller et al., 2019). Instead, the C-terminal region of NdhS has been suggested to contribute to Fd binding via a fly-catching mechanism (Schuller et al., 2019). NdhV is another component of SubED and at least transiently interacts with SubA in cyanobacteria (Gao et al., 2016) and Arabidopsis (Fan, Zhang, Li, & Peng, 2015). Association of NdhV with the NDH complex modifies the structure of SubA and may alter the mode of Fd binding (Pan et al., 2020; Zhang et al., 2020). In Arabidopsis, two additional subunits with a transmembrane domain, NdhT/CRRJ, and NdhU/CRRL are required for the stability of NdhS (Yamamoto et al., 2011). Because NdhT and NdhS have J and J-like domains respectively, these subunits may also involve the assembly of SubED.

The NDH complex of angiosperms contains additional two subcomplexes, subcomplex B (SubB) and lumenal subcomplex (SubL), which are absent in the cyanobacterial NDH-1 complex (Ifuku et al., 2011). Both subcomplexes are likely related to the supercomplex formation with PSI. I will explain their function in the following section focusing on the supercomplex formation.

4.2 Structure of the NDH-PSI supercomplex

Blue-native (BN) gel electrophoresis is a powerful tool to separate large protein complexes localized to the thylakoid membrane (Järvi, Suorsa, Paakkarinen, & Aro, 2011). NDH subunits were detected at the top of the BN gel as a green band, leading to the discovery of the NDH-PSI supercomplex (Peng, Shimizu, & Shikanai, 2008). Because the NDH complex interacts with two copies of the PSI supercomplex consisting of a PSI core and four light-harvesting chlorophyll protein complex I (LHCI) molecules (Lhca1–Lhca4), the NDH-PSI supercomplex is a super-super complex to be more precise. The mass spectrometry analysis of the NDH-PSI supercomplex clarified two linker proteins, Lhca5 and Lhca6, as well as subunits of the NDH complex and the PSI supercomplex (Peng, Fukao, Fujiwara, Takami, & Shikanai, 2009). Lhca5 and Lhca6 are members of the LHCI family. The Arabidopsis genome encodes six members of the family and

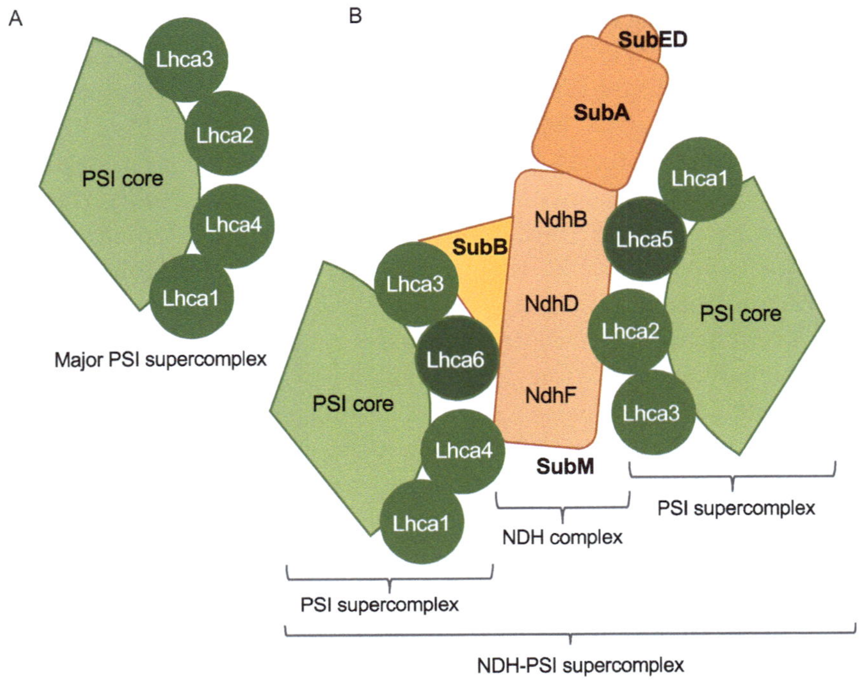

Fig. 5 Structural model of the NDH-PSI supercomplex. (A) More than 95% of the PSI core forms the major PSI supercomplex consisting of a core and four LHCI molecules (Lhca1–Lhca4). (B) The NDH complex interacts with two copies of the PSI supercomplex to form the NDH-PSI supercomplex. The NDH-PSI supercomplex is viewed from the stroma side and SubL is on the opposite side of this figure. In the left copy of the PSI supercomplex, Lhca6 substitutes for Lhca2. In the right copy, Lhca5 substitutes for Lhca4. SubB likely provides the binding site for Lhca6. The position of SubED was predicted by the positions of NdhS in the cryo-electron microscopy structure of the cyanobacterial NDH-1.

Lhca1/Lhca4 and Lhca2/Lhca3 form a heterodimer of heterodimers present in the major PSI supercomplex (Qin, Suga, Kuang, & Shen, 2019) (Fig. 5A). However, the exact molecular function of minor members of Lhca5 and Lhca6 has not been clarified. They are preferentially detected in the NDH-PSI supercomplex and their main function is the intermediation of the supercomplex formation with a single copy of the PSI supercomplex, respectively. In the Arabidopsis mutant defective in Lhca5 or Lhca6, the mutant versions of the NDH-PSI supercomplex migrated slightly faster than the intact supercomplex in the BN gel. In the *lhca5 lhca6* double mutant, however, NDH subunits were detected at the position of the NDH complex which was unassociated with any PSI supercomplex (Peng et al., 2009). Our initial model is based on this genetic information, hypothesizing the

NDH complex to be sandwiched between two copies of the PSI supercomplex (Fig. 5B). The model was confirmed by a visualization of the single particles of the NDH-PSI supercomplex (Kouřil et al., 2014).

A question was on the exact position of Lhca5 and Lhca6 in the NDH-PSI supercomplex. Although Kouřil et al. (2014) proposed their position at the one side of the NDH complex (Fig. 5B, at the position of SubB), this model is inconsistent with our genetic results indicating that Lhca5 and Lhca6 are responsible for binding a single copy of the PSI supercomplex, respectively (Peng et al., 2011). On the basis of the high sequence similarity, our hypothesis was the substitution of Lhca2 by Lhca6 in a copy of the PSI supercomplex. In the *lhca5* mutant defective in another copy of the PSI supercomplex, the mutant version of the NDH-PSI supercomplex contained Lhca6 instead of Lhca2 (Otani, Kato, & Shikanai, 2018). By the similar analysis using the *lhca6* mutant, we also showed that Lhca5 substituted for Lhca4 in another copy of the NDH-PSI supercomplex. When the NDH complex is viewed from the stroma side and placing SubA at the top, Lhca2 is substituted by Lhca6 in the left copy of the PSI supercomplex and Lhca4 is substituted by Lhca5 in the right copy of the PSI supercomplex (Fig. 5B).

Lhca6 clearly evolved from Lhca2 during the evolutional history of angiosperms (Peng et al., 2009). Lhca6 is not conserved in *Physcomitrella* and the NDH complex interacts with a single copy of the PSI supercomplex via Lhca5 (Kato, Odahara, Fukao, & Shikanai, 2018). The Lhca6-dependent supercomplex formation has been acquired during the evolution of angiosperms for stabilizing the NDH complex especially under stressful light conditions (Peng & Shikanai, 2011). How did Lhca6 acquire the linker function in the evolution from Lhca2? Swapping of the stromal loop between Lhca2 and Lhca6 was necessary and sufficient for the linker function (Otani, Yamamoto, & Shikanai, 2017). Chimeric Lhca6 with the stromal loop of Lhca2 interacts with the PSI core but not with the NDH complex, suggesting the stromal loop of Lhca6 to be necessary for the interaction with the NDH complex (Otani et al., 2018). In the structure of the PSI supercomplex, the stromal loop of Lhca2 extrudes to the outside of the PSI supercomplex (Qin et al., 2019). The stromal loop of Lhca6 should also extrude to the outside of the PSI supercomplex so that the loop can provide the contact site with the NDH complex to form the NDH-PSI supercomplex (Otani et al., 2018).

How was the NDH complex-side modified for the Lhca6-dependent supercomplex formation? SubB is specific to the chloroplast NDH complex and consists of five subunits (Ifuku et al., 2011). Although PnsL3 (PsbQ-like protein) was originally classified into SubL, the stability of PnsL3 is more tightly linked to the SubB subunits than other SubL subunits (Yabuta et al., 2010). We proposed to reclassify PnsL3 into SubB (Kato, Sugimoto, & Shikanai, 2018; Table 1). The CRR3 protein was identified in the Arabidopsis mutant defective in the accumulation of the NDH complex and is required for the assembly of SubB (Kato, Sugimoto, & Shikanai, 2018; Muraoka, Okuda, Kobayashi, & Shikanai, 2006). The assembly intermediate including CRR3 likely consists of SubB subunits and a single copy of the PSI supercomplex including Lhca6 instead of Lhca2. Most likely, SubB is a contact site with Lhca6. In the single-particle images, an assigned protein density was observed at the shoulder of the Lhca3/Lhca6 heterodimer (Kouřil et al., 2014). Probably, this protein density corresponds to SubB (Otani et al., 2018) (Fig. 5B). Some of SubB subunits are likely to present in *Marchantia* (Ueda et al., 2012). Modification of SubB was probably necessary for the interaction with Lhca6 in the evolution of angiosperms.

SubL consists of four subunits excluding PnsL3 which is reclassified into SubB (Table 1). PnsL1 and PnsL2 are PsbP-like and PsbQ-like protein, respectively (Ishihara et al., 2007; Peng et al., 2009; Suorsa et al., 2010; Yabuta et al., 2010). PnsL4 is FKBP16–2 (Majeran et al., 2008; Peng et al., 2009), whereas PnsP5 is AtCYP20–2 (Majeran et al., 2008; Sirpiö, Holmström, Battchikova, & Aro, 2009). In contrast to PnsL3, the stability of PnsL1, PnsL2, PnsL4, and PnsL5 was more tightly linked to the stability of SubA (Peng et al., 2009). On the basis of this observation, we place PnsL3 at the bottom (lumenal side) of SubB, whereas other SubL subunits are at the bottom of SubA (Peltier, Aro, & Shikanai, 2016). The exact molecular function of SubL is still unclear, but SubL is specific to angiosperms, in which the NDH complex interacts with two copies of the PSI supercomplex. SubL may stabilize the Lhca6-dependent interaction with a copy of the PSI supercomplex from the lumenal side. This idea is consistent with the discovery that PsbP is necessary for the stability of the PSII supercomplex (Ido et al., 2009).

4.3 Physiological function of the chloroplast NDH complex

Although the exact molecular function of PGR5/PGRL1 proteins has been unclear, it has been experimentally shown that the NDH complex mediates Fd-dependent PQ reduction (Schuller et al., 2019). At least, the NDH complex is able to mediate CET around PSI in the activity. Although the direct evidence is still lacking, the movement of two electrons in the complex to oxidize a molecule of PQ is likely coupled with the translocation of four H^+ across the thylakoid membrane (Strand, Fisher, & Kramer, 2017). The PGR5-dependent CET solely depends on the Q cycle for H^+ translocation, indicating the $2H^+/e^-$ ratio in CET (Fig. 1). In contrast, the NDH complex additionally pumps four H^+, consequently increasing the H^+/e^- ratio to 4. The efficiency of H^+ translocation is double in the NDH-dependent CET than PGR5-dependent CET. However, on the basis of the mutant phenotypes, the contribution of the NDH complex to the *pmf* formation is much lower than that of PGR5-dependent CET (Wang, Yamamoto, & Shikanai, 2015). This observation is consistent with lower activity of Fd-dependent PQ reduction in ruptured chloroplasts and the subtle mutant phenotype in photosynthesis (Munekage et al., 2004). A question is on the physiological significance of such minor activity of NDH-dependent CET in photosynthesis.

The most evident impact of the Arabidopsis NDH-less (*chlororespiratory reduction, crr*) mutations has been observed in the *pgr5* mutant background (Munekage et al., 2004; Nakano, Yamamoto, & Shikanai, 2019). Although the regulation of photosynthesis is severely impaired in the *pgr5* mutant, it can grow like WT plants under the constant low light conditions of a growth chamber. However, plant growth and photosynthesis are severely impaired in the double mutant defective in both pathways of CET. The similar phenotype was also observed in the rice double mutants defective in both CET pathways (Wada et al., 2018). On the basis of these observations, a possible idea is the involvement of the NDH complex in the alleviation of oxidative stresses in chloroplasts. The severe mutant phenotypes in the regulation of photosynthesis suggest the oxidative stress in the chloroplasts of the *pgr5* mutant. Impacts of a series of environmental stresses including high light (Endo, Shikanai, Takabayashi, Asada, & Sato, 1999) and low temperature (Yamori, Sakata, Suzuki, Shikanai, & Makino, 2011) have been tested in the Arabidopsis, tobacco, and rice NDH-less mutants. The mutants are slightly sensitive to the stresses but the mechanism as to how the chloroplast NDH complex alleviates the stresses has been unclear. In the original idea, the oxidative stresses may induce chloroplast conditions similar to those in the *pgr5* mutant in the WT plants. However, we may have overlooked the fact that the size of *pmf* is lower in the *pgr5* mutant background than in WT plants. Because the NDH complex pumps H^+ across the thylakoid membrane coupled with electron transport, it may not be active in the presence of large *pmf*. If this were the case, we might have to focus on the low light conditions, where the size of *pmf* is not saturated. In fact, electron transport is affected at the low light intensity in the rice mutant defective in the chloroplast NDH complex (Yamori, Shikanai, & Makino, 2015). A similar mutant phenotype was also observed in the *Marchantia* mutant defective in the *ndhB* gene (Ueda et al., 2012). In angiosperms, the NDH-less mutants were first generated in tobacco (*Nicotiana tabacum*) (Burrows et al., 1998; Shikanai et al., 1998) and this characterization was followed by the analysis of a series of the Arabidopsis *crr* mutants (Hashimoto, Endo, Peltier, Tasaka, & Shikanai, 2003). Contribution of the NDH complex to total CET activity is relatively low in Arabidopsis and tobacco in angiosperms and this is one of the reasons why we have overlooked the significant contribution of the NDH complex at a low light intensity. In fact, careful inspection of the Arabidopsis *crr2-2* mutant defective in the expression of the *ndhB* gene clarified the slightly highly reduced state of the PQ pool at low light intensity (Nakano et al., 2019). The chloroplast NDH complex may significantly function when the size of *pmf* is not large at a low light intensity. The rice mutant defective in the chloroplast NDH complex is also sensitive to fluctuating light intensity (Yamori, Makino, & Shikanai, 2016). Immediately after the sift from low light, in which the size of *pmf* is low, to high light, the NDH-dependent CET also acidifies the thylakoid lumen to induce the donor-side regulation rapidly, as well as the PGR5-dependent CET. A question is still remaining on the physiological function of the NDH complex in angiosperms.

In C_4 photosynthesis, extra ATP is required for energizing the CO_2 concentration mechanism (Munekage, 2016). Although the ATP/NADPH ratio required by the Calvin-Benson cycle is 1.5 in the absence of photorespiration in C_3 plants, it is increased to 2.5 in C_4 plants for recycling phosphoenolpyruvate. In plants operating NADP-malic enzyme (ME)-type C_4 photosynthesis like maize (*Zea mays*) and *Flaveria bidentis*, extra ATP is synthesized by the NDH-dependent CET in bundle-sheath cells (Ishikawa et al., 2016; Munekage et al., 2010; Peterson, Schultes, McHale, & Zelitch, 2016; Takabayashi, Kishine, Asada, Endo, & Sato, 2005). Because the NDH complex is more efficient machinery for the *pmf* formation than the PGR5-dependent CET, it would be reasonable to use the NDH complex in C_4 photosynthesis requiring more ATP. As discussed above, however, the NDH complex may not be able to efficiently operate in the presence of large *pmf*. How do NADP-ME-type C_4 plants utilize the NDH complex as machinery for supplying ATP during steady-state photosynthesis? In the NADP-ME-type C_4 plants, the level of PSII is extremely low in the bundle-sheath cells (Majeran, Cai, Sun, & van Wijk, 2005). It is probably unnecessary to control LET by the qE induction and the donor-side regulation via the lumenal acidification in the bundle-sheath cells. Consequently, it may be possible to keep the low level of *pmf* during the steady-state photosynthesis in the bundle-sheath cells. More research is required to clarify how the electron transport is regulated in a cell-type-specific manner in C_4 plants.

5. Regulation of the *pmf* components

The *pmf* contributes to both ATP synthesis and downregulation of electron transport. How are two conflicting mechanisms of the accelerator (ATP synthesis) and the brakes (qE and the donor-side regulation) regulated by a single factor? The *pmf* is composed of two components, ΔpH and ΔΨ (Fig. 2). Although both components contribute to ATP synthesis, the only ΔpH downregulates electron transport by acidifying the thylakoid lumen (Fig. 3). It is possible to control the operation of the accelerator and the brakes by fine-tuning the partitioning of ΔpH and ΔΨ in *pmf*. In fact, at moderate light intensity (100–200 μmol photons m^{-2} s^{-1} for Arabidopsis plants cultured in a growth chamber), the size of *pmf* is already saturated but NPQ (qE) is not fully induced. This is achieved by a larger contribution of ΔΨ to *pmf* at a moderate light intensity, although ΔpH preferentially contributes to *pmf* at higher light intensity (Yamamoto & Shikanai, 2019). This regulation is important not to dissipate absorbed light energy, when light intensity limits the rate of photosynthesis.

How is the partitioning of two components regulated? Photosynthetic electron transport is coupled with H^+ translocation across the thylakoid membrane. But the thylakoid lumen is not immediately acidified by the H^+ uptake because H^+ are absorbed by negative buffering charges which are abundant in the thylakoid lumen (Fig. 6A). As a result, the energy of the H^+ translocation is stored in the form of ΔΨ (Kramer, Cruz, & Kanazawa, 2003). To acidify the thylakoid lumen by releasing H^+ from the buffering anions, the partitioning of ΔΨ should be reduced, because there is a limitation in the size of *pmf* (ΔpH + ΔΨ). This is accomplished by the function of ion channels localized to the thylakoid membrane (Dukic et al., 2019) (Fig. 6B). TPK3 is a potassium (K^+) channel and has been proposed to induce NPQ by leaking K^+ from the thylakoid lumen (Carraretto et al., 2013), although TPK3 has been recently proposed to

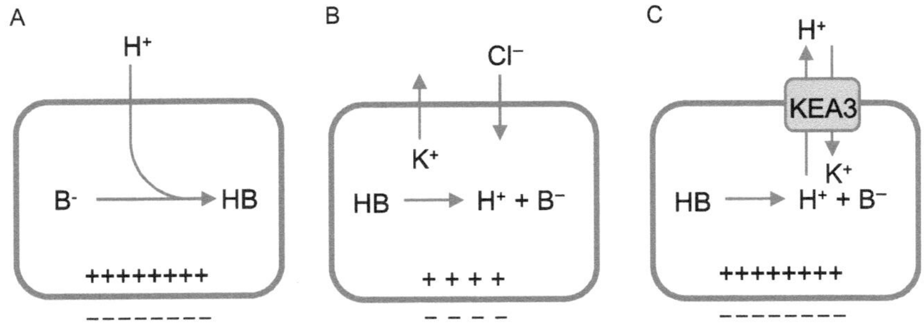

Fig. 6 Regulation of *pmf* partitioning. (A) Photosynthetic electron transport is coupled with H^+ uptake by the thylakoid membranes. The thylakoid lumen is not immediately acidified because buffering anions (B^-) absorbed H^+. The energy of H^+ translocation is stored as ΔΨ. (B) To acidify the thylakoid lumen, ΔΨ is relaxed with the aid of ion channels. The lumenal acidification downregulates electron transport, as described in this chapter. (C) To relax downregulation of electron transport, KEA3 probably antiport H^+ and K^+ to substitute ΔpH by ΔΨ.

localize to the vacuole (Höhner et al., 2019). Two chloride (Cl^-) channels, VCCN1 and CLCe also contribute to a similar process by incorporating Cl^- into the thylakoid lumen (Duan et al., 2016; Herdean, Nziengui, et al., 2016; Herdean, Teardo, et al., 2016). In the absence of these channels, ΔΨ contributes more to *pmf* than in the WT, leading to the reduced induction of the downregulation of electron transport (Dukic et al., 2019). It is still unclear how the activity of channels is regulated to balance the operation of the accelerator and the brakes.

In contrast to the function of ion channels, a putative H^+/K^+ antiporter KEA3 relaxes ΔpH by substituting it by ΔΨ (Armbruster et al., 2014; Kunz et al., 2014). During the induction of photosynthesis, NPQ is transiently induced even at low-to-moderate light intensity because the Calvin-Benson cycle is not fully activated. PGR5-dependent CET is necessary to induce this transient NPQ by recycling electrons especially when the acceptor from PSI is limited (Munekage et al., 2002). In contrast, KEA3 is necessary to immediately relax this transient NPQ once the CO_2 fixation starts operating by exchanging H^+ by K^+ (Fig. 6C). In the *kea3* mutant, absorbed light energy is dissipated as heat under the conditions where the light intensity limits the rate of photosynthesis. The collaboration of the PGR5-dependent CET and KEA3 is required for inducing photosynthesis efficiently.

The Arabidopsis *disturbed proton gradient regulation* (*dpgr*) mutant was isolated on the basis of its high chlorophyll fluorescence at the low light intensity in the artificial air (CO_2-free air including 5% O_2) (Wang et al., 2017). The *dpgr* mutant has a single amino acid alteration in the gene encoding KEA3. The *dpgr* mutation does not affect the stability of KEA3 but enhances or modifies its activity (Tsujii et al., 2019; Wang & Shikanai, 2019). The *dpgr* mutant could not induce transient NPQ during the induction of photosynthesis especially after overnight dark adaptation, in which the Calvin-Benson cycle enzymes are inactivated (Wang et al., 2017). On the basis of this observation, we concluded that KEA3 should be inactivated during the early induction of photosynthesis.

Two routes of CET around PSI play a central role in the regulation of the size of *pmf*. To balance the strength of the accelerator and the breaks, the partitioning of *pmf* components (ΔpH and $\Delta\Psi$) is regulated by the function of ion channels and transporters localized to the thylakoid membrane. It is necessary to coordinate the regulation of size and partitioning but the information is still very limited on the cross talk between the size and partitioning regulation. The introduction of the Flv-dependent pseudo-CET complements the function of PGR5-dependent CET in the size of *pmf* (Yamamoto, Takahashi, et al., 2016). But the qE induction is only partly restored at moderate light intensity because of the leger contribution of $\Delta\Psi$ to *pmf*. This observation suggests the unknown cross talk between the PGR5-dependent CET and the regulatory system for the *pmf* partitioning. The exogenous Flv system is unlikely linked to the regulation of *pmf* partitioning. Especially on the regulatory system for the partitioning, our molecular knowledge has been still very limited. We are understanding that multiple steps of tradeoff are optimized in the metabolic network of photosynthesis. To improve the efficiency of photosynthesis, it is probably necessary to reoptimize this network of regulation to the environments used in agriculture.

6. Conclusion and future prospective

For the optimization of photosynthesis under fluctuating light conditions, it is necessary to precisely regulate the size of *pmf* and the partitioning of *pmf* components. Two CET pathways depending on PGR5/PGRL1 proteins and the NDH complex play a central role in the regulation of the size of *pmf*. In contrast, ion channels and transporters including KEA3 dynamically regulate the partitioning of the *pmf* components. It is still unclear how the activity of each regulatory machinery is regulated. It is also necessary to understand the cross talk between the regulatory systems. It is still a mystery why angiosperms have lost Flvs during the evolution, although they could function as an excellent safety valve in transgenic Arabidopsis and rice (Wada et al., 2018; Yamamoto, Takahashi, et al., 2016). The basic machinery of photosynthesis is highly conserved among O_2-evolving phototrophs and it may not be simple to artificially improve the performance. In contrast, the regulatory systems are rather divergent among the plant lineage. The systemic biological approach may be able to answer the question why angiosperms have selected the current strategy and whether there have been alternative approaches to adapt different light environments.

Acknowledgment

This work was supported by the Japanese Society for the Promotion of Science KAKENHI (16H06553 and 19H00992).

References

Allahverdiyeva, Y., Mustila, H., Ermakova, M., Bersanini, L., Richaud, P., Ajlani, G., et al. (2013). Flavodiiron proteins Flv1 and Flv3 enable cyanobacterial growth and photosynthesis under fluctuating light. *Proceeding of National Academy of Sciences of the United States of America*, *110*, 4111–4116.

Allahverdiyeva, Y., Suorsa, M., Tikkanen, M., & Aro, E.-M. (2015). Photoprotection of photosystems in fluctuating light intensities. *Journal of Experimental Botany*, *66*, 2427–2436.

Allen, J. F. (2002). Photosynthesis of ATP—Electrons, proton pumps, rotors, and poise. *Cell*, *110*, 273–276.

Armbruster, U., Carrillo, L. R., Venema, K., Pavlovic, L., Schmidtmann, E., Kornfeld, A., et al. (2014). Ion antiport accelerates photosynthetic acclimation in fluctuating light environments. *Nature Communications*, *5*, 5439.

Arnon, D. I., Allen, M. B., & Whatley, F. R. (1954). Photosynthesis by isolated chloroplasts. *Nature*, *174*, 394–396.

Asada, K. (1999). The water-water cycle in chloroplasts: Scavenging of active oxygens and dissipation of excess photons. *Annual Review of Plant Physiology and Plant Molecular Biology*, *50*, 601–639.

Avenson, T. J., Cruz, J. A., Kanazawa, A., & Kramer, D. M. (2005). Regulating the proton budget of higher plant photosynthesis. *Proceeding of National Academy of Sciences of the United States of America*, *102*, 9709–9713.

Bendall, D. S., & Manasse, R. S. (1995). Cyclic photophosphorylation and electron transport. *Biochimica et Biophysica Acta*, *1229*, 23–38.

Burrows, P. A., Sazanov, L. A., Svab, Z., Maliga, P., & Nixon, P. J. (1998). Identification of a functional respiratory complex in chloroplasts through analysis of tobacco mutants containing disrupted plastid *ndh* genes. *The EMBO Journal*, *17*, 868–876.

Carraretto, L., Formentin, E., Teardo, E., Checchetto, V., Tomizioli, M., Morosinotto, T., et al. (2013). A thylakoid-located two-pore K^+ channel controls photosynthetic light utilization in plants. *Science*, *342*, 114–118.

Chaux, F., Burlacot, A., Mekhalfi, M., Auroy, P., Blangy, S., Richaud, P., et al. (2017). Flavodiiron proteins promote fast and transient O_2 photoreduction in *Chlamydomonas*. *Plant Physiology*, *174*, 1825–1836.

de Wolf, F. A., Krab, K., Visschers, R. W., de Waard, J. H., & Kraayenhof, R. (1998). Studies on well-coupled photosystem 1-enriched subchloroplast vesicles -characteristics and reinterpretation of single-turnover cyclic electron transfer. *Biochimica et Biophysica Acta-Bioenergetics, 936*, 487–503.

Duan, Z., Kong, F., Zhang, L., Li, W., Zhang, J., & Peng, L. (2016). A bestrophin-like protein modulates the proton motive force across the thylakoid membrane in Arabidopsis. *Journal of Integrative Plant Biology, 58*, 848–858.

Dukic, E., Herdean, A., Cheregi, O., Sharma, A., Nziengui, H., Dmitruk, D., et al. (2019). K^+ and Cl^- channels/transporters independently fine-tune photosynthesis in plants. *Scientific Reports, 9*, 8639.

Endo, T., Shikanai, T., Takabayashi, A., Asada, K., & Sato, F. (1999). The role of chloroplastic NAD(P)H dehydrogenase in photoprotection. *FEBS Letters, 457*, 5–8.

Fan, X., Zhang, J., Li, W., & Peng, L. (2015). The NdhV subunit is required to stabilize the chloroplast NADH dehydrogenase-like complex in Arabidopsis. *The Plant Journal, 82*, 221–231.

Gao, F., Zhao, J., Wang, X., Qin, S., Wei, L., & Ma, W. (2016). NdhV is a subunit of NADPH dehydrogenase essential for cyclic electron transport in *Synechocystis* sp. strain PCC 6803. *Plant Physiology, 170*, 752–760.

Grieco, M., Tikkanen, M., Paakkarinen, V., Kangasjärvi, S., & Aro, E.-M. (2012). Steady-state phosphorylation of light-harvesting complex II proteins preserves photosystem I under fluctuating white light. *Plant Physiology, 160*, 1896–1910.

Hashimoto, M., Endo, T., Peltier, G., Tasaka, M., & Shikanai, T. (2003). A nucleus-encoded factor, CRR2, is essential for the expression of chloroplast *ndhB* in *Arabidopsis*. *The Plant Journal, 36*, 541–549.

Herdean, A., Nziengui, H., Zsiros, O., Solymosi, K., Garab, G., Lundin, B., et al. (2016). The Arabidopsis thylakoid chloride channel AtCLCe functions in chloride homeostasis and regulation of photosynthetic electron transport. *Frontiers in Plant Science*, 7, 115.

Herdean, A., Teardo, E., Nilsson, A. K., Pfeil, B. E., Johansson, O. N., Ünnep, R., et al. (2016). A voltage-dependent chloride channel fine-tunes photosynthesis in plants. *Nature Communications*, 7, 11654.

Hertle, A. P., Blunder, T., Wunder, T., Pesaresi, P., Pribil, M., Armbruster, U., et al. (2013). PGRL1 is the elusive ferredoxin-plastoquinone reductase in photosynthetic cyclic electron flow. *Molecular Cell, 49*, 511–523.

Höhner, R., Galvis, V. C., Strand, D. D., Völkner, C., Krämer, M., Messer, M., et al. (2019). Photosynthesis in Arabidopsis is unaffected by the function of the vacuolar K^+ channel TPK3. *Plant Physiology, 180*, 1322–1335.

Ido, K., Ifuku, K., Yamamoto, Y., Ishihara, S., Murakami, A., Takabe, K., et al. (2009). Knockdown of the PsbP protein does not prevent assembly of the dimeric PSII core complex but impairs accumulation of photosystem II supercomplexes in tobacco. *Biochimica Biophysica Acta-Bioenergetics, 1787*, 873–881.

Ifuku, K., Endo, T., Shikanai, T., & Aro, E.-M. (2011). Structure of the chloroplast NADH dehydrogenase-like complex: Nomenclature for nuclear-encoded subunits. *Plant Cell Physiology, 52*, 1560–1568.

Ilik, P., Pavlovič, A., Kouril, R., Alboresi, A., Morosinotto, T., Allahverdieyeva, Y., et al. (2017). Alternative electron transport mediated by flavodiiron proteins is operational in organisms from cyanobacteria up to gymnosperms. *New Phytologist, 214*, 967–972.

Ishihara, S., Takabayashi, A., Ido, K., Endo, T., Ifuku, K., & Sato, F. (2007). Distinct functions for the two PsbP-like proteins PPL1 and PPL2 in the chloroplast thylakoid lumen of Arabidopsis. *Plant Physiology, 145*, 668–679.

Ishikawa, N., Takabayashi, A., Noguchi, K., Tazoe, Y., Yamamoto, H., von Caemmerer, S., et al. (2016). NDH-mediated cyclic electron flow around photosystem I is crucial for C_4 photosynthesis. *Plant Cell Physiology, 57*, 2020–2028.

Jahns, P., Graf, M., Munekage, Y., & Shikanai, T. (2002). Single point mutation in the Rieske iron-sulfur subunit of cytochrome b6/f leads to an altered pH dependence of plastoquinol oxidation in Arabidopsis. *FEBS Letters*, *519*, 99–102.

Järvi, S., Suorsa, M., Paakkarinen, V., & Aro, E.-M. (2011). Optimized native gel systems for separation of thylakoid protein complexes: Novel super- and mega-complexes. *Biochemical Journal*, *439*, 207–214.

Kato, Y., Odahara, M., Fukao, Y., & Shikanai, T. (2018). Stepwise evolution of supercomplex formation with photosystem I is required for stabilization of chloroplast NADH dehydrogenase-like complex: Lhca5-dependent supercomplex formation in *Physcomitrella patens*. *The Plant Journal*, *96*, 937–948.

Kato, Y., Sugimoto, K., & Shikanai, T. (2018). NDH-PSI supercomplex assembly precedes full assembly of the NDH complex in chloroplast. Plant Physiol*ogy*, *176*, 1728–1738.

Kouřil, R., Strouhal, O., Nosek, L., Lenobel, R., Chamrád, I., Boekema, E. J., et al. (2014). Structural characterization of a plant photosystem I and NAD(P)H dehydrogenase supercomplex. *The Plant Journal*, *77*, 568–576.

Kramer, D. M., Cruz, J. A., & Kanazawa, A. (2003). Balancing the central roles of the thylakoid proton gradient. *Trends in Plant Science*, *8*, 27–32.

Külheim, C., Agren, J., & Jansson, S. (2002). Rapid regulation of light harvesting and plant fitness in the field. *Science*, *297*, 91–93.

Kunz, H. H., Gierth, M., Herdean, A., Satoh-Cruz, M., Kramer, D. M., Spetea, C., et al. (2014). Plastidial transporters KEA1, −2, and −3 are essential for chloroplast osmoregulation, integrity, and pH regulation in Arabidopsis. *Proceeding of National Academy of Sciences of the United States of America*, *111*, 7480–7485.

Laughlin, T. G., Bayne, A. N., Trempe, J. F., Savage, D. F., & Davies, K. M. (2019). Structure of the complex I-like molecule NDH of oxygenic photosynthesis. *Nature*, *566*, 411–414.

Li, X. P., Björkman, O., Shih, C., Grossman, A. R., Rosenquist, M., Jansson, S., et al. (2000). A pigment-binding protein essential for regulation of photosynthetic light harvesting. *Nature*, *403*, 391–395.

Majeran, W., Cai, Y., Sun, Q., & van Wijk, K. J. (2005). Functional differentiation of bundle sheath and mesophyll maize chloroplasts determined by comparative proteomics. *The Plant Cell*, *17*, 3111–3140.

Majeran, W., Zybailov, B., Ytterberg, A. J., Dunsmore, J., Sun, Q., & van Wijk, K. J. (2008). Consequences of C_4 differentiation for chloroplast membrane proteomes in maize mesophyll and bundle sheath cells. *Molecular & Cellular Proteomics*, 7, 1609–1638.

Matsubayashi, T., Wakasugi, T., Shinozaki, K., Yamaguchi-Shinozaki, K., Zaita, N., Hidaka, T., et al. (1987). Six chloroplast genes (*ndhA-F*) homologous to human mitochondrial genes encoding components of the respiratory chain NADH dehydrogenase are actively expressed: Determination of the splice sites in *ndhA* and *ndhB* pre-mRNAs. *Molecular and General Genetics*, *210*, 385–393.

Munekage, Y. N. (2016). Light harvesting and chloroplast electron transport in NADP-malic enzyme type C_4 plants. *Current Opinion in Plant Biology*, *31*, 9–15.

Munekage, Y. N., Eymery, F., Rumeau, D., Cuiné, S., Oguri, M., Nakamura, N., et al. (2010). Elevated expression of PGR5 and NDH-H in bundle sheath chloroplasts in C_4 *Flaveria* species. *Plant Cell Physiology*, *51*, 664–668.

Munekage, Y., Hashimoto, M., Miyake, C., Tomizawa, K., Endo, T., Tasaka, M., et al. (2004). Cyclic electron flow around photosystem I is essential for photosynthesis. *Nature*, *429*, 579–582.

Munekage, Y., Hojo, M., Meurer, J., Endo, T., Tasaka, M., & Shikanai, T. (2002). *PGR5* is involved in cyclic electron flow around photosystem I and is essential for photoprotection in Arabidopsis. *Cell*, *110*, 361–371.

Munekage, Y., Takeda, S., Endo, T., Jahns, P., Hashimoto, T., & Shikanai, T. (2001). Cytochrome b_6f mutation specifically affects thermal dissipation of absorbed light energy in Arabidopsis. *The Plant Journal*, *28*, 351–359.

Muraoka, R., Okuda, K., Kobayashi, Y., & Shikanai, T. (2006). A eukaryotic factor required for accumulation of the chloroplast NAD(P)H dehydrogenase complex in Arabidopsis. Plant Physiol*ogy*, *142*, 1683–1689.

Nakano, H., Yamamoto, H., & Shikanai, T. (2019). Contribution of NDH-dependent cyclic electron transport around photosystem I to the generation of proton motive force in the weak mutant allele of *pgr5*. *Biochimica Biophysica Acta-Bioenergetics*, *1860*, 369–374.

Nandha, B., Finazzi, G., Joliot, P., Hald, S., & Johnson, G. N. (2007). The role of PGR5 in the redox poising of photosynthetic electron transport. *Biochimica Biophysica Acta-Bioenergetics*, *1767*, 1252–1259.

Nishio, J. N., & Whitmarsh, J. (1993). Dissipation of the proton electrochemical potential in intact chloroplasts (II. The pH gradient monitored by cytochrome f reduction kinetics). *Plant Physiology*, *101*, 89–96.

Ogawa, T. (1991). A gene homologous to the subunit-2 gene of NADH dehydrogenase is essential to inorganic carbon transport of Synechocystis PCC6803. *Proceeding of National Academy of Sciences of the United States of America*, *88*, 4275–4279.

Otani, T., Kato, Y., & Shikanai, T. (2018). Specific substitutions of light-harvesting complex I proteins associated with photosystem I are required for supercomplex formation with chloroplast NADH dehydrogenase-like complex. *The Plant Journal*, *94*, 122–130.

Otani, T., Yamamoto, H., & Shikanai, T. (2017). Stromal loop of Lhca6 is responsible for the linker function required for the NDH-PSI supercomplex formation. Plant Cell Physiol*ogy*, *58*, 851–861.

Pan, X., Cao, D., Xie, F., Xu, F., Su, X., Mi, H., et al. (2020). Structural basis for electron transport mechanism of complex I-like photosynthetic NAD(P)H dehydrogenase. *Nature Communications*, *11*, 610.

Parey, K., Wirth, C., Vonck, J., & Zickermann, V. (2020). Respiratory complex I—Structure, mechanism and evolution. *Current Opinion in Structural Biology*, *63*, 1–9.

Peltier, G., Aro, E.-M., & Shikanai, T. (2016). NDH-1 and NDH-2 plastoquinone reductases in oxygenic photosynthesis. *Annual Review of Plant Biology*, *67*, 55–80.

Peng, L., Cai, W., & Shikanai, T. (2010). Chloroplast stromal proteins, CRR6 and CRR7, are required for assembly of the NAD(P)H dehydrogenase subcomplex A in Arabidopsis. *The Plant Journal*, *63*, 203–211.

Peng, L., Fukao, Y., Fujiwara, M., & Shikanai, T. (2012). Multistep assembly of chloroplast NADH dehydrogenase-like subcomplex a requires several nucleus-encoded proteins, including CRR41 and CRR42, in Arabidopsis. *The Plant Cell*, *24*, 202–214.

Peng, L., Fukao, Y., Fujiwara, M., Takami, T., & Shikanai, T. (2009). Efficient operation of NAD(P)H dehydrogenase requires supercomplex formation with photosystem I via minor LHCI in *Arabidopsis*. *The Plant Cell*, *21*, 3623–3640.

Peng, L., Fukao, Y., Myouga, F., Motohashi, R., Shinozaki, K., & Shikanai, T. (2011). A chaperonin subunit with unique structures is essential for folding of a specific substrate. *PLoS Biology*, *9*, e1001040.

Peng, L., & Shikanai, T. (2011). Supercomplex formation with photosystem I is required for the stabilization of the chloroplast NADH dehydrogenase-like complex in Arabidopsis. Plant Physiol*ogy*, *155*, 1629–1639.

Peng, L., Shimizu, H., & Shikanai, T. (2008). The chloroplast NAD(P)H dehydrogenase complex interacts with photosystem I in Arabidopsis. *Journal of Biological Chemistry*, *283*, 34873–34879.

Peterson, R. B., Schultes, N. P., McHale, N. A., & Zelitch, I. (2016). Evidence for a role for NAD(P)H dehydrogenase in concentration of CO_2 in the bundle sheath cell of *Zea mays*. *Plant Physiology*, *171*, 125–138.

Qin, X., Suga, M., Kuang, T., & Shen, J.-R. (2019). Structural basis for energy transfer pathways in the plant PSI-LHCI supercomplex. *Science, 348*, 989–995.

Rantala, S., Lempiäinen, T., Gerotto, C., Tiwari, A., Aro, E.-M., & Tikkanen, M. (2020). PGR5 and NDH-1 systems do not function as protective electron acceptors but mitigate the consequences of PSI inhibition. *Biochimica Biophysica Acta-Bioenergetics, 1861*, 148154.

Ruban, A. V. (2018). Light harvesting control in plants. *FEBS Letters, 592*, 3030–3039.

Rumberg, B., & Siggel, U. (1969). pH changes in the inner phase of the thylakoids during photosynthesis. *Naturwissenschaften, 56*, 130–132.

Rumeau, D., Becuwe-Linka, N., Beyly, A., Louwagie, M., Garin, J., & Peltier, G. (2005). New subunits NDH-M, -N, and -O, encoded by nuclear genes, are essential for plastid Ndh complex functioning in higher plants. *The Plant Cell, 17*, 219–232.

Schuller, J. M., Birrell, J. A., Tanaka, H., Konuma, T., Wulfhorst, H., Cox, N., et al. (2019). Structural adaptations of photosynthetic complex I enable ferredoxin-dependent electron transfer. *Science, 363*, 257–260.

Shikanai, T. (2007). Cyclic electron transport around photosystem I; genetic approaches. *Annual Review of Plant Biology, 58*, 199–217.

Shikanai, T., Endo, T., Hashimoto, T., Yamada, Y., Asada, K., & Yokota, A. (1998). Directed disruption of the tobacco ndhB gene impairs cyclic electron flow around photosystem I. *Proceeding of National Academy of Sciences of the United States of America, 95*, 9705–9709.

Shikanai, T., & Yamamoto, H. (2017). Contribution of cyclic and pseudo-cyclic electron transport to the formation of proton motive force in chloroplasts. *Molecular Plant, 10*, 20–29.

Sirpiö, S., Holmström, M., Battchikova, N., & Aro, E.-M. (2009). AtCYP20-2 is an auxiliary protein of the chloroplast NAD(P)H dehydrogenase complex. *FEBS Letters, 583*, 2355–2358.

Strand, D. D., Fisher, N., & Kramer, D. M. (2017). The higher plant plastid NAD(P)H dehydrogenase-like complex (NDH) is a high efficiency proton pump that increases ATP production by cyclic electron flow. *Journal of Biological Chemistry, 292*, 11850–11860.

Sugimoto, K., Okegawa, Y., Tohri, A., Long, T. A., Sarah, F. S., Hisabori, T., et al. (2013). A single amino acid alteration in PGR5 confers resistance to antimycin A in cyclic electron transport around PSI. Plant Cell Physio*logy, 54*, 1525–1534.

Suorsa, M., Järvi, S., Grieco, M., Nurmi, M., Pietrzykowska, M., Rantala, M., et al. (2012). PROTON GRADIENT REGULATION5 is essential for proper acclimation of Arabidopsis photosystem I to naturally and artificially fluctuating light conditions. *The Plant Cell, 24*, 2934–2948.

Suorsa, M., Sirpiö, S., Paakkarinen, V., Kumari, N., Holmström, M., & Aro, E.-M. (2010). Two proteins homologous to PsbQ are novel subunits of the chloroplast NAD(P)H dehydrogenase. *Plant Cell Physiology, 51*, 877–883.

Tagawa, K., Tsujimoto, H. Y., & Arnon, D. I. (1963). Role of chloroplast ferredoxin in the energy conversion process of photosynthesis. *Proceeding of National Academy of Sciences of the United States of America, 49*, 567–572.

Takabayashi, A., Kishine, M., Asada, K., Endo, T., & Sato, F. (2005). Differential use of two cyclic electron flows around photosystem I for driving CO2-concentration mechanism in C4 photosynthesis. *Proceeding of National Academy of Sciences of the United States of America, 102*, 16898–16903.

Takagi, D., & Miyake, C. (2018). PROTON GRADIENT REGULATION 5 supports linear electron flow to oxidize photosystem I. *Physiologia Plantarum, 164*, 337–348.

Tikhonov, A. N. (2014). The cytochrome b_6f complex at the crossroad of photosynthetic electron transport pathways. *Plant Physiology and Biochemistry, 81*, 163–183.

Tikhonov, A. N., Khomutov, G. B., Ruuge, E. K., & Blumenfeld, L. A. (1981). Electron transport control in chloroplasts: Effects of photosynthetic control monitored by the intrathylakoid pH. *Biochimica et Biophysica Acta*, *637*, 321–333.

Tikkanen, M., Grieco, M., Kangasjärvi, S., & Aro, E.-M. (2010). Thylakoid protein phosphorylation in higher plant chloroplasts optimizes electron transfer under fluctuating light. *Plant Physiology*, *152*, 723–735.

Tiwari, A., Mamedov, F., Grieco, M., Suorsa, M., Jajoo, A., Styring, S., et al. (2016). Photodamage of iron-Sulphur clusters in photosystem I induces non-photochemical energy dissipation. *Nature Plants*, *2*, 16035.

Tsujii, M., Kera, K., Hamamoto, S., Kuromori, T., Shikanai, T., & Uozumi, N. (2019). Evidence for potassium transport activity of Arabidopsis KEA1-KEA6. *Scientific Reports*, *9*, 10040.

Ueda, M., Kuniyoshi, T., Yamamoto, H., Sugimoto, K., Ishizaki, K., Kohchi, T., et al. (2012). Composition and physiological function of the chloroplast NADH dehydrogenase-like complex in *Marchantia polymorpha*. *The Plant Journal*, *72*, 683–693.

Vicente, J. B., Gomes, C. M., Wasserfallen, A., & Teixeira, M. (2002). Module fusion in an A-type flavoprotein from the cyanobacterium *Synechocystis* condenses a multiple-component pathway in a single polypeptide chain. *Biochemical Biophysical Research Communications*, *294*, 82–87.

Wada, S., Yamamoto, H., Suzuki, Y., Yamori, Y., Shikanai, T., & Makino, A. (2018). Flavodiiron protein substitutes for cyclic electron flow without competing CO_2 assimilation in rice. Plant Physiol*ogy*, *176*, 1509–1518.

Wang, C., & Shikanai, T. (2019). Modification of activity of the thylakoid H^+/K^+ antiporter KEA3 disturbs ΔpH-dependent regulation of photosynthesis. Plant Physiol*ogy*, *181*, 762–773.

Wang, C., Takahashi, H., & Shikanai, T. (2018). PROTON GRADIENT REGULATION 5 contributes to ferredoxin-dependent cyclic phosphorylation in ruptured chloroplasts. *Biochimica Biophysica Acta-Bioenergetics*, *1859*, 1173–1179.

Wang, C., Yamamoto, H., Narumiya, F., Munekage, Y. N., Finazzi, G., Szabo, I., et al. (2017). Fine-tuned regulation of the K^+/H^+ antiporter KEA3 is required to optimize photosynthesis during induction. *The Plant Journal*, *89*, 540–553.

Wang, C., Yamamoto, H., & Shikanai, T. (2015). Role of cyclic electron transport around photosystem I in regulating proton motive force. *Biochimica Biophysica Acta-Bioenergetics*, *1847*, 931–938.

Yabuta, S., Ifuku, K., Takabayashi, A., Ishihara, S., Ido, K., Ishikawa, N., et al. (2010). Three PsbQ-like proteins are required for the function of the chloroplast NAD(P)H dehydrogenase complex in Arabidopsis. *Plant Cell Physiology*, *51*, 866–876.

Yamamoto, H., Fan, X., Sugimoto, K., Fukao, Y., Peng, L., & Shikanai, T. (2016). CHLORORESPIRATORY REDUCTION 9 is a novel factor required for formation of subcomplex A of the chloroplast NADH dehydrogenase-like complex. Plant Cell Physiol*ogy*, *57*, 2122–2132.

Yamamoto, H., Peng, L., Fukao, Y., & Shikanai, T. (2011). An Src homology 3 domain-like fold protein forms a ferredoxin-binding site for the chloroplast NADH dehydrogenase-like complex in *Arabidopsis*. *The Plant Cell*, *23*, 1480–1493.

Yamamoto, H., & Shikanai, T. (2013). *In planta* mutagenesis of Src homology 3 domain-like fold of NdhS, a ferredoxin-binding subunit of the chloroplast NADH dehydrogenase-like complex in Arabidopsis. A conserved Arg-193 plays a critical role in ferredoxin binding. *The Journal of Biological Chemistry*, *288*, 36328–36337.

Yamamoto, H., & Shikanai, T. (2019). PGR5-dependent cyclic electron flow protects PSI under fluctuating light at donor and acceptor sides. *Plant Physiology*, *179*, 588–600.

Yamamoto, H., Takahashi, S., Badger, M. R., & Shikanai, T. (2016). Artificial remodeling of alternative electron flow by flavodiiron proteins in Arabidopsis. *Nature Plants*, *2*, 16012.

Yamori, W., Makino, A., & Shikanai, T. (2016). A physiological role of cyclic electron transport around photosystem I in sustaining photosynthesis under fluctuating light in rice. *Scientific Reports*, *6*, 20147.
Yamori, W., Sakata, N., Suzuki, Y., Shikanai, T., & Makino, A. (2011). Cyclic electron flow around photosystem I via chloroplast NAD(P)H dehydrogenase (NDH) complex performs a significant physiological role during photosynthesis and plant growth at low temperature in rice. *The Plant Journal*, *68*, 966–976.
Yamori, W., & Shikanai, T. (2016). Physiological functions of cyclic electron transport around photosystem I in sustaining photosynthesis and plant growth. *Annual Review of Plant Biology*, *67*, 81–106.
Yamori, W., Shikanai, T., & Makino, A. (2015). Photosystem I cyclic electron flow via chloroplast NADH dehydrogenase-like complex performs a physiological role for photosynthesis at low light. *Scientific Reports*, *5*, 13908.
Zhang, C., Shuai, J., Ran, Z., Zhao, J., Wu, Z., Liao, R., et al. (2020). Structural insights into NDH-1 mediated cyclic electron transfer. *Nature Communications*, *11*, 888.

CHAPTER SEVEN

Regulation of the biogenesis of chloroplast ATP synthase

Lin Zhang[a], Jean-David Rochaix[b], and Lianwei Peng[a,*]

[a]Shanghai Key Laboratory of Plant Molecular Sciences, College of Life Sciences, Shanghai Normal University, Shanghai, China

[b]Departments of Molecular Biology and Plant Biology, University of Geneva, Geneva, Switzerland

*Corresponding author: e-mail address: penglianwei@shnu.edu.cn

Contents

Abstract

Chloroplast ATP synthase belongs to the F-type ATP synthases. It is localized in the thylakoid membrane and consists of nine subunits with a stoichiometry of $\alpha_3\beta_3\gamma_1\varepsilon_1\delta_1a_1b_1b'_1c_{14}$. Since its subunits are encoded by both nuclear and chloroplast genomes, its biogenesis requires an intricate coordination of gene expression between these two genetic systems as well as a tight regulation of assembly by various chloroplast protein chaperones. Isolation of mutants defective in the biogenesis of chloroplast ATP synthase and detailed characterization of the corresponding genes revealed the existence of several pentatricopeptide repeat (PPR) proteins that are required for proper expression of chloroplast *atp* genes at various post-transcriptional steps including RNA splicing, RNA editing, RNA stabilization as well as protein translation. As other F-type ATP synthases, assembly of the chloroplast ATP synthase occurs in a stepwise manner through the formation of several transient intermediates. Several chaperones facilitate the assembly of chloroplast ATP synthase by interacting with specific subunits at different stages. In particular, structural studies of some assembly factors have provided many new insights into the assembly of the chloroplast ATP synthase at

Advances in Botanical Research, Volume 96
ISSN 0065-2296
https://doi.org/10.1016/bs.abr.2020.06.001

© 2020 Elsevier Ltd
All rights reserved.

the molecular level. Because the structure and the assembly pathway of the F-type ATP synthase are highly conserved among most species, some of the assembly mechanisms of chloroplast ATP synthase are likely to be valid for other ATP synthases.

1. Introduction

The light reactions of oxygenic photosynthesis are driven mainly by four protein complexes embedded in the thylakoid membrane of chloroplasts, which are involved in photosynthetic electron transport and ATP synthesis. These complexes include photosystem II (PSII), cytochrome b_6f complex (Cytb_6f), PSI, and ATP synthase. During photosynthesis, light is captured by the light-harvesting-complexes LHCII and LHCI and the energy is then transferred to the reaction centers of the two photosystems for charge separation. PSII uses light energy to split water giving rise to electrons and molecular oxygen. The electrons are then transferred in series through plastoquinone, Cytb_6f, and plastocyanin to PSI, where the electrons on the acceptor side of this complex are used to generate NADPH by the ferredoxin-NADP$^+$-reductase (FNR). Besides this linear electron transport chain, plant chloroplasts also perform cyclic electron transport that is solely driven by PSI. PSI cyclic electron transport involves PGR5/PGRL1- and NDH (NAD(*P*)H dehydrogenase-like complex)-dependent pathways (Yamori & Shikanai, 2016). Electrons from ferredoxin are transferred to plastoquinone via thylakoid-localized PGR5/PGRL1 or NDH complexes and then recycled back to PSI through Cytb_6f for reduction of ferredoxin. Both linear and cyclic electron transport pathways generate a proton gradient across the thylakoid membrane to drive ATP synthesis by chloroplast ATP synthase.

Chloroplast ATP synthase belongs to the F-type ATP synthases ubiquitously present in bacteria, mitochondria, and chloroplasts (Junge & Nelson, 2015; Yoshida, Muneyuki, & Hisabori, 2001). Recently, a high-resolution structure of intact spinach chloroplast ATP synthase was obtained by cryo-electron microscopy (Hahn, Vonck, Mills, Meier, & Kühlbrandt, 2018). The structure reveals that this complex is highly similar to other F-type ATP synthases and it is composed of two subcomplexes, CF_o and CF_1 (Fig. 1). While CF_o subunits are partially or entirely embedded in the thylakoid membrane, the CF_1 subcomplex protrudes from thylakoids into the chloroplast stroma. CF_1 catalyzes ATP synthesis and consists of five

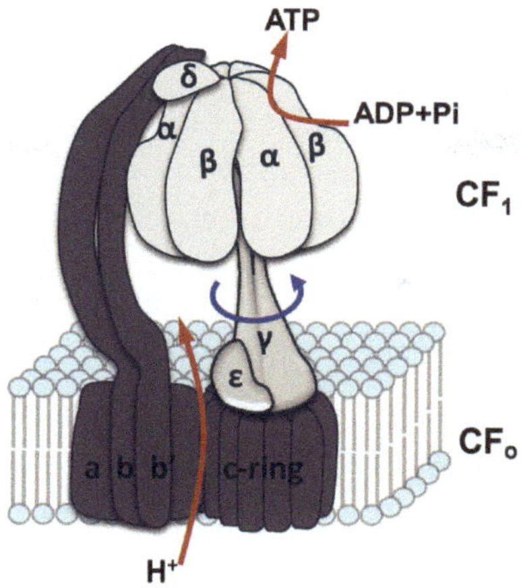

Fig. 1 Structural and working model of chloroplast ATP synthase. Chloroplast ATP synthase consists of two parts, CF_1 and CF_o, whose subunits are shown with white and gray background, respectively. While CF_1 and a part of the b and b′ subunits protrude into the stroma, CF_o subunits are embedded in the thylakoid membrane. Three subunits of α and of β form together a spherical $\alpha_3\beta_3$ hexamer. One end of the central stalk γ/ε inserts into the $\alpha_3\beta_3$ hexamer and the other end is attached to the c-ring subcomplex. The peripheral stalk containing b, b′ and $CF_1\delta$ stabilizes the $\alpha_3\beta_3$ hexamer in the stroma and clamps the a subunit in its position next to the c-ring in the thylakoid membrane. During photosynthesis, protons are pumped into the thylakoid lumen generating the proton motive force for driving chloroplast ATP synthase. Efflux of protons across thylakoids drives the rotation of the c-ring rotor and of the γ/ε stalk, leading to conformational changes in the $\alpha_3\beta_3$ hexamer that ultimately result in ATP synthesis (Hahn et al., 2018).

subunits with the stoichiometry $\alpha_3\beta_3\gamma_1\varepsilon_1\delta_1$. The three α and three β subunits are alternately arranged forming a spherical $\alpha_3\beta_3$ hexamer with a cavity. The γ subunit forms a long α-helix with a coiled-coil domain and is inserted in the $\alpha_3\beta_3$ hexamer forming a central stalk with the ε subunit. CF_o is composed of four subunits with a stoichiometry of $a_1b_1b'_1c_{14}$. A total of 14 c subunits form a rotor ring which acts as a nanomotor. While subunit a is localized next to the c-ring rotor, the peripheral stalk subunits b and b′ clamp the a subunit in its position next to the c-ring. The γ/ε stalk is fixed on the *c*-ring of CF_o and transfers the energy to the $\alpha_3\beta_3$ hexamer, and is maintained by the peripheral stalk subunits b and b′ and the $CF_1\delta$ subunit. During photosynthesis, efflux of protons from thylakoid lumen to stroma drives the rotation of the c-ring rotor and the γ/ε stalk. Rotation of the γ subunit coiled-coil domain inside the $\alpha_3\beta_3$ induces conformational changes in the α/β surface, resulting in the biosynthesis of ATP (Hahn et al., 2018).

Chloroplast ATP synthase comprises nine subunits encoded by both the nuclear and chloroplast genomes (Fig. 1). Subunits $CF_1\gamma$ (*atpC*), $CF_1\delta$ (*atpD*), and b′ (*atpG*) are encoded by the nuclear genome and they are expressed in the cytosol and then imported into chloroplasts through the TOC/TIC translocon in the chloroplast envelope. In Arabidopsis and other dicot species, two genes *atpC1* and *atpC2* encoding $CF_1\gamma$ have been identified (Kohzuma et al., 2012). While the $CF_1\gamma1$-containing ATP synthase represents the dominant ATP synthase in chloroplasts (Dal Bosco et al., 2004), the $CF_1\gamma2$-containing ATP synthase is predominantly expressed in roots or other non-photosynthetic tissues and probably functions in catalyzing ATP hydrolysis-driven proton translocation and maintains a proper proton gradient required for some physiological processes (Kohzuma et al., 2012). Knockout of the *atpC1*, *atpD*, or *atpG* genes results in the loss of chloroplast ATP synthase indicating that these subunits are essential for the formation of the complex (Dal Bosco et al., 2004; Kong, Wang, Yang, & Mi, 2013; Maiwald et al., 2003). In land plants, six of nine subunits are encoded by the chloroplast genome in two polycistronic chloroplast transcription units, the large (*atpI/H/F/A*) and the small (*atpB/E*) *atp* operons. Chloroplasts are semi-autonomous organelles. Although they contain more than 3000 proteins, their genome encodes only about 100 proteins. Synthesis of chloroplast proteins requires coordinated expression of two genomes, and many nucleus-encoded proteins are involved in the precise control of chloroplast gene expression. In the past decade, genetic approaches have revealed several proteins required for expression of the chloroplast *atp* genes. In addition, assembly of the chloroplast ATP synthase is a complicated process that occurs in a stepwise manner and involves several assembly factors isolated and characterized recently.

Here, we summarize recent progress in understanding the molecular mechanisms of chloroplast ATP synthase biogenesis in land plants. We discuss the function of the factors involved in the expression of the chloroplast *atp* genes and in the assembly of this complex. Based on these data we propose a model for the assembly of the chloroplast ATP synthase.

2. Expression of the chloroplast *atp* genes

2.1 Expression of the *atp* synthase operons

In chloroplasts of higher plants, the large ATP synthase operon consists of four genes *atpI/H/F/A*, which encodes the CF_oa, CF_oc, CF_ob, and $CF_1\alpha$ subunits, respectively (Fig. 2A). Transcription of this operon is performed

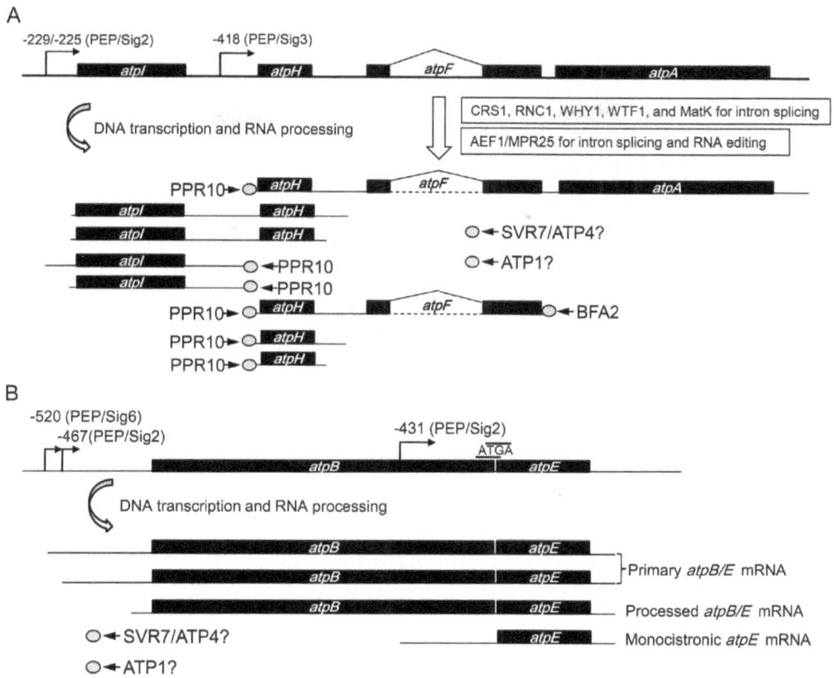

Fig. 2 Scheme of chloroplast *atp* gene expression in Arabidopsis. (A) Model for the expression of the large ATP synthase operon *atpI/H/F/A*. After transcription initiation from the −229/−225 *atpI* and −418 *atpH* promoters performed by PEP/Sig2 and PEP/Sig3, respectively, the primary *atpI/H/F/A* transcripts are processed into mono-, di-, or tri-cistronic transcripts for subsequent translation. Only major transcripts are shown. Several splicing factors including CRS1, WHY1, RNC1, WTF1, and MatK are involved in *atpF* intron splicing. PPR protein AEF1/MPR25 is required for *atpF* intron splicing and RNA editing. Binding of PPR10 to the 5′ end of *atpH* and 3′ end of *psaJ* mRNA as well as binding of BFA2 to the 3′ end of *atpH/F* protects the target RNA from 5′ or 3′ exonucleases, respectively. Binding of PPR10 to the 5′ end of *atpH* also enhances the translation of *atpH*. The targets of the SVR7/ATP4 and unidentified maize ATP1 proteins are still unknown and labeled with question marks. Specific RNA structures in the transcript termini proposed by Malik Ghulam et al. (2013) are not shown in the model. (B) Model for the expression of the small ATP synthase operon *atpB/E*. This operon is transcribed by PEP/Sig6 and PEP/Sig2 holoenzyme at the −520/−467 *atpB* and −431 *atpE* promoters. Two primary *atpB/E* transcripts are further processed producing the processed *atpB/E* transcript. SVR7/ATP4 and unidentified maize ATP1 are proposed to be involved in *atpB/E* translation initiation but the binding sites are not known.

by the plastid-encoded RNA polymerase (PEP) and, in Arabidopsis, it starts at two promoters, the −225/−229 *atpI* promoter at the 5′ end of the large *atp* operon and the −418 *atpH* promoter upstream of the *atpH* gene (Malik Ghulam, Zghidi-Abouzid, Lambert, Lerbs-Mache, & Merendino, 2012).

While expression from the −225/−229 *atpI* promoter is dependent on the Sigma factor 2 (Sig2), that from the −418 *atpH* promoter is specifically dependent on Sig3. It has been proposed that Sig2 is responsible for the basal transcription of the *atp* genes and Sig3 might modulate *atp* expression under certain physiological and environmental conditions (Malik Ghulam et al., 2012). It is currently unclear whether the −418 *atpH* promoter is operational in maize. After transcription, the primary transcripts are processed to mono-, di, or tri-cistronic transcripts for subsequent translation (Fig. 2A). The 5′ or 3′ end of these transcripts either binds specific proteins or forms stem-loop structures for protection against degradation by exoribonucleases (Malik Ghulam et al., 2013). During the past decade, several PPR proteins were identified and shown to associate with these transcripts for their stabilization or activation of translation.

A single group II intron is present in the *atpF* gene that is widely conserved across angiosperms except for some species of Malpighiales (Daniell et al., 2008). In maize, splicing of the *atpF* intron requires several proteins including CRS1, WHY1, RNC1, WTF1, and MatK (Kroeger, Watkins, Friso, van Wijk, & Barkan, 2009; Prikryl, Watkins, Friso, van Wijk, & Barkan, 2008; Till, Schmitz-Linneweber, Williams-Carrier, & Barkan, 2001; Watkins et al., 2007; Zoschke et al., 2010). In addition, there is one RNA editing site at position 92 of the *atpF* transcript (C to T substitution) in Arabidopsis. However, in monocots such as rice and maize, this nucleotide is a T in the chloroplast genome and no editing is required (Yap et al., 2015).

The small ATP synthase operon consists of two genes *atpB/E*, which encodes the $CF_1\beta$ and $CF_1\varepsilon$ subunits, respectively (Fig. 2B). In Arabidopsis, there are two transcription start sites for *atpB*, −520 and −467 nucleotides upstream of the *atpB* AUG start codon and they are specifically recognized by the PEP/Sig6 and PEP/Sig2 holoenzyme, respectively (Malik Ghulam et al., 2012). After transcription, the primary *atpB/E* transcripts are cleaved at the −84 position to produce the processed dicistronic *atpB/E* transcript (Fig. 2B). In contrast, there are five transcription start sites for *atpB* in tobacco and the processing site is at −90 from the *atpB* AUG (Suzuki, Kuroda, Yukawa, & Sugiura, 2011). For *atpE*, there is only one transcription start site, −431 from the *atpE* AUG in both Arabidopsis and tobacco and this site is recognized by PEP/Sig2 holoenzyme to produce the monocistronic *atpE* transcript (Malik Ghulam et al., 2012; Suzuki et al., 2011). In the dicistronic *atpB/E* transcripts, there is an overlap (AUGA) between the *atpB* stop codon (UGA) and the *atpE* start codon (AUG), which is a typical feature of a translational coupling mechanism. However, in vitro translation assays showed that the efficiency of *atpB* translation is higher than that of *atpE*, and translation of *atpE* is independent of that of *atpB* (Suzuki et al., 2011). The *cis*-element(s) located ~25 nt upstream of the *atpE* start codon might play essential roles for decoupling the expression of *atpE* from *atpB* (Suzuki et al., 2011). In addition, the monocistronic *atpE* transcript can be efficiently translated (Hirose & Sugiura, 2004). These facts suggest that translation of *atpE* is not significantly affected when a highly dynamic regulation of *atpB* translation is required during chloroplast ATP synthase assembly or under some specific conditions.

2.2 Biogenesis factors involved in *atp* gene expression

2.2.1 PPR10

PPR10 is a chloroplast P-class PPR protein with 19 PPR motifs (Pfalz, Bayraktar, Prikryl, & Barkan, 2009; Yin et al., 2013). Fine mapping of the PPR10-binding sites revealed that it binds to a similar sequence in the *atpI-atpH* and *psaJ-rpl33* intercistronic regions (Pfalz et al., 2009). This binding defines the 5′ and 3′ termini of *atpH* and *psaJ* mRNA, respectively, by protecting the target RNA from 5′ or 3′ exonucleases (Fig. 2A). Moreover, studies on *atpH* translation further revealed that binding of PPR10 to the *atpH* 5′ end prevents the formation of an RNA structure that masks the *atpH* ribosome-binding site, thereby enhancing translation of *atpH* (Prikryl, Rojas, Schuster, & Barkan, 2011). Thus, PPR10 has a dual function for the expression of *atpH*: stabilization of *atpH* mRNA and enhancement of its translation. *atpH* encodes the CF_oc subunit. Each chloroplast ATP synthase contains 14 CF_oc subunits. PPR10-mediated translational enhancement of *atpH* may be important for producing larger amounts of CF_oc subunit compared with the other subunits of the CF_o complex to achieve this uneven subunit stoichiometry,

2.2.2 BFA2

BFA2 (biogenesis factor required for ATP synthase2) is also a chloroplast stroma-localized P-class PPR protein with 22 PPR motifs (Zhang et al., 2019). P-class PPR proteins usually act as barrier proteins to prevent RNA degradation by exoribonucleases from the 5′- and 3′-ends of transcripts (Barkan & Small, 2014). Bioinformatics and EMSA analysis showed that BFA2 binds to the consensus sequence of the *atpF-atpA* intergenic region in a sequence-specific manner. Consistently, absence of BFA2 leads to the loss of the dicistronic *atpH/F* transcript and other minor transcripts overlapping the intergenic region of *atpF-atpA*. Thus, BFA2 mainly binds to the 3′ end of *atpH/F* and acts as barrier to prevent *atpH/F* transcript degradation by exoribonucleases (Fig. 2A). Hence, the accumulation of dicistronic *atpH/F* transcript requires two distinct PPR proteins, PPR10 and BFA2, which bind, respectively, to the 5′- and 3′-UTR of *atpH/F* to protect the mRNA against exoribonucleases.

2.2.3 AEF1/MPR25

Arabidopsis AEF1 (ATPF editing factor1) and its rice ortholog MPR25 belong to the PLS-PPR protein subfamily with an E domain in their C-terminus (Toda, Fujii, Noguchi, Kazama, & Toriyama, 2012; Yap et al., 2015). Absence of AEF1 leads to a loss of RNA editing of *atpF* as well as a clear decrease in splicing of the *atpF* transcripts in Arabidopsis (Yap et al., 2015). Although the *atpF* transcript is not edited in rice, splicing of *atpF* transcript is less efficient in the rice *mpr25* mutant. These results indicate that AEF1/MPR25 has a conserved function in splicing of the *atpF* transcript across angiosperms and a specific function in editing *atpF* in dicots (Fig. 2A). In addition to chloroplasts, AEF1/MPR25 is also localized in mitochondria and required for the RNA editing of *aad5* (Toda et al., 2012; Yap et al., 2015).

2.2.4 ATP4/SVR7/OsPPR676

ATP4 is a maize PPR protein with a small MutS-related (SMR) domain in its C-terminus (Zoschke et al., 2012). ATP4 can bind to the 5′ UTR of the *atpB/E* mRNA and polysome association of this transcript is disrupted in the absence ATP4, indicating that ATP4 facilitates translation initiation of this dicistronic mRNA (Fig. 2B). In addition, polysome association with *atpA* transcript is altered and the level of the *atpH/F* dicistronic mRNA is also significantly reduced in the *atp4* mutant. Together with the facts that ATP4 weakly associates with the *atpF/atpA* intergenic region, these results suggests that ATP4 is also required for *atpA* translation and stabilization of *atpH/F* dicistronic mRNA (Zoschke et al., 2012). Recently, the SMR motif was shown to have RNA endonuclease activity in vitro (Zhang & Lu, 2019; Zhou et al., 2017). Thus it is possible that ATP4 is required for *atpF* metabolism by cleavage in the *atpF/atpA* intergenic region to produce the *atpH/F* dicistronic mRNA. In addition to the *atp* transcripts, ATP4 is also essential for the normal accumulation of the *psaJ* mRNA, which encodes a nonessential PSI subunit (Zoschke et al., 2012), indicating that ATP4 has multiple functions.

SVR7 (suppressor of variegation7) is the ortholog of ATP4 in Arabidopsis (Liu, Yu, & Rodermel, 2010). The *svr7* mutant was initially isolated in a suppressor screen of *var2* variegation, in which the thylakoid membrane-localized FtsH2 metalloprotease is absent, leading to the production of variegated leaves with green and yellow sectors (Liu et al., 2010). Similar to ATP4, SVR7 also has multiple roles in gene expression (Fig. 2). Absence of SVR7 leads to accumulation of the precursors of 16S and 23S rRNA, and of the 23S–4.5S rRNA dicistronic precursor as well as a decrease of mature 23S rRNA, suggesting its involvement in chloroplast rRNA processing (Liu et al., 2010). In the case of the chloroplast *atp* genes, accumulation of the *atpH/F* dicistronic transcript was half-reduced and polysome association with *atpB/E* mRNA was decreased in the *svr7* mutant (Zoschke, Qu, Zubo, Börner, & Schmitz-Linneweber, 2013). This suggests that SVR7 and its maize ortholog ATP4 may play similar roles in the expression of the *atp* genes. However, the decrease of polysome association with *rbcL* mRNA and normal accumulation of *psaJ* transcripts in the *svr7* mutant suggests that some of the functions of SVR7 and ATP4 have diverged during evolution. Consistent with this idea, their rice ortholog OsPPR676 was shown to promote *atpB/E* translation in chloroplasts as well as to interact with the β subunit of the nascent polypeptide-associated complex, Osj10gBTF3, for pollen development (Liu et al., 2017; Wang et al., 2012).

2.2.5 ATP1

ATP1 was discovered by characterizing the maize mutant *atp1-1*, in which the level of chloroplast ATP synthase is specifically reduced (McCormac & Barkan, 1999). Loss of ATP1 leads to a clear defect in the association of *atpB/E* mRNA with polysomes and synthesis of $CF_1\beta$ subunit, suggesting that ATP1 is also involved in the translation initiation or early elongation of *atpB* mRNA. In addition, the level of *atpH/F* mRNA was also reduced in the *atp1* mutant. Although the *ATP1* gene has not yet been isolated, genetic analysis showed that ATP1 is not allelic to ATP4 (Zoschke et al., 2012). Thus, ATP1 may represent a novel factor required for *atpB/E* translation and/or *atpH/F* mRNA accumulation (Fig. 2).

Based on the above information, we propose a working model for the regulation of expression of the chloroplast *atp* operons (Fig. 2). After transcription from their promoters, the primary transcripts are further processed to mono-, di, or tri-cistronic transcripts. PPR protein AEF1/MPR25 and several proteins are required for the splicing of the *atpF* intron. Transcripts are protected from exonucleases by binding of PPR10 and

BFA2 or forming specific RNA structures in their 5′ or 3′ termini (Malik Ghulam et al., 2013). Binding of PPR10 to the 5′ end of *atpH* also enhances its translation. However, it is noteworthy that accumulation of dicistronic *atpH/F* transcript is highly variable and polysome association with processed dicistronic *atpB/E* transcript is also less efficient in some Arabidopsis mutants lacking chloroplast ATP synthase assembly factors (see below for more details). This observation suggests that the level of *atpH/F* transcript and translation initiation of *atpB/E* mRNA may also be controlled by assembly intermediates of chloroplast ATP synthase. Thus, some defects in *atp* gene expression observed in the *atp4/svr7*, *atp1*, and *aef1/mpr25* mutants may be secondary effects resulting from deficiencies of the primary targets of ATP4, ATP1, and AEF1/MPR25. Further investigations are needed to distinguish the primary and secondary targets of these biogenesis factors.

3. Assembly of the chloroplast ATP synthase

3.1 Auxiliary factors for assembly of the chloroplast ATP synthase

Radiolabeling of chloroplast proteins has revealed that intact chloroplast ATP synthase and the CF_1 subcomplex can be rapidly assembled in chloroplasts and other assembly intermediates are barely detected. This makes it difficult to study the assembly steps of the chloroplast ATP synthase by biochemical approaches. By analogy to the bacterial and mitochondrial ATP synthase assembly model, assembly of the chloroplast ATP synthase depends very likely on a variety of protein interactions between subunits and between subunits and assembly chaperones (reviewed in Rühle & Leister, 2015; Schöttler, Tóth, Boulouis, & Kahlau, 2015). Without assembly factors, assembly of chloroplast ATP synthase might be blocked at certain stages resulting in the accumulation of assembly intermediates or unassembled subunits. Detailed characterization of the chloroplast-specific assembly factors and their corresponding mutants has provided novel insights into the regulatory mechanisms of chloroplast ATP synthase assembly.

3.1.1 PAB

PAB (protein in chloroplast ATPase biogenesis) is a chloroplast stromal protein without any known motifs. Protein-protein interaction analysis showed that PAB specifically interacts with the $CF_1\gamma$ subunits (Mao et al., 2015). Among the nine subunits of chloroplast ATP synthase, three subunits ($CF_1\gamma$, $CF_1\delta$, and CF_ob') are encoded by the nuclear genome of land plants.

After synthesis in the cytosol, these subunits are post-translationally imported into chloroplasts in an unfolded state through the translocon in the envelope. In the chloroplast stroma, several types of chaperones are required for folding the imported or newly synthesized ATP synthase subunits encoded by chloroplast genes, such as Cpn60 and Hsp70/40 (Chen & Jagendorf, 1994). Chaperonin Cpn60 consists of 14 subunits forming two stacked heptameric rings with a central cavity in each ring where folding occurs (Zhao et al., 2019). Together with the co-chaperonin Cpn20, Cpn60 assists the folding of a large number of proteins in the chloroplast stroma (Apuya et al., 2001; Suzuki et al., 2009). Among these proteins, unfolded $CF_1\gamma$ subunit is one of the substrates of Cpn60 (Mao et al., 2015). In vitro pull-down analysis indicates that PAB likely associates with the folded rather than the unfolded form of $CF_1\gamma$ (Mao et al., 2015). Thus, unfolded $CF_1\gamma$ subunit is first captured by the Cpn60/Cpn20 complex for protein folding upon its import into chloroplasts. After folding to its native conformation, $CF_1\gamma$ is released from the Cpn60 complex and further interacts with the PAB protein. The interaction between $CF_1\gamma$ and PAB may lead to a conformational change of $CF_1\gamma$ which is required for its association with $CF_1\alpha/\beta$ heterodimers and for further assembly of the CF_1 subcomplex. Thus, PAB protein represents a dedicated assembly chaperone that directly interacts with $CF_1\gamma$ and functions downstream of Cpn60 to promote the assembly of $CF_1\gamma$ into the catalytic CF_1 subcomplex (Fig. 3; Mao et al., 2015).

3.1.2 BFA3/CGLD11

BFA3/CGLD11 was discovered in a mutant screen (BFA3, Zhang, Duan, Zhang, & Peng, 2016) and through bioinformatic analysis of the GreenCut set of proteins (CGLD11, Grahl et al., 2016). It is located in the chloroplast stroma and is conserved in higher plants, green algae and a few eukaryotic algae (Zhang et al., 2016). Absence of BFA3/CGLD11 results in a reduced assembly efficiency of the CF_1 subcomplex of chloroplast ATP synthase. No known domain was found in the BFA3/CGLD11 protein and it specifically interacts with the $CF_1\beta$ subunit (Grahl et al., 2016; Zhang et al., 2016). Except for $CF_1\beta$, no other ATP synthase subunits can be co-purified with BFA3/CGLD11, suggesting that BFA3/CGLD11 is involved in the assembly of the CF_1 subcomplex at a very early stage, probably before the formation of $CF_1\alpha/\beta$ heterodimers (Fig. 3). Although the detailed function and the mode of action of BFA3/CGLD11 are unclear, its binding site in $CF_1\beta$ was mapped to a region in the catalytic

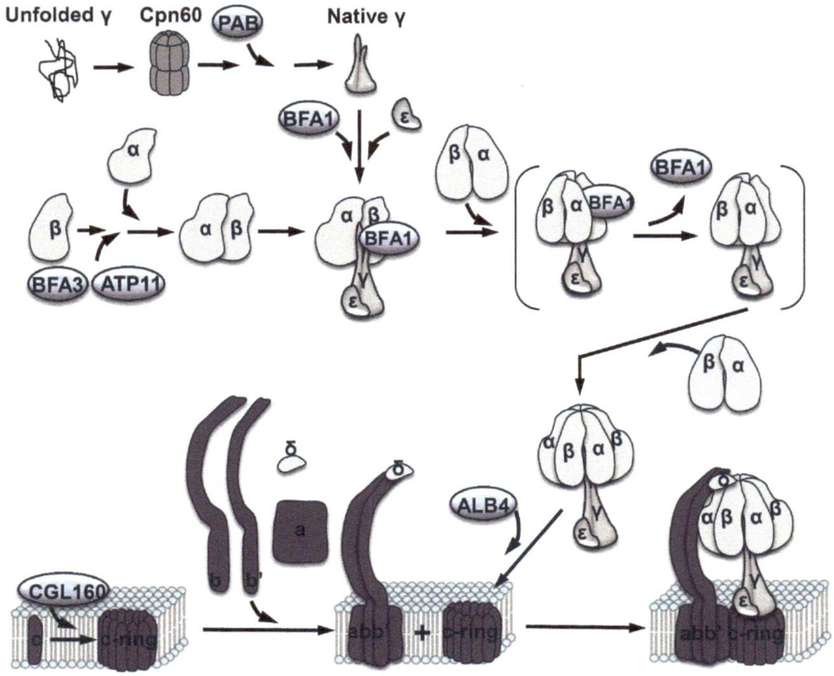

Fig. 3 Scheme of the assembly of the chloroplast ATP synthase. After its import into chloroplasts, unfolded $CF_1\gamma$ is folded by the Cpn60 chaperonin and then interacts with PAB to produce the native $CF_1\gamma$ subunit. Simultaneously, the interaction of $CF_1\beta$ with BFA3 and ATP11 is likely required for $CF_1\alpha/\beta$ heterodimer formation. Afterward, BFA1 associates with $CF_1\gamma$ and $CF_1\alpha/\beta$ forming the transient $CF_1\alpha/\beta/\gamma$-BFA1 assembly intermediate. Since BFA1 also interacts with $CF_1\varepsilon$ and an *E. coli* γ/ε heterodimer has been purified (Rodgers & Wilce, 2000), we propose that $CF_1\varepsilon$ assembly occurs at this step. Other $CF_1\alpha/\beta$ heterodimers bind around $CF_1\gamma$ and promote the release of BFA1 and finally form the $CF_1\alpha/\beta/\gamma/\varepsilon$ subcomplex. In thylakoids, CGL160 facilitates assembly of c subunits into the c-ring. Since $CF_1\delta$ comigrates with b and b′ in BN-PAGE (Zhang et al., 2018), we propose that $CF_1\delta$ is first associated with the peripheral stalk, which is likely required for the recruitment of the $CF_1\alpha/\beta/\gamma/\varepsilon$ subcomplex on the CF_o subcomplex. ALB4 may facilitate the final step of assembly of the mature complex. The two putative pathways for CF_1 assembly are not illustrated in the model.

site of $CF_1\beta$ (Zhang et al., 2016). The region of the $CF_1\beta$ catalytic site is hydrophobic and has a strong tendency to aggregate (Ackerman, 2002). Binding of BFA3/CGLD11 to this region of $CF_1\beta$ may prevent aggregation of unassembled $CF_1\beta$ until the $CF_1\alpha$ subunit replaces BFA3/CGLD11 to form the $CF_1\alpha/\beta$ heterodimer. It is also possible that association of BFA3/CGLD11 with $CF_1\beta$ can induce some structural changes in $CF_1\beta$ and enhance the assembly efficiency of the $CF_1\alpha/\beta$ intermediate (Zhang et al., 2016).

In addition to chloroplasts, BFA3/CGLD11 was also localized in the mitochondria of Arabidopsis (Grahl et al., 2016). However, the level of mitochondrial ATP synthase is not reduced in the *cgld11* mutant indicating that BFA3/CGLD11 is not essential for the biogenesis of mitochondrial ATP synthase.

3.1.3 BFA1

BFA1 is also a chloroplast stroma protein which is required for efficient assembly of the CF_1 subcomplex (Zhang et al., 2018). In contrast to BFA3 that specifically interacts with $CF_1\beta$, BFA1 transiently interacts with the $CF_1\beta$, $CF_1\gamma$, and $CF_1\varepsilon$ subunits. Moreover, $CF_1\alpha$ can also be co-purified with BFA1 and $CF_1\beta/\gamma$, suggesting that BFA1 facilities CF_1 assembly after $CF_1\alpha/\beta$ heterodimer formation. The crystal structure of BFA1 revealed that it contains two interacting distinct β-barrels oriented at right angles to each other. A detailed yeast-two hybrid and in silico molecular docking analysis showed that both $CF_1\beta$ and $CF_1\gamma$ could fit on the surface of BFA1 without steric conflicts. The relative positions of one $CF_1\alpha/\beta$ heterodimer and $CF_1\gamma$ in the $CF_1\alpha/\beta/\gamma$-BFA1 model are similar to their final positions in the assembled CF_1 subcomplex (Zhang et al., 2018). This suggests that the $CF_1\alpha/\beta$ heterodimer and $CF_1\gamma$ are transiently associated by BFA1 nearly in their final conformation in CF_1 during the assembly process. It is very likely that, in the next step, two additional $CF_1\alpha/\beta$ heterodimers are recruited to the $CF_1\alpha/\beta/\gamma$-BFA1 intermediate around the coiled-coil N/C-terminus of the $CF_1\gamma$ subunit and BFA1 is subsequently released (Fig. 3).

CF_1 is the catalytic domain for ATP synthesis and it is mainly composed of a $\alpha_3\beta_3$ hexamer surrounding a γ/ε stalk. Given the importance of this core catalytic domain in ATP synthesis and its potential usage in the development of nanomotors for nanomedicine by synthetic biology, its assembly pathway has been extensively investigated. In vitro reconstitution assays using overexpressed CF_1 subunits showed that $CF_1\alpha$ and $CF_1\beta$ subunits are firstly assembled into a heterodimer with the assistance of several molecular chaperonins and chaperones such as Cpn60, Cpn20, and Hsp70. Three $CF_1\alpha/\beta$ heterodimers are then assembled to form a $\alpha_3\beta_3$ hexamer through some unknown pathway. In the final step, the α-helical coiled-coil structure formed by the N- and C-termini of $CF_1\gamma$ is inserted into the $\alpha_3\beta_3$ hexamer to form the functional CF_1 catalytic core subcomplex (Chen & Jagendorf, 1994; Gao, Lipscomb, Wu, & Richter, 1995). However, this assembly pathway has never been demonstrated in vivo. Moreover, the structural model of the $CF_1\alpha/\beta/\gamma$-BFA1 intermediate complex does not support this assembly pathway of CF_1 because the BFA1 protein occupies the position of at least one of the three $CF_1\alpha/\beta$ heterodimers in the precursor of the $\alpha_3\beta_3\gamma$ subcomplex (Zhang et al., 2018). Thus it is more likely that three $CF_1\alpha/\beta$ heterodimers are assembled into $\alpha_3\beta_3\gamma$ around the coiled-coil N/C-termini of $CF_1\gamma$ with the help of BFA1, although the detailed mechanism is still unclear (Zhang et al., 2018).

BFA1 is also interacting with $CF_1\varepsilon$, a small subunit binding to the central domain of $CF_1\gamma$, suggesting that BFA1 is also involved in the integration of $CF_1\varepsilon$ into CF_1 (Hahn et al., 2018; Zhang et al., 2018). In vitro reconstitution studies showed that $CF_1\varepsilon$ does not bind to the $CF_1\alpha/\beta$ heterodimers but weakly associates with $CF_1\gamma$ (Gao et al., 1995), in accordance with the findings that the $F_1\gamma/\varepsilon$ subcomplex from *E. coli* and the photosynthetic cyanobacterium *Thermosynechococcus elongatus* BP-1 could be purified (Murakami et al., 2018; Rodgers & Wilce, 2000). This suggests that $CF_1\varepsilon$ is assembled into the CF_1 subcomplex together with $CF_1\gamma$ with the help of the BFA1 protein. The $F_1\varepsilon$ subunit serves as an intrinsic inhibitor of ATP hydrolysis activity in various species and two different states of $F_1\varepsilon$ were found in *E. coli* (Yagi et al., 2007). Based on the proposed BFA1-$CF_1\beta/\gamma$ structural model, BFA1 may facilitate the assembly of the $CF_1\varepsilon$ subunit into CF_1 subcomplex in its extended state, although the details of this process have not yet been elucidated (Fig. 3; Zhang et al., 2018).

3.1.4 Atp11 and Atp12

Atp11 and Atp12 represent the members of two large chaperone families present in most eukaryotes. They play a conserved role in the assembly of mitochondrial ATP synthase by binding to the β and α subunits, respectively (Ackerman & Tzagoloff, 2005; Pícková, Potocký, & Houštěk, 2005; Rak, Gokova, & Tzagoloff, 2011; Wang & Ackerman, 2000; Wang, Sheluho, Gatti, & Ackerman, 2000). Structural analysis of Atp11p from *Candida glabrata* and Atp12p from *Paracoccus denitrificans* showed that formation of Atp11-β and Atp12-α heterodimers is likely to act as decoys preventing the aggregation of α/β subunits (Ludlam et al., 2009). Homologs of Atp11 and Atp12 are present in all photosynthetic eukaryotes that harbor two organellar ATP synthases in the chloroplast and mitochondria (Duan et al., unpublished data). Loss of either Atp11 or Atp12 is lethal in Arabidopsis. As their homologs in yeast, Arabidopsis Atp11 and Atp12 specifically interact with the β and α subunits of mitochondrial ATP synthase, suggesting their conserved function in the assembly of this complex. While Atp12 is only localized in mitochondria, Atp11 is present both in chloroplasts and mitochondria, raising the possibility that Atp11 is also required for assembly of the chloroplast ATP synthase (Fig. 3). Although Atp11 interacts with the β subunit of the chloroplast ATP synthase, the binding sites for Atp11 in the β subunit of chloroplast and mitochondrial ATP synthase are slightly different. Thus, Atp11 may fulfill a conserved task during assembly of two organellar ATP synthases in higher plants but their mechanisms of action may have evolved in different ways. Although Atp11 interacts with $CF_1\beta$ as the two assembly factors BFA3 and BFA1, they bind to different sites of the β subunit (Duan et al., unpublished data).

3.1.5 CGL160

CGL160 (conserved only in the green lineage160) is a 34-kDa thylakoid membrane protein with four transmembrane domains at its C-terminus (Fristedt et al., 2015; Rühle et al., 2014). CGL160 is conserved in photosynthetic eukaryotes and its C-terminal transmembrane domains display moderate similarity to prokaryotic Atp1/UncI in *Escherichia coli* and *Synechocystis* sp. PCC6803. In bacteria, the Atp1/UncI protein is involved in the oligomerization of the c subunits into the c-ring (Ozaki, Suzuki, Kuruma, Ueda, & Yoshida, 2008; Suzuki, Ozaki, Sone, Feniouk, & Yoshida, 2007). Consistently, yeast-two hybrid assays showed that CGL160 interacts with the c subunit and the Atp1 protein from *Synechocystis* can replace the C-terminal transmembrane domains in CGL160 for the interaction with the c subunit in yeast. These results indicate that the function of CGL160 and Atp1/UncI is conserved for c-ring formation (Fig. 3; Rühle et al., 2014). Besides the c subunit, CGL160 also interacts with the b subunit of chloroplast ATP synthase in yeast-two hybrid assays. Co-migration of CGL160 with a subcomplex containing CF_ob in blue native PAGE supports this conclusion. Thus, CGL160 may be also required for the incorporation of CF_ob into the c-ring subcomplex in thylakoids (Fristedt et al., 2015; Rühle et al., 2014).

3.1.6 ALB4

ALB4 belongs to the YidC/Oxa1/Alb3 protein family that functions in membrane protein integration and complex assembly (Gerdes et al., 2006). Higher plant chloroplasts contain two homologs, ALB3 and ALB4, localized in the thylakoid membrane. ALB3 was proposed to be essential for incorporation of LHCPs into thylakoids and interacts with the chloroplast Sec translocon for mediating co-translational integration of membrane proteins into thylakoids. Analysis of the *alb4* mutant showed that ALB4 is required for the assembly or stability of chloroplast ATP synthase (Benz et al., 2009). Physical interactions between ALB4 and $CF_1\beta$/CF_ob' subunits raise the possibility that ALB4 facilitates the incorporation of the CF_1 subcomplex into the membrane-embedded CF_o subcomplex at the final step of assembly of chloroplast ATP synthase (Fig. 3; Benz et al., 2009).

3.2 Assembly of the chloroplast ATP synthase

Compared to other photosynthetic complexes like PSII and PSI, our knowledge about chloroplast ATP synthase assembly is relatively limited and arises mainly from the analogy to the bacterial and mitochondrial ATP synthase assembly model which suggests that several distinct modules are assembled separately and then incorporated to form the intact ATP synthase (reviewed in Rühle & Leister, 2015; Schöttler et al., 2015). Blue native PAGE analysis showed that, besides the intact chloroplast ATP synthase, a CF_1 subcomplex containing $CF_1\alpha$, $CF_1\beta$, $CF_1\gamma$, and $CF_1\varepsilon$ can be detected in thylakoids and this subcomplex can be rapidly assembled after protein pulse-labeling (Rühle et al., 2014; Takabayashi et al., 2013; Zhang et al., 2016, 2018), indicating that this subcomplex constitutes a distinct module, which is also consistent with the assembly model in bacteria and mitochondria (Deckers-Hebestreit, 2013; Rak et al., 2011). In vitro reconstitution experiments using purified CF_1 tobacco subunits synthesized in *E. coli* have demonstrated that formation of $CF_1\alpha/\beta$ heterodimers is the first critical step for the assembly of CF_1 (Chen & Jagendorf, 1994). However, the details of the subsequent steps are still unclear. Based on the characterization of PAB, BFA1, BFA3/CGLD11, and ATP11, we propose a model for the assembly of chloroplast ATP synthase in land plants (Fig. 3). While BFA3 and ATP11 interact with $CF_1\beta$ and facilitate formation of $CF_1\alpha/\beta$ heterodimer, Cpn60 assists the folding of $CF_1\gamma$ and PAB further interacts with native $CF_1\gamma$ for subsequent assembly steps. Afterwards, BFA1 acts as a chaperone to assemble the $CF_1\alpha/\beta$ heterodimer and $CF_1\gamma$ together. Other $CF_1\alpha/\beta$ heterodimers may be further assembled by surrounding $CF_1\gamma$. During these early steps of CF_1 assembly, $CF_1\varepsilon$ may be assembled together with $CF_1\gamma$, since BFA1 also interacts with $CF_1\varepsilon$ and an *E. coli* γ/ε heterodimer has been purified (Rodgers & Wilce, 2000; Zhang et al., 2018).

For the CF_o module, BN-PAGE and immunoblot analyses detected two subcomplexes containing CF_oc subunits (Rühle et al., 2014). From their molecular masses, the smallest subcomplex likely represents the c-ring module (about 110 kDa). This implies that the c-ring and other parts of CF_o are independently assembled in thylakoids, a process similar to the assembly of their bacterial and mitochondrial analogs. Thylakoid protein CGL160 may play important roles in the assembly of the c-ring in chloroplasts (Fig. 3). CF_ob, CF_ob', and $CF_1\delta$ co-fractionate in BN-PAGE forming a subcomplex slightly smaller than the CF_1 subcomplex containing $CF_1\alpha$, $CF_1\beta$, $CF_1\gamma$, and $CF_1\varepsilon$ (Zhang et al., 2018). The large subcomplex containing CF_oc was detected at a similar position (Rühle et al., 2014). This suggests that CF_ob, CF_ob', and the c-ring (probably as well as CF_oa) form a CF_o subcomplex in thylakoids and $CF_1\delta$ binds to it rather than to the CF_1 module (Fig. 3). Details for the formation of $CF_oa/b/b'$ are not known currently. However, by analogy to the function of δ during bacterial ATP synthase assembly (Deckers-Hebestreit, 2013), $CF_1\delta$ in chloroplasts is also likely a linker for recruitment of the $CF_1\alpha/\beta/\gamma/\varepsilon$ subcomplex on the CF_o subcomplex in the final step of assembly of the mature ATP synthase complex. ALB4 protein may also assist the association of the $CF_1\alpha/\beta/\gamma/\varepsilon$ subcomplex with CF_o (Fig. 3).

3.3 Two assembly pathways for CF_1?

Pulse-chase labeling of Arabidopsis leaves showed that the CF_1 subcomplex containing $CF_1\alpha$, $CF_1\beta$, $CF_1\gamma$, and $CF_1\varepsilon$ is rapidly assembled in the thylakoid membrane. Simultaneously, another CF_1 subcomplex containing all five CF_1 subunits is assembled in the chloroplast stroma (Zhang et al., 2016, 2018). Formation of this subcomplex is independent on the CF_ob subunit (Zhang et al., 2019). At the current stage, we cannot rule out the possibility that the stroma-localized CF_1 subcomplex is washed off from isolated thylakoids. However, stroma-localized CF_1 subcomplex contains $CF_1\delta$ whereas thylakoid-localized CF_1 subcomplex does not (Zhang et al., 2018). Moreover, the level of labeled stroma-localized CF_1 subcomplex is not reduced during a 1 h chase but accumulates to a certain level in both WT and mutants, implying that it is not incorporated into thylakoids (Zhang et al., 2016, 2018). These observations do not support the notion that CF_1 subcomplex assembly occurs only in thylakoids or only in the stroma but raise the possibility of two assembly pathways of CF_1. The major pathway occurs in the thylakoid membrane; probably the $CF_1\gamma/\varepsilon$ stalk is firstly anchored in the thylakoids via some unknown complex (maybe c-ring) and then the $CF_1\alpha/\beta$ heterodimers are assembled around the coiled-coil N/C-terminus of $CF_1\gamma$ with the help of BFA1. Another pathway occurs in the chloroplast stroma, where an intact CF_1 subcomplex was found to assemble and accumulate to some extent even in the absence of

BFA1 and BFA3 (Zhang et al., 2016, 2018). This raises the question of the significance of these two assembly pathways. One possibility is that the less efficient stromal pathway is a salvage pathway to maintain ATP synthesis under conditions when the thylakoid pathway is compromised, for example under certain stress conditions. It is also possible that the stroma-localized CF_1 subcomplex plays some unknown physiological role for chloroplast development and function.

4. Regulation linking chloroplast *atp* gene expression and chloroplast ATP synthase assembly

The most prominent feature of the ATP synthase that distinguishes it from other thylakoid complexes is its uneven subunit stoichiometry. Thus, tight coordination in gene expression in the nucleus and plastids is required. To ensure the stoichiometric accumulation of the ATP synthase subunits, chloroplasts not only evolved gene-specific regulatory factors but also use a mechanism called CES (control by epistasy of synthesis) in *Chlamydomonas* (Drapier, Rimbault, Vallon, Wollman, & Choquet, 2007). In this process, synthesis of its subunits is tightly coupled to the assembly of the photosynthetic complex. In the case of the ATP synthase, the nucleus-encoded γ subunit is required for the translation of the chloroplast-encoded β subunit, which in turn activates translation initiation of the chloroplast-encoded α subunit. Moreover, α/β heteromers can bind to the 5′ UTR of *atpB* mRNA and repress its translation, a process which in turn decreases the translation of the α subunit. Once the γ subunit associates with the α/β heteromer to form the $\alpha_3\beta_3\gamma$ complex, repression of translation of the β subunit by the α/β heteromer is relieved (Drapier et al., 2007). This mechanism prevents the accumulation of excess amounts of α/β relative to the γ subunit. However, whether the CES process occurs in land plants is controversial (Rühle & Leister, 2015). It is still unclear how chloroplasts in land plants achieve the accumulation of subunits with a different stoichiometry during the biogenesis of the ATP synthase.

We observed that some of the *atp* transcripts accumulate to rather variable levels in different chloroplast ATP synthase mutants. In *bfa3–1* and *svr7* mutant plants, the amount of *atpH/F* transcript is reduced to less than half compared with WT but is slightly higher than in WT in the *bfa1* and *cgl160* mutants (Rühle et al., 2014; Zhang et al., 2016, 2018; Zoschke et al., 2013). In the case of the *atpB/E* transcript, while the *bfa3-1* mutant accumulates high levels of processed dicistronic *atpB/E* transcript, the amount of this transcript is very low in *bfa1-1*, *bfa2-1*, and *cgl160* (Fristedt et al., 2015;

Rühle et al., 2014; Zhang et al., 2016, 2018, 2019). These results show that accumulation of dicistronic *atpH/F* and processed dicistronic *atpB/E* transcripts is highly variable in mutants defective in ATP synthase assembly. In addition, polysome association with primary dicistronic *atpB/E* transcript is affected in the *bfa1-1*, *bfa2-1*, and *cgl160* mutants, but polysome association with *atpH/F/A* transcripts is enhanced in the *bfa1-1* mutant, resulting in enhanced synthesis of $CF_1\alpha$ subunits (Zhang et al., 2018, 2019).

The above facts imply that the levels of *atpH/F* and processed *atpB/E* transcripts as well as translation initiation of some subunits are controlled by some unknown mechanisms, probably involving assembly intermediates during ATP synthase biogenesis. As an example processing of *atpB/E* transcript and assembly of the *c*-ring of the CF_o subcomplex is impaired in *cgl160* whereas *atpH/F* RNA is absent in *bfa2* (Fristedt et al., 2015; Rühle et al., 2014; Zhang et al., 2019). In these mutants the level of CF_o and CF_1 assembly intermediate(s) may also be increased. In contrast, processed *atpB/E* transcript was significantly increased in the *bfa3* mutants (Zhang et al., 2016). BFA3 specifically interacts with $CF_1\beta$ at its catalytic site and functions as a molecular chaperone to facilitate association of unassembled $CF_1\beta$ with $CF_1\alpha$ during assembly of the CF_1 subcomplex (Zhang et al., 2016). Thus, assembly of CF_1 may be blocked at a certain stage and the assembly intermediate(s) in *bfa3*, different from those in *cgl160* and *bfa2*, could play critical roles in the up-regulation of processed *atpB/E* transcript and, at the same time, in the down regulation of *atpH/F* transcript (Zhang et al., 2016). These regulatory mechanisms may be important for allowing subunits of this complex to accumulate with a different stoichiometry.

5. Conclusions and perspectives

Although several auxiliary proteins were discovered recently, the number of biogenesis factors for chloroplast ATP synthase is still relatively modest compared with other photosynthetic complexes. Unlike the other three thylakoid complexes, chloroplast ATP synthase does not contain redox-active cofactors. Thus, its biogenesis does not require auxiliary proteins for the insertion of cofactors into the complex. This may be one reason for the small number of assembly factors of chloroplast ATP synthase. More assembly factors are likely to be required, for example the assembly factors specifically interacting with $CF_1\alpha$. Atp12 is localized in mitochondria and interacts with $CF_1\alpha$ but it is not present in chloroplasts. However, it is difficult to screen for nonessential biogenesis factors required for chloroplast

ATP synthase assembly or biogenesis because the activity of chloroplast ATP synthase is regulated in a highly dynamic manner. Photosynthetic proprieties are largely unaffected even with a moderate reduction in the level of chloroplast ATP synthase in tobacco (Rott et al., 2011). Thus, a novel strategy for forward genetic screening is needed to identify auxiliary proteins that are nonessential for the accumulation of chloroplast ATP synthase. Screening rice or maize mutant grown in the field using chlorophyll fluorescence may be a good strategy because these crop plants are often subjected to more severe environmental stress conditions that increase the turnover rate of chloroplast ATP synthase than when they are grown in a greenhouse.

Further complications in mutant screens could arise because of the remarkable plasticity of compensatory mechanisms that can alleviate deficiencies in ATP synthesis. A striking example is provided by the Chlamydomonas mutant *Fud50* which lacks the chloroplast gene *atpB* (Woessner et al., 1984). This mutant therefore lacks the chloroplast ATP synthase and is unable to grow under phototrophic conditions. It was however possible to isolate a *Fud50* suppressed strain able to grow photoautotrophically although it is unable to synthesize ATP in the chloroplast (Lemaire, Wollman, & Bennoun, 1988). One possible explanation is that mitochondrial ATP is shuttled from the mitochondria to the chloroplast through a process which is still unclear.

Another challenging topic is the feedback regulation of chloroplast *atp* gene expression, which is closely related to the problem of how chloroplasts synthesize subunits with a different stoichiometry. At least two different mechanisms could be involved in this task. The simplest would be that all subunits made in excess that are not assembled into the mature ATP synthase complex are rapidly degraded by proteases. The second mechanism could be CES in which some of the chloroplast-encoded ATP synthase subunits are subject to assembly-dependent translational regulation. Thus, one key question is whether the CES process elucidated in Chlamydomonas also operates in land plants for ATP synthase assembly. At this time there is no convincing evidence for CES in this particular case although experimental evidence for CES has been provided in land plants for the regulation of RubisCO assembly (Wostrikoff & Stern, 2007).

Acknowledgments

This work was supported by funds from the National Science Foundation of China (31700202), the Shanghai Engineering Research Center of Plant Germplasm Resources (17DZ2252700), and the Science and Technology Commission of Shanghai Municipality (18DZ2260500).

References

Ackerman, S. H. (2002). Atp11p and Atp12p are chaperones for F_1-ATPase biogenesis in mitochondria. *Biochimica et Biophysica Acta, 1555*, 101–105.

Ackerman, S. H., & Tzagoloff, A. (2005). Function, structure, and biogenesis of mitochondrial ATP synthase. *Progress in Nucleic Acid Research and Molecular Biology, 80*, 95–133.

Apuya, N. R., Yadegari, R., Fischer, R. L., Harada, J. J., Zimmerman, J. L., & Goldberg, R. B. (2001). The Arabidopsis embryo mutant schlepperless has a defect in the *chaperonin-60α* gene. *Plant Physiology, 126*, 717–730.

Barkan, A., & Small, I. (2014). Pentatricopeptide repeat proteins in plants. *Annual Review of Plant Biology, 65*, 415–442.

Benz, M., Bals, T., Gügel, I. L., Piotrowski, M., Kuhn, A., Schünemann, D., et al. (2009). Alb4 of Arabidopsis promotes assembly and stabilization of a non chlorophyll-binding photosynthetic complex, the CF_1CF_o-ATP synthase. *Molecular Plant, 2*, 1410–1424.

Chen, G. G., & Jagendorf, A. T. (1994). Chloroplast molecular chaperone-assisted refolding and reconstitution of an active multisubunit coupling factor core. *Proceedings of the National Academy of Sciences of the United States of America, 91*, 11497–11501.

Dal Bosco, C., Lezhneva, L., Biehl, A., Leister, D., Strotmann, H., Wanner, G., et al. (2004). Inactivation of the chloroplast ATP synthase gamma subunit results in high non-photochemical fluorescence quenching and altered nuclear gene expression in *Arabidopsis thaliana*. *Journal of Biological Chemistry, 279*, 1060–1069.

Daniell, H., Wurdack, K. J., Kanagaraj, A., Lee, S. B., Saski, C., & Jansen, R. K. (2008). The complete nucleotide sequence of the cassava (Manihot esculenta) chloroplast genome and the evolution of *atpF* in Malpighiales: RNA editing and multiple losses of a group II intron. *Theoretical and Applied Genetics, 116*, 723.

Deckers-Hebestreit, G. (2013). Assembly of the Escherichia coli F_oF_1 ATP synthase involves distinct subcomplex formation. *Biochemical Society Transactions, 41*, 1288–1293.

Drapier, D., Rimbault, B., Vallon, O., Wollman, F. A., & Choquet, Y. (2007). Intertwined translational regulations set uneven stoichiometry of chloroplast ATP synthase subunits. *EMBO Journal, 26*, 3581–3591.

Fristedt, R., Martins, N. F., Strenkert, D., Clarke, C. A., Suchoszek, M., Thiele, W., et al. (2015). The thylakoid membrane protein CGL160 supports CF_1CF_o ATP synthase accumulation in *Arabidopsis thaliana*. *PLoS One, 10*, e0121658.

Gao, F., Lipscomb, B., Wu, I., & Richter, M. L. (1995). In vitro assembly of the core catalytic complex of the chloroplast ATP synthase. *Journal of Biological Chemistry, 270*, 9763–9769.

Gerdes, L., Bals, T., Klostermann, E., Karl, M., Philippar, K., Hünken, M., et al. (2006). A second thylakoid membrane-localized Alb3/OxaI/YidC homologue is involved in proper chloroplast biogenesis in *Arabidopsis thaliana*. *Journal of Biological Chemistry, 281*, 16632–16642.

Grahl, S., Reiter, B., Gügel, I. L., Vamvaka, E., Gandini, C., Jahns, P., et al. (2016). The Arabidopsis protein CGLD11 is required for chloroplast ATP synthase accumulation. *Molecular Plant, 9*, 885–899.

Hahn, A., Vonck, J., Mills, D. J., Meier, T., & Kühlbrandt, W. (2018). Structure, mechanism, and regulation of the chloroplast ATP synthase. *Science, 360*, eaat4318.

Hirose, T., & Sugiura, M. (2004). Multiple elements required for translation of plastid atp B m RNA lacking the Shine–Dalgarno sequence. *Nucleic Acids Research, 32*, 3503–3510.

Junge, W., & Nelson, N. (2015). ATP synthase. *Annual Review of Biochemistry, 84*, 631–657.

Kohzuma, K., Dal Bosco, C., Kanazawa, A., Dhingra, A., Nitschke, W., Meurer, J., et al. (2012). Thioredoxin-insensitive plastid ATP synthase that performs moonlighting functions. *Proceedings of the National Academy of Sciences of the United States of America, 109*, 3293–3298.

Kong, M., Wang, F., Yang, Z., & Mi, H. (2013). ATPG is required for the accumulation and function of chloroplast ATP synthase in *Arabidopsis*. *Chinese Science Bulletin, 58*, 3224–3232.

Kroeger, T. S., Watkins, K. P., Friso, G., van Wijk, K. J., & Barkan, A. (2009). A plant-specific RNA-binding domain revealed through analysis of chloroplast group II intron splicing. *Proceedings of the National Academy of Sciences of the United States of America*, *106*, 4537–4542.

Lemaire, C., Wollman, F. A., & Bennoun, P. (1988). Restoration of phototrophic growth in a mutant of Chlamydomonas reinhardtii in which the chloroplast *atpB* gene of the ATP synthase has a deletion: An example of mitochondria-dependent photosynthesis. *Proceedings of the National Academy of Sciences of the United States of America*, *85*, 1344–1348.

Liu, Y., Liu, X., Chen, H., Zheng, P., Wang, W., Wang, L., et al. (2017). A plastid-localized pentatricopeptide repeat protein is required for both pollen development and plant growth in rice. *Scientific Reports*, *7*, 1–12.

Liu, X., Yu, F., & Rodermel, S. (2010). An Arabidopsis pentatricopeptide repeat protein, SUPPRESSOR OF VARIEGATION7, is required for FtsH-mediated chloroplast biogenesis. *Plant Physiology*, *154*, 1588–1601.

Ludlam, A., Brunzelle, J., Pribyl, T., Xu, X., Gatti, D. L., & Ackerman, S. H. (2009). Chaperones of F_1-ATPase. *Journal of Biological Chemistry*, *284*, 17138–17146.

Maiwald, D., Dietzmann, A., Jahns, P., Pesaresi, P., Joliot, P., Joliot, A., et al. (2003). Knock-out of the genes coding for the Rieske protein and the ATP-synthase delta-subunit of Arabidopsis. Effects on photosynthesis, thylakoid protein composition, and nuclear chloroplast gene expression. *Plant Physiology*, *133*, 191–202.

Malik Ghulam, M., Courtois, F., Lerbs-Mache, S., & Merendino, L. (2013). Complex processing patterns of mRNAs of the large ATP synthase operon in Arabidopsis chloroplasts. *PLoS One*, *8*, e78265.

Malik Ghulam, M., Zghidi-Abouzid, O., Lambert, E., Lerbs-Mache, S., & Merendino, L. (2012). Transcriptional organization of the large and the small ATP synthase operons, *atpI/H/F/A* and *atpB/E*, in *Arabidopsis thaliana* chloroplasts. *Plant Molecular Biology*, *79*, 259–272.

Mao, J., Chi, W., Ouyang, M., He, B., Chen, F., & Zhang, L. (2015). PAB is an assembly chaperone that functions downstream of chaperonin 60 in the assembly of chloroplast ATP synthase coupling factor 1. *Proceedings of the National Academy of Sciences of the United States of America*, *112*, 4152–4157.

McCormac, D. J., & Barkan, A. (1999). A nuclear gene in maize required for the translation of the chloroplast *atpB/E* mRNA. *The Plant Cell*, *11*, 1709–1716.

Murakami, S., Kondo, K., Katayama, S., Hara, S., Sunamura, E. I., Yamashita, E., et al. (2018). Structure of the γ–ε complex of cyanobacterial F_1-ATPase reveals a suppression mechanism of the γ subunit on ATP hydrolysis in phototrophs. *Biochemical Journal*, *475*, 2925–2939.

Ozaki, Y., Suzuki, T., Kuruma, Y., Ueda, T., & Yoshida, M. (2008). UncI protein can mediate ring-assembly of c-subunits of F_oF_1-ATP synthase in vitro. *Biochemical and Biophysical Research Communications*, *367*, 663–666.

Pfalz, J., Bayraktar, O. A., Prikryl, J., & Barkan, A. (2009). Site-specific binding of a PPR protein defines and stabilizes 5′ and 3′ mRNA termini in chloroplasts. *EMBO Journal*, *28*, 2042–2052.

Pícková, A., Potocký, M., & Houštěk, J. (2005). Assembly factors of F_1F_o-ATP synthase across genomes. *Proteins: Structure Function and Bioinformatics*, *59*, 393–402.

Prikryl, J., Rojas, M., Schuster, G., & Barkan, A. (2011). Mechanism of RNA stabilization and translational activation by a pentatricopeptide repeat protein. *Proceedings of the National Academy of Sciences of the United States of America*, *108*, 415–420.

Prikryl, J., Watkins, K. P., Friso, G., van Wijk, K. J., & Barkan, A. (2008). A member of the whirly family is a multifunctional RNA- and DNA-binding protein that is essential for chloroplast biogenesis. *Nucleic Acids Research*, *36*, 5152.

Rak, M., Gokova, S., & Tzagoloff, A. (2011). Modular assembly of yeast mitochondrial ATP synthase. *EMBO Journal*, *30*, 920–930.

Rodgers, A. J., & Wilce, M. C. (2000). Structure of the γ–ε complex of ATP synthase. *Nature Structural Biology*, 7, 1051–1054.

Rott, M., Martins, N. F., Thiele, W., Lein, W., Bock, R., Kramer, D. M., et al. (2011). ATP synthase repression in tobacco restricts photosynthetic electron transport, CO_2 assimilation, and plant growth by overacidification of the thylakoid lumen. *Plant Cell*, *23*, 304–321.

Rühle, T., & Leister, D. (2015). Assembly of F_1F_o-ATP synthases. *Biochimica et Biophysica Acta*, *1847*, 849–860.

Rühle, T., Razeghi, J. A., Vamvaka, E., Viola, S., Gandini, C., Kleine, T., et al. (2014). The Arabidopsis protein CONSERVED ONLY IN THE GREEN LINEAGE160 promotes the assembly of the membranous part of the chloroplast ATP synthase. *Plant Physiology*, *165*, 207–226.

Schöttler, M. A., Tóth, S. Z., Boulouis, A., & Kahlau, S. (2015). Photosynthetic complex stoichiometry dynamics in higher plants: Biogenesis, function, and turnover of ATP synthase and the cytochrome b_6f complex. *Journal of Experimental Botany*, *66*, 2373–2400.

Suzuki, H., Kuroda, H., Yukawa, Y., & Sugiura, M. (2011). The downstream *atpE* cistron is efficiently translated via its own cis-element in partially overlapping *atpB-atpE* dicistronic mRNAs in chloroplasts. *Nucleic Acids Research*, *39*, 9405–9412.

Suzuki, K., Nakanishi, H., Bower, J., Yoder, D. W., Osteryoung, K. W., & Miyagishima, S. (2009). Plastid chaperonin proteins Cpn60α and Cpn60β are required for plastid division in *Arabidopsis thaliana*. *BMC Plant Biology*, *9*, 38.

Suzuki, T., Ozaki, Y., Sone, N., Feniouk, B. A., & Yoshida, M. (2007). The product of uncI gene in F_1F_o-ATP synthase operon plays a chaperone-like role to assist c-ring assembly. *Proceedings of the National Academy of Sciences of the United States of America*, *104*, 20776–20781.

Takabayashi, A., Kadoya, R., Kuwano, M., Kurihara, K., Ito, H., Tanaka, R., et al. (2013). Protein co-migration database (PCo M-DB) for thylakoids and cells. *Springerplus*, *2*, 148.

Till, B., Schmitz-Linneweber, C., Williams-Carrier, R., & Barkan, A. (2001). CRS1 is a novel group II intron splicing factor that was derived from a domain of ancient origin. *RNA*, 7, 1227–1238.

Toda, T., Fujii, S., Noguchi, K., Kazama, T., & Toriyama, K. (2012). Rice *MPR25* encodes a pentatricopeptide repeat protein and is essential for RNA editing of *nad5* transcripts in mitochondria. *The Plant Journal*, *72*, 450–460.

Wang, Z. G., & Ackerman, S. H. (2000). The assembly factor Atp11p binds to the β-subunit of the mitochondrial F_1-ATPase. *Journal of Biological Chemistry*, *275*, 5767–5772.

Wang, Z. G., Sheluho, D., Gatti, D. L., & Ackerman, S. H. (2000). The α-subunit of the mitochondrial F_1 ATPase interacts directly with the assembly factor Atp12p. *EMBO Journal*, *19*, 1486–1493.

Wang, Y., Zhang, X., Lu, S., Wang, M., Wang, L., Wang, W., et al. (2012). Inhibition of a basal transcription factor 3-like gene Osj10gBTF3 in rice results in significant plant miniaturization and typical pollen abortion. *Plant and Cell Physiology*, *53*, 2073–2089.

Watkins, K. P., Kroeger, T. S., Cooke, A. M., Williams-Carrier, R. E., Friso, G., Belcher, S. E., et al. (2007). A ribonuclease III domain protein functions in group II intron splicing in maize chloroplasts. *Plant Cell*, *19*, 2606–2623.

Woessner, J. P., Masson, A., Harris, E. H., Bennoun, P., Gillham, N. W., & Boynton, J. E. (1984). Molecular and genetic analysis of the chloroplast ATPase of *Chlamydomonas*. *Plant Molecular Biology*, *3*, 177–190.

Wostrikoff, K., & Stern, D. (2007). Rubisco large-subunit translation is autoregulated in response to its assembly state in tobacco chloroplasts. *Proceedings of the National Academy of Sciences of the United States of America*, *104*, 6466–6471.

Yagi, H., Kajiwara, N., Tanaka, H., Tsukihara, T., Kato-Yamada, Y., Yoshida, M., et al. (2007). Structures of the thermophilic F_1-ATPase ε subunit suggesting ATP-regulated arm motion of its C-terminal domain in F_1. *Proceedings of the National Academy of Sciences of the United States of America, 104*, 11233–11238.

Yamori, J. W., & Shikanai, T. (2016). Physiological functions of cyclic electron transport around photosystem I in sustaining photosynthesis and plant growth. *Annual Review of Plant Biology, 67*, 81–106.

Yap, A., Kindgren, P., Colas Des Francs-Small, C., Kazama, T., Tanz, S. K., Toriyama, K., et al. (2015). AEF1/MPR25 is implicated in RNA editing of plastid *atpF* and mitochondrial *nad5*, and also promotes *atpF* splicing in Arabidopsis and rice. *Plant Journal, 81*, 661–669.

Yin, P., Li, Q., Yan, C., Liu, Y., Liu, J., Yu, F., et al. (2013). Structural basis for the modular recognition of single-stranded RNA by PPR proteins. *Nature, 504*, 168–171.

Yoshida, M., Muneyuki, E., & Hisabori, T. (2001). ATP synthase—A marvellous rotary engine of the cell. *Nature Reviews Molecular Cell Biology, 2*, 669–677.

Zhang, L., Duan, Z., Zhang, J., & Peng, L. (2016). Biogenesis factor required for ATP synthase 3 facilitates assembly of the chloroplast ATP synthase complex. *Plant Physiology, 171*, 1291–1306.

Zhang, Y., & Lu, C. (2019). The enigmatic roles of PPR-SMR proteins in plants. *Advanced Science, 6*, 1900361.

Zhang, L., Pu, H., Duan, Z., Li, Y., Liu, B., Zhang, Q., et al. (2018). Nucleus-encoded protein BFA1 promotes efficient assembly of the chloroplast ATP synthase coupling factor 1. *Plant Cell, 30*, 1770–1788.

Zhang, L., Zhou, W., Che, L., Rochaix, J. D., Lu, C., Li, W., et al. (2019). PPR protein BFA2 is essential for the accumulation of the *atpH/F* transcript in chloroplasts. *Frontiers in Plant Science, 10*, 446.

Zhao, Q., Zhang, X., Sommer, F., Ta, N., Wang, N., Schroda, M., et al. (2019). Hetero-oligomeric CPN60 resembles highly symmetric group-I chaperonin structure revealed by cryo-EM. *The Plant Journal, 98*, 798–812.

Zhou, W., Lu, Q., Li, Q., Wang, L., Ding, S., Zhang, A., et al. (2017). PPR-SMR protein SOT1 has RNA endonuclease activity. *Proceedings of the National Academy of Sciences of the United States of America, 114*, E1554–E1563.

Zoschke, R., Kroeger, T., Belcher, S., Schottler, M. A., Barkan, A., & Schmitz-Linneweber, C. (2012). The pentatricopeptide repeat-SMR protein ATP4 promotes translation of the chloroplast *atpB/E* mRNA. *Plant Journal, 72*, 547–558.

Zoschke, R., Nakamura, M., Liere, K., Sugiura, M., Börner, T., & Schmitz-Linneweber, C. (2010). An organellar maturase associates with multiple group II introns. *Proceedings of the National Academy of Sciences of the United States of America, 107*, 3245–3250.

Zoschke, R., Qu, Y., Zubo, Y. O., Börner, T., & Schmitz-Linneweber, C. (2013). Mutation of the pentatricopeptide repeat-SMR protein SVR7 impairs accumulation and translation of chloroplast ATP synthase subunits in *Arabidopsis thaliana*. *Journal of Plant Research, 126*, 403–414.